First Principles of Meteorology and Air Pollution

ENVIRONMENTAL POLLUTION

VOLUME 19

Editors

Brain J. Alloway, *Department of Soil Science, The University of Reading, U.K.*
Jack T. Trevors, *School of Environmental Sciences, University of Guelph, Ontario, Canada*

Editorial Board

I. Colbeck, *Interdisciplinary Centre for Environment and Society, Department of Biological Sciences, University of Essex, Colchester, U.K.*
R.L. Crawford, *Food Research Center (FRC) 204, University of Idaho, Moscow, Idaho, U.S.A.*
W. Salomons, *GKSS Research Center, Geesthacht, Germany*

For other titles published in this series, go to
www.springer.com/series/5929

Mihalis Lazaridis

First Principles of Meteorology and Air Pollution

 Springer

Mihalis Lazaridis
Technical University of Crete
Department of Environmental Engineering
Polytechneioupolis
73100 Chania
Greece

ISSN 1566-0745
ISBN 978-94-007-3418-0   ISBN 978-94-007-0162-5 (eBook)
DOI 10.1007/978-94-007-0162-5
Springer Dordrecht New York Heidelberg London

© Springer Science+Business Media B.V. 2010
Softcover reprint of the hardcover 1st edition 2010
No part of this work may be reproduced, stored in a retrieval system, or transmitted in any form or by any means, electronic, mechanical, photocopying, microfilming, recording or otherwise, without written permission from the Publisher, with the exception of any material supplied specifically for the purpose of being entered and executed on a computer system, for exclusive use by the purchaser of the work.

Cover illustration: Cover Image © 2010 JupiterImages Corporation

Printed on acid-free paper

Springer is part of Springer Science+Business Media (www.springer.com)

Contents

1	**Description of the Earth's Atmosphere**...............................		1
1.1	Introduction to Atmospheric Structure and Composition		2
	1.1.1	Emissions of Air Pollutants in the Atmosphere	2
	1.1.2	The Earth's Atmosphere	4
	1.1.3	Origin and Evolution of the Atmosphere....................	7
1.2	Atmosphere's Characteristics..		10
1.3	Lower Atmosphere's Composition		11
	1.3.1	Dry Atmospheric Air...	11
	1.3.2	Water in the Atmosphere	11
	1.3.3	Atmospheric Aerosols	13
1.4	Vertical Division of the Atmosphere – Temperature Change.......		14
	1.4.1	Troposphere...	14
	1.4.2	Boundary Layer...	16
	1.4.3	Stratosphere...	20
	1.4.4	Mesosphere..	21
	1.4.5	Thermosphere...	21
	1.4.6	Exosphere	21
	1.4.7	Ionosphere – Magnetosphere	22
1.5	Change of Meteorological Parameters with Height		22
	1.5.1	Temperature Inversion	25
	1.5.2	Air Density Variation with Height............................	27
	1.5.3	Change of Atmospheric Pressure with Height...............	28
1.6	Model of the Standard Atmosphere		29
	1.6.1	Units of Chemical Components in the Atmosphere.........	30
	1.6.2	Unit Conversion of Concentration of Component I ($\mu g/m^3$) to Volume Concentration (Ppm)....................	31
1.7	Radiation in the Atmosphere		31
	1.7.1	Laws of Radiation..	33
	1.7.2	Sun's Radiation	36
	1.7.3	Earth's Radiation..	36

		1.7.4	Factors That Affect the Sun's Radiation Flux to Earth	37

　
      1.7.4 Factors That Affect the Sun's Radiation Flux to Earth 37
      1.7.5 Interaction of the Sun's Radiation in the Atmosphere 38
      1.7.6 Greenhouse Effect 45
      1.7.7 Energy Balance of Earth and its Atmosphere 47
      1.7.8 Distribution of Sun's Radiation at the System
             Atmosphere-Surface 49
      1.7.9 The Earth's Climate 50
  1.8 Examples 54
  1.9 Ambient Air Quality Standards 58
  1.10 Appendixes 59
      1.10.1 Appendix 1: The Hydrostatic Equation 59
  References 64

2 First Principles of Meteorology 67

  2.1 General Aspects of Meteorology 68
  2.2 Vertical Structure of the Temperature and Conditions
       of Atmospheric Stability 70
      2.2.1 Dry Vertical Temperature Lapse Rate 71
      2.2.2 Wet Vertical Temperature Lapse Rate 74
      2.2.3 Temperature Inversion 75
  2.3 Atmospheric Variability – Air Masses – Fronts 77
      2.3.1 Air Masses 77
      2.3.2 Classification of Air Masses 78
      2.3.3 Fronts 79
      2.3.4 Wave Cyclone 85
  2.4 Turbulence – Equations for the Mean Values 86
  2.5 Statistical Properties of Turbulence 87
  2.6 Atmospheric Temperature 92
      2.6.1 Temperature Season Variability 92
      2.6.2 Temperature Daily Variability 96
      2.6.3 Heating of the Earth's Surface and Heat Conduction 98
      2.6.4 Distribution of Temperature in the Air 100
  2.7 Humidity in the Atmosphere 101
      2.7.1 Mathematical Expressions of Humidity
             in the Atmosphere 102
      2.7.2 Dew Point 104
      2.7.3 Clouds in the Atmosphere 106
      2.7.4 Precipitation 107
      2.7.5 Study of Precipitation Scavenging 109
  2.8 Applications and Examples 112
  References 117

3 Atmospheric Circulation ... 119
- 3.1 Atmospheric Pressure and Pressure Gradient Systems ... 120
- 3.2 Atmospheric Pressure Changes ... 120
  - 3.2.1 Vertical Pressure Changes ... 120
  - 3.2.2 Non-canonical Pressure Changes ... 121
  - 3.2.3 Canonical Pressures Changes ... 121
- 3.3 Transfer of the Pressure to Its Mean Value at Sea Level ... 122
- 3.4 Isobaric Curves: Pressure Gradient Systems ... 122
- 3.5 Pressure Gradient Force ... 124
- 3.6 Movement of Air: Wind ... 126
  - 3.6.1 Forces Which Affect the Movement of Air ... 128
  - 3.6.2 Transport Equations of Air Masses in the Atmosphere . 131
  - 3.6.3 Wind Categories ... 132
- 3.7 General Circulation in the Atmosphere ... 137
  - 3.7.1 Single and Three Cell Models ... 138
  - 3.7.2 Continuous Winds ... 140
  - 3.7.3 Periodic Winds ... 141
  - 3.7.4 Sea and Land Breeze ... 141
  - 3.7.5 Mountain and Valley Breezes ... 143
- 3.8 Vertical Structure of Pressure Gradient Systems ... 144
- 3.9 Equations of Atmospheric Circulation ... 145
  - 3.9.1 Equations of Circulation for a Compressible Fluid ... 145
- References ... 149

4 Atmospheric Chemistry ... 151
- 4.1 Chemical Components in the Atmosphere ... 151
- 4.2 Chemistry of the Troposphere ... 152
  - 4.2.1 Sulphur Components ... 155
  - 4.2.2 Nitrogen Components ... 156
  - 4.2.3 Carbon Components ... 156
  - 4.2.4 Halogen Components ... 157
- 4.3 Particulate Matter ... 157
- 4.4 Photochemistry in the Free Troposphere ... 158
  - 4.4.1 Photochemical Cycle of Ozone and Nitrogen Oxides ... 158
  - 4.4.2 Chemistry of Carbon Dioxide ... 160
  - 4.4.3 Chemistry of Hydrocarbons ... 161
  - 4.4.4 Chemistry of Sulphur Compounds ... 161
- 4.5 Components of Aquatic Chemistry in the Atmosphere ... 162
- 4.6 Chemistry of the Stratosphere - Ozone ... 163
- References ... 167

5 Atmospheric Aerosols ... 169
- 5.1 Introduction ... 169
- 5.2 Size Distribution of Aerosols ... 171
- 5.3 Chemical Composition of Aerosols ... 180

	5.4	Organic Aerosols..	182
		5.4.1 Elemental Carbon- Primary Organic Carbon............	184
		5.4.2 Secondary Organic Matter Formation (Secondary Organic Carbon)	185
	5.5	Dynamics of Atmospheric Particulate Matter	186
		5.5.1 New Particle Formation.................................	186
		5.5.2 Condensation and Evaporation	193
		5.5.3 Coagulation	196
	5.6	Bioaerosols – Definition	197
		References...	198
6	**Atmospheric Dispersion: Gaussian Models**		201
	6.1	Theories of Atmospheric Diffusion.............................	202
	6.2	Euler Description..	202
	6.3	Lagrange Description	203
	6.4	Equations Describing the Concentration of Pollutants at Turbulent Conditions...	204
		6.4.1 Diffusion Equation in Euler Description	204
		6.4.2 Diffusion Equation in Lagrange Description...........	205
		6.4.3 Solution of the Diffusion Equation for a Continuous Source with the Euler Methodology........	207
	6.5	Gaussian Model	207
		6.5.1 Limitations of the Gaussian Model.....................	208
		6.5.2 Calculation of the σ_y and σ_z Coefficients. Stability Methodology	209
		6.5.3 Plume Rise	212
		6.5.4 Atmospheric Stability – Application to the Gaussian Models...	216
	6.6	Analytical Solutions of the Atmospheric Diffusion Equation...	217
	6.7	Two-Dimensional, Time Independent Line-Continuous Source with Changing Values of Velocity and Diffusion Coefficient	220
	6.8	Characteristics of Plume Dispersion – Stability Conditions	223
	6.9	Examples and Applications	226
	6.10	Appendix 6.1: The Continuity Equation	230
		References...	232
7	**Atmospheric Models: Emissions of Pollutants**		233
	7.1	Introduction	233
	7.2	Dispersion Equations for Pollutant Transport at the Euler and Lagrange Coordinating Systems	235
		7.2.1 Model of a Single Volume in the Euler System.........	236
		7.2.2 Three Dimensional Models of Atmospheric Pollution ..	237
	7.3	Statistical Evaluation of Atmospheric Models	238

	7.4	Emissions of Atmospheric Pollutants............................	239
	7.5	Emissions from the Biosphere	240
		7.5.1 Emissions of Volatile Organic Compounds from Vegetation..	243
		7.5.2 Calculation of Biogenic Emissions	243
		7.5.3 Sea Salt Emissions	245
		7.5.4 Emissions of Air Pollutants from the Earth's Surface..	246
		7.5.5 Emissions of Pollutants from Forest Fires...............	247
	7.6	Examples and applications.......................................	249
		References...	253
8	**Indoor Air Pollution**		**255**
	8.1	Introduction to Indoor Air Quality..............................	256
	8.2	Ozone...	262
	8.3	Nitrogen Oxides...	265
	8.4	Volatile Organic Compounds	267
	8.5	Chemistry of Organic Compounds Indoors.........................	268
	8.6	Radon...	275
		8.6.1 Radiative Decay of Radon Isotopes	277
		8.6.2 Exposure and Dose of Radon in Indoor Environment...	279
		8.6.3 Examples...	282
	8.7	Carbon Monoxide	283
	8.8	Asbestos..	284
	8.9	Heavy Metals..	285
	8.10	Formaldehyde	287
	8.11	Pesticides	290
	8.12	Polycyclic Aromatic Hydrocarbons (PAH)	291
	8.13	Polychloric Biphenyls (Pcbs)	291
	8.14	Tobacco Smoke	293
	8.15	Bioaerosols	294
	8.16	Microenvironmental Models	296
	8.17	Air Exchange Rate by Infiltration...............................	299
	8.18	Emission Models	300
	8.19	Deposition Models	301
		8.19.1 Examples..	301
		References...	303
9	**Human Exposure and Health Risk from Air Pollutants**		**305**
	9.1	Human Exposure and Doses from Air Pollutants	306
	9.2	Exposure Pathways...	309
		9.2.1 Dermal Absorption.......................................	309
		9.2.2 Inhalation Exposure.....................................	310
	9.3	Calculation of Dose–Response Functions	315

		9.3.1	Dose Calculation Through Intake........................	315
		9.3.2	Internal Dose Calculation Through Dermal Absorption	316
		9.3.3	Internal Dose Calculation Through Inhalation and Food Intake..	318
		9.3.4	Functions of Dose–Response	319
	9.4	Particulate Matter Dose Through Inhalation		322
		9.4.1	Deposition of Particles in the Respiratory Tract	322
		9.4.2	Classification of Particles Based on Their Ability to Penetrate the Respiratory Tract...............	328
		9.4.3	Calculation of Particle Deposition in the Respiratory Tract...	331
		9.4.4	Particle Clearance in the Human Respiratory Tract.....	333
		9.4.5	Particle Deposition Measurements......................	338
	9.5	Application: Internal Dose from Radon Inhalation..............		341
	9.6	Health Effects from Air Pollutants		346
	9.7	Health Effects from Exposure to Particulate Matter		350
		References...		354
Appendix A				355
Index...				359

Chapter 1
Description of the Earth's Atmosphere

Contents

1.1	Introduction to Atmospheric Structure and Composition	2
	1.1.1 Emissions of Air Pollutants in the Atmosphere	2
	1.1.2 The Earth's Atmosphere	4
	1.1.3 Origin and Evolution of the Atmosphere	7
1.2	Atmosphere's Characteristics	10
1.3	Lower Atmosphere's Composition	11
	1.3.1 Dry Atmospheric Air	11
	1.3.2 Water in the Atmosphere	11
	1.3.3 Atmospheric Aerosols	13
1.4	Vertical Division of the Atmosphere – Temperature Change	14
	1.4.1 Troposphere	14
	1.4.2 Boundary Layer	16
	1.4.3 Stratosphere	20
	1.4.4 Mesosphere	21
	1.4.5 Thermosphere	21
	1.4.6 Exosphere	21
	1.4.7 Ionosphere – Magnetosphere	22
1.5	Change of Meteorological Parameters with Height	22
	1.5.1 Temperature Inversion	25
	1.5.2 Air Density Variation with Height	27
	1.5.3 Change of Atmospheric Pressure with Height	28
1.6	Model of the Standard Atmosphere	29
	1.6.1 Units of Chemical Components in the Atmosphere	30
	1.6.2 Unit Conversion of Concentration of Component I ($\mu g/m^3$) to Volume Concentration (Ppm)	31
1.7	Radiation in the Atmosphere	31
	1.7.1 Laws of Radiation	33
	1.7.2 Sun's Radiation	36
	1.7.3 Earth's Radiation	36
	1.7.4 Factors That Affect the Sun's Radiation Flux to Earth	37
	1.7.5 Interaction of the Sun's Radiation in the Atmosphere	38
	1.7.6 Greenhouse Effect	45
	1.7.7 Energy Balance of Earth and its Atmosphere	47
	1.7.8 Distribution of Sun's Radiation at the System Atmosphere-Surface	49
	1.7.9 The Earth's Climate	50
1.8	Examples	54
1.9	Ambient Air Quality Standards	58

1.10	Appendixes	59
	1.10.1 Appendix 1: The Hydrostatic Equation	59
References		64

Abstract The atmosphere's dynamics have a direct effect on weather conditions and consequently on the climate and human existence. Emissions of air pollutants affect the public health, the ecosystem and result in climatic changes. The first chapter examines in depth the meteorology and the dynamics of the atmosphere. It examines the atmospheric structure, the chemical composition of the atmosphere and the change of meteorological parameters with height. There is also an examination of the Sun's radiation effect on the Earth's surface, on the atmosphere and finally the greenhouse effect.

1.1 Introduction to Atmospheric Structure and Composition

1.1.1 Emissions of Air Pollutants in the Atmosphere

During the last 50 industry years has produced over 60,000 new chemical compounds which have accumulated in the environment and some of them bioaccumulated in the human body. Emissions of air pollutants disperse in the atmosphere and are transported at long distances from their sources. The medium in which the air pollutants are transported is the atmosphere, and its dynamics determines their lifetime and their impact on the environment and humans.

Many of these anthropogenic and xenobiotic chemical compounds have been transported to the biosphere and are characterized as toxic or potentially harmful. These chemical components in their gaseous or particulate matter form have the potential, besides their direct implication for human health, to produce changes in the environment and even to the climate of our planet. The concentration of these chemical species is usually less than one part per million parts of air by volume (1 ppm) which are called trace gasses. During the last century the concentration of ozone (O_3), nitrogen oxides (NO_x), sulphur and organic particulate matter, as well as carbon dioxide (CO_2), has increased considerably with potential implications for the climate.

We often assume that air pollution is a modern phenomenon, and that it has become worse in recent times. However since the dawn of history, mankind has been burning biological and fossil fuel to produce heat. The walls of caves, inhabited millennia ago, are covered with layers of soot and many of the lungs of mummified bodies from Palaeolithic times have a black tone (McNeill 2001). Zimmerman (2004) concluded that continual exposure of early man to campfires in enclosed areas contributed to increased incidences of nasal cancer. The nature of air pollution has changed over the millennia. No longer is it predominantly smoke and sulphur related but now it is associated with nitrogen oxides, volatile organic compounds and particulate matter connected with growing vehicle traffic (Morawska et al. 2008; Fenger 2009)

The term climate is related to the atmosphere's condition for a long time interval (minimum time interval of 30 years) and contains data on the average values of atmosphere's temperature, relative humidity, precipitation, cloud cover and wind. However, the term weather refers to the state of the atmosphere at a specific time and place and it is restricted to small time intervals (days to a week) containing information for the same atmospheric properties as the climate but for small time intervals.

Figure 1.1 presents measurements of carbon dioxide (CO_2) at the observatory of Mauna Loa at Hawaii which show an increase of its concentration the last 40 years. This figure represents a total increase of the carbon dioxide concentration over 15% from 1958 until 2004. The higher concentrations are observed during the winter period due to the reduced photosynthesis of plants. During the spring and autumn the carbon dioxide concentrations are lower since the plants absorb large quantities of carbon dioxide from the atmosphere and digest it into organic compounds. The increase of the carbon dioxide concentration in the atmosphere is a result of combustion of liquid and solid fuels and of the destruction of forests in the planet.

Since there is a prognosis for an increase in industrial production and economic growth in general, the emissions of air pollutants in the atmosphere will increase. Therefore there is an urgent need for the study of air pollution characteristics and dynamics and its transport in the atmosphere at various spatial and time scales.

The spatial and temporal variability of air pollutants is an important subject of study, since long-range transport of air pollutants contributes significantly to air pollution levels. Important is also the air pollution in urban agglomerations, since observations have shown high concentrations of gaseous and particulate matter

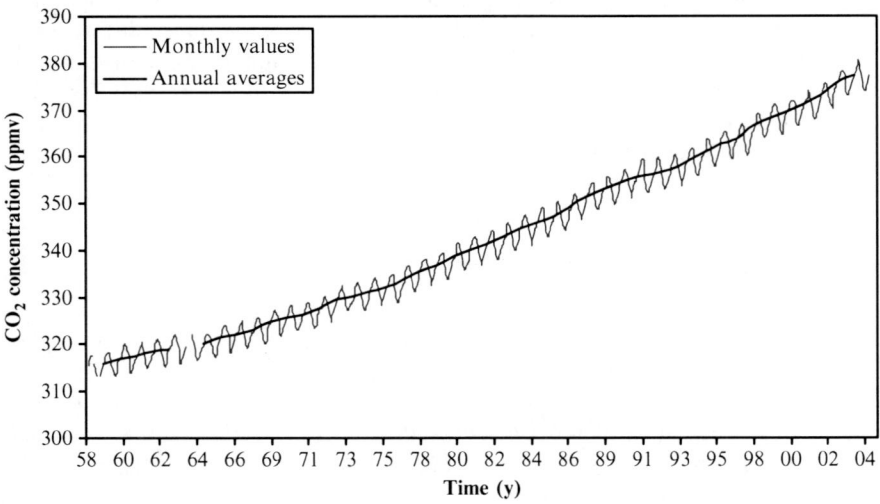

Fig. 1.1 Observation of the increase of CO_2 (parts per million - ppm) at the Mauna Loa observatory in Hawaii during the period from 1958 to 2004

pollutants such as nitrogen dioxide (NO_2). An emerging issue is the air pollution in large urban agglomerates which are called megacities. The world's urban population increased over the past 50 years with an average rate of 2.7% $year^{-1}$. By contrast, the total world population increased with an average rate of only 1.8% $year^{-1}$. This illustrates the increasing global importance of megacities, thus their growing significance for the world's air quality and climate problems. Air quality degradation in large urban agglomerations has significant effects on human health leading to increased morbidity and mortality, to lower urban and regional air quality and consequently to the earth's climate.

Gaseous and particulate matter chemical and physical properties have been found (IPCC 2007) to have a strong direct and indirect impact on earth's radiation budget. The latter originates from modification of cloud microphysical and optical properties and of the earth's hydrological cycle. To quantify the impact of gaseous pollutants and aerosols on climate and to assess, in turn, the feedback of climate change on aerosols requires a thorough understanding of the physico-chemical aerosol processes on a micro-scale and aerosol evolution in the context of regional and global scale circulation.

In the current chapter the structure and the composition of the atmosphere will be studied. There is also an interaction between the atmosphere and the earth's surface and consequently with the surface water (oceans, lakes) and even with groundwater availability. Interactions between the oceans and the atmosphere, as well as, between the air and the Earth's surface inside the boundary layer will be examined in this chapter. However, a more extensive study of air pollutant deposition and further intrusion in ground waters will not be examined here.

A more general conceptual picture of the effects that air pollutants have on ecosystems and humans is shown in Fig. 1.2. The emissions and transport of pollutants from various sources influence air pollution and climate. The importance of this transport varies between different chemical compounds due to differences in their atmospheric residence times. The multiplicity of linkages demonstrates the complexity of interactions among different pollutants at different receptors. This picture relates the different air pollutants together with their sources and their effects including the different international protocols related to specific compounds.

1.1.2 The Earth's Atmosphere

The term atmosphere refers to the gaseous cover of Earth (Fig. 1.3) and participates in the Earth's motion.

The Earth is rotating around its axis with a constant speed. The atmosphere is rotating also together with the Earth and tends to move towards empty space due to centrifugal motion. This is the reason that the atmosphere above the equator is higher compared to the atmosphere above the poles.

The atmosphere is invisible, odorless and has various properties that favor the survival of plants and living organisms in the planet. The atmosphere contains a

1.1 Introduction to Atmospheric Structure and Composition

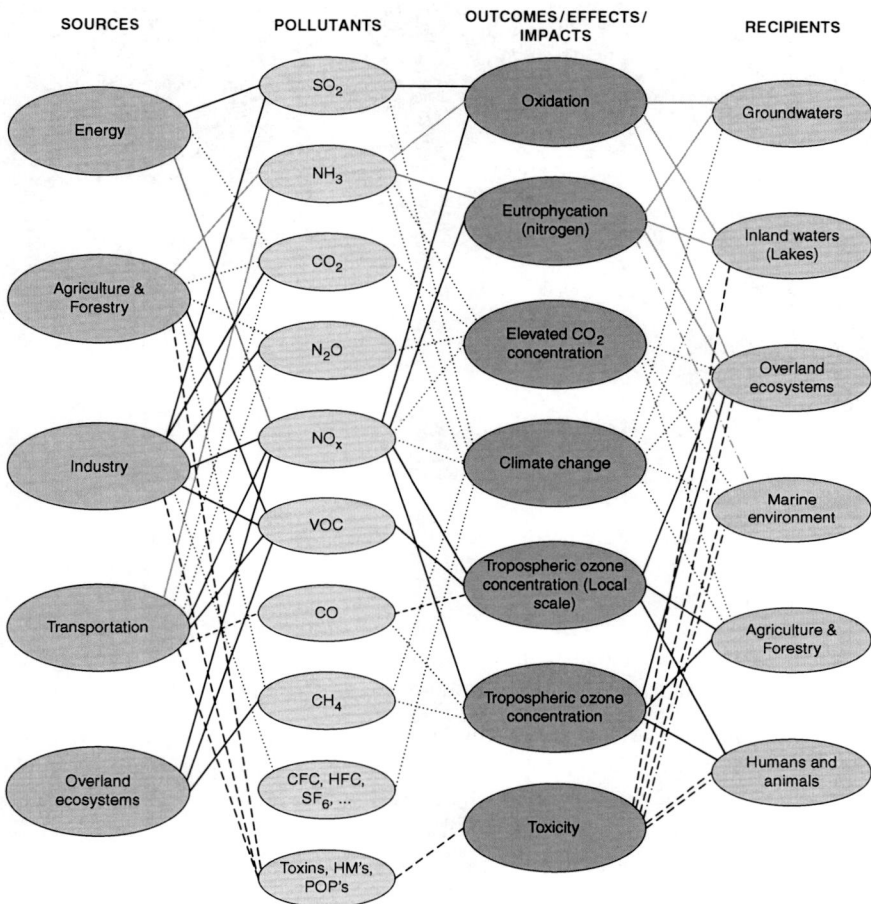

Fig. 1.2 Conceptual linkages between the sources of trace gas emissions, environmental issues and receptors of concern. The line styles represent the main international policy structure dealing with each link. (Adapted from Sutton et al. 2003; Erisman et al. 2001; Grennfelt and Schjoldager 1984)

mixture of gaseous chemical species which are known as atmospheric air and remains close to the Earth's surface due to the force of gravity.

The movement of air masses inside the atmosphere is called atmospheric circulation and is mainly a result of temperature differences between the tropics and the poles and at the same time is affected by the Earth's rotation. Conversion of the Sun's and Earth's radiation to other energy forms inside the atmosphere is occurring (e.g. heat and kinetic energy). Therefore the atmosphere can be considered as a medium of complex thermodynamic and mechanical processes which are responsible for various phenomena. These phenomena which occur in the atmosphere and are observed by the humans (direct or indirect with the help of instruments) are called meteorological phenomena.

Fig. 1.3 The Earth's atmosphere (Adapted from Ahrens 1994)

The atmosphere's height is not accurately known since its upper layers have a very small density and it is difficult to distinguish the limit between the atmosphere and interstellar space.

Estimation of the atmosphere's height is based on measurements or estimation of the height at which specific phenomena take place and need a satisfactory air concentration. Specific examples of the estimation of the atmosphere's height are given:

- From the duration of dawn and twilight, which are caused by multiple reflection and flux of the Sun's radiation at the upper layers of the atmosphere, it can be concluded that the atmosphere's height reaches 80 Km.
- From observation of shooting stars, it can be concluded that the Earth's atmosphere reaches 250 km, whereas from the Northern Lights (Aurora Borealis) and Southern Lights (Aurora Australis) there is evidence that atmospheric air exists above 1,000 km.
- Finally, observed irregularities in satellite orbits imply that the atmosphere's height reaches 3.500 km.

The atmosphere's dynamics is one very interesting subject to study. Intensive weather conditions affect human daily life. For example, differences in meteorological conditions between northern Europe and Mediterranean countries are reflected in social life and building architecture.

Furthermore, the periodic appearance of natural phenomena, such as volcanic eruptions, affects the climate through the ejection of large quantities of particles into the atmosphere that reflect the Sun's radiation (Fig. 1.4). Figure 1.5 shows a microscope photograph of particles produced from combustion processes. More specifically it shows electron microscope pictures of coal fly ash and grain oil char particles emitted from coal and oil combustion respectively. The complex nature and size of the resulting particles indicate the wide range of particles present in the atmosphere.

Fig. 1.4 Eruption from the St. Helen volcano in May 1980 and electron microscope pictures of emitted volcanic ash

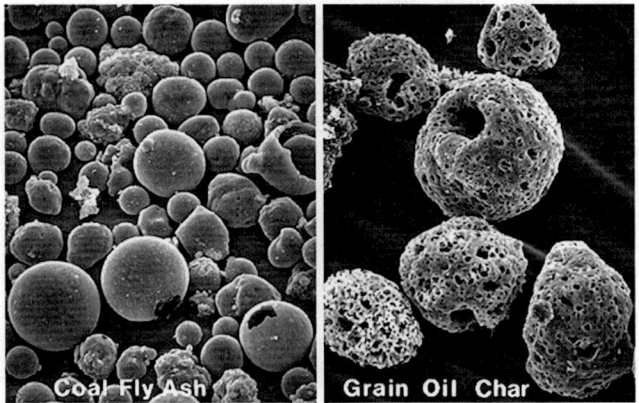

Fig. 1.5 Electron microscope pictures of coal fly ash and grain oil char particles

The atmosphere's dynamics influences the concentration and spatial and temporal patterns of pollutants which arise from natural and anthropogenic sources.

1.1.3 Origin and Evolution of the Atmosphere

Earth belongs to our solar system which comprises nine planets and the Sun (Fig. 1.6). The Sun is an average size star that is located at the edge of our galaxy which is named the Milky Way. The universe comprises a billion galaxies and each of them contains billions of stars. Stars are hot balls of gas which produce energy with the transformation of hydrogen to helium.

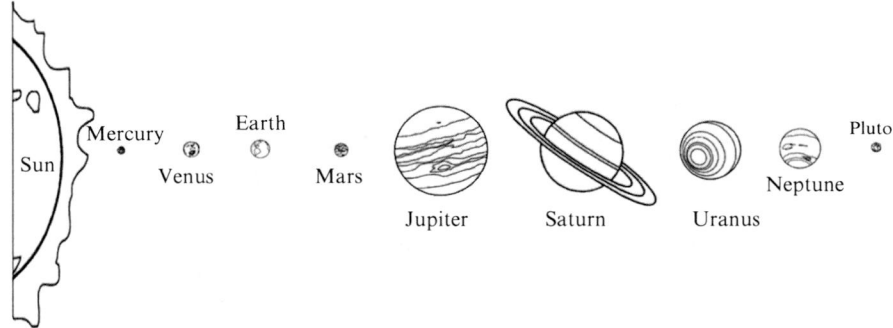

Fig. 1.6 Relative size and position of the planets in the solar system (positions are not to scale)

The initial atmosphere of the Earth about 4.6 billion years ago had a totally different composition than today's atmosphere. This initial atmosphere contained large quantities of hydrogen, helium and other hydrogen compounds such as ammonia, and methane is transported to space due to the flux of the solar wind, which is a huge flux of particles emitted from the Sun. The loss of this initial atmosphere occurred also due to the extreme high temperatures which occurred in the planet. Gradually the formation of a denser atmosphere developed as the Earth's temperature decreased. The escape of gasses which were dissolved in the molten rocks in the Earth created this second atmosphere. The whole process is called outgassing. The escape of gasses from the Earth's interior occurs also today from hundreds of active volcanoes. The main gasses which are emitted from volcanoes today are water vapour (80%) and carbon dioxide (10%).

Geologists argue from chemical analysis of older rocks in Earth that volcanoes outpour the same gasses today as they did in the past for the formation of the first atmosphere. In the early atmosphere no oxygen was present (reduced atmosphere) and there were gasses such as H_2, CH_4, N_2, NH_3, H_2O, CO and CN^-. As the cooling of the planet proceeded, the water vapor started to condense into clouds and extreme weather conditions occurred. At the beginning the water from the rain did not reach the Earth's surface since it was evaporated from the hot air in the atmosphere. This process gradually resulted to the acceleration of cooling of the Earth's surface.

When the earth's surface was cooled below the temperature of boiling water, intense rain filled in the huge cavities on the surface and formed the oceans. This dynamic procedure caused a drastic reduction in the water vapor (H_2O) and carbon dioxide (CO_2) concentration in the atmosphere. Huge quantities of CO_2 drifted to the earth's surface as rain, leaving in the atmosphere only large quantities of nitrogen (N_2). At the same time the atmosphere became overwhelmed from ultraviolet radiation and lightings. The coexistence of energy sources and of the first atmosphere resulted in the creation of a primitive form of life inside water which, through photosynthesis, enriched the atmosphere with oxygen.

The oxygen in the atmosphere is a necessary component for the creation and the maintenance of life on Earth. It is consumed and produced through biological and

1.1 Introduction to Atmospheric Structure and Composition

chemical processes which take place in the biosphere and the lithosphere. The sources and the production processes of oxygen inside the atmosphere are the following:

- The photo-dissociation of water vapor under the influence of ultraviolet radiation of Sun, a process which takes place in the upper atmosphere through the chemical reaction:

$$2H_2O + hv \rightarrow 2H_2 + O_2 \quad (1.1)$$

- The photosynthesis through which live organisms (plants, algal and Cyanobacterial) compose organic material from H_2O and CO_2 with the action of Sun's radiation through the reaction:

$$H_2O + CO_2 + hv \rightarrow CH_2O + O_2 \quad (1.2)$$

From the above two pathways of oxygen production, photosynthesis is the dominant mechanism in relation to the photo-dissociation process. If only the photo-dissociation was present in the atmosphere the production of oxygen would correspond to concentration levels between 10^{-6} and 10^{-4} lower than its present level in the atmosphere.

The exit of primitive life from the water was achieved with shielding of the atmosphere from the ultraviolet radiation of the Sun. This protection was created by the protective layer of ozone at the upper layers of the atmosphere. The ozone (O_3) is produced from oxygen through the photosynthesis of micro-organisms in the water. The ozone is able to absorb the ultraviolet radiation which is harmful for living organisms. Theoretical calculations with an atmospheric model have shown that the formation of this protective layer of ozone required at least 10^{-3} of the current oxygen concentration for its formation. With the formation and further enhancement of this ozone layer, the living organisms were able to exit from the water and live on land. Life started to use the bountiful constituents of the environment, producing extended forest areas on Earth.

Photosynthesis was the main source of oxygen in the atmosphere and it is calculated that one billion years ago the atmosphere contained the same levels of oxygen as today. The total quantity of oxygen in the atmosphere corresponds only to 10% of the total oxygen produced from plants through photosynthesis. The remaining 90% of oxygen was undertaken for the production of several oxides on the Earth's crust (e.g. $CaCO_3$). Especially important is the formation of carbonate compounds due to the participation of carbon dioxide which is consequently removed from the atmosphere. In the atmosphere exists a dynamic equilibrium between the photosynthesis, the bonding of CO_2 in carbonaceous rocks, the burning processes and the emission of gasses from volcanoes. Therefore, the concentration of CO_2 and O_2 in the atmosphere remained almost constant during the Earth's recent history of evolution.

Had water not been present on Earth, then the photosynthesis would not have occurred and the Earth's atmosphere would have resembled the atmosphere of Venus where 95% of it consists of CO_2 in small quantities of water and nitrogen. The nitrogen (N_2) which also originates from the Earth's interior remains a dominant constituent of the Earth's atmosphere due to its chemical inactivity and its very small solubility in water (close to 1/70 of the CO_2 solubility).

The Earth's atmosphere reached today's composition close to 0.5 billion years ago. The presence of man on Earth occurred after that of the first animal in the equatorial zone. Changes in atmospheric composition occur continuously and these changes in the atmospheric composition may result in changes in the ecosystems and the environment.

1.2 Atmosphere's Characteristics

The mass of the atmosphere can be assessed even though its height is not known exactly. It is supposed that Earth can be approximated by a sphere with radius $R_e = 6.371 \times 10^3$ m, its surface is flat with an average surface pressure $p_0 = 1,013.25$ mb $= (1,013.25 \times 10^2$ Nm$^{-2})$ and gravity acceleration g $= 9.81$ ms^{-2}. The expression of the atmospheric pressure concludes that:

$$p_o = \frac{F}{S} = \frac{M_a g}{4\pi R_e^2} \Rightarrow M_a = \frac{4\pi R_e^2 p_o}{g} = \frac{4\pi (6.371 \times 10^3)^2 \times (1013.25 \times 10^2)}{9.81},$$

$$M_a = 5.27 \times 10^{18} kg, \tag{1.3}$$

where F is the applied force, S the surface and M_a the atmosphere's mass. For comparison the total mass of oceans is close to 1.35×10^{21} kg and the total mass of the solid crust of Earth is close to 5.98×10^{21} kg.

Calculations have shown that 50% of the total mass of the atmosphere is included in the layer between the Earth's surface and the height of 5.5 km, whereas, 75%, 95% and 99% of the mass is contained up to the height of 10, 20 and 40 km respectively. In addition, the density of air is changing inversely proportional to the height. This result is due to the gravity force which acts on the air molecules and concentrates them close to the Earth's surface. The air density at sea level is close to 1.225 g/m^3.

The atmosphere is following the Earth's rotation with a velocity which ranges between 300 and 600 km/h. The atmospheric air contains large quantities of air pollutants including aerosols. Aerosols can be introduced from natural sources. For example a single sand storm may introduce over 500 million ton of sand into the atmosphere. The atmosphere contains also huge quantities of water since only a small cloud contains 100–1,000 t of humidity. As another example it can be shown that evaporation of water from the Gulf of Mexico can reach levels close to 20 billion liters per hour.

From an energy point of view the atmosphere is considered as an inefficient engine which uses only 3% of the energy received from the Sun to transfer it to kinetic energy. It is possible for a summer storm to consume the energy equivalent of 10 atomic bombs (Hiroshima type), whereas, on a daily basis, about 45,000 storms occur on our planet. For comparison a typhoon releases the same energy as a storm during one second.

1.3 Lower Atmosphere's Composition

The Earth's atmosphere in the lower layers is composed of

(a) A mixture of gasses which comprise the dry air.
(b) Water in all three phases (vapor, liquid and solid).
(c) Aerosols (solid or liquid particles suspended in air).

1.3.1 Dry Atmospheric Air

The mixture of gasses which compose the "dry" atmospheric air does not contain water and aerosols. Table 1.1 depicts the composition of "dry" air.

The "dry" air contains also several other gasses in small and variable concentrations. These include nitrogen oxides (NO, N_2O_5, NO_2), sulphur dioxide (SO_2), carbon monoxide (CO), hydrogen peroxide (H_2O_2), ammonia (NH_3), nitric acid (HNO_3), sulphuric acid (H_2SO_4), radon (Rn) and iodine (I_2).

The composition of "dry" air remains constant inside the atmosphere until the height of 85 km. Even though the density of air is reduced with the height from the Earth's surface, the composition ratio remains the same. This concludes that the atmosphere is well mixed. Above the height of 85 km the air composition is changing constantly. The atmosphere is divided in two regions in respect to the air composition. The region from the Earth's surface until the height of 85 km is called the homosphere (constant atmospheric composition) and the region above the height of 85 km is called the heterosphere (variable atmospheric composition).

The atmosphere is directly connected with human existence since the absence of air for a few minutes would result in the disappearance of life on our planet. Life occurs in the first layer of the atmosphere which is close to the Earth's surface (troposphere). The troposphere's height is close to 12 km but above 8 km height the low air density and the low temperatures make impossible the human presence.

1.3.2 Water in the Atmosphere

The atmosphere, except for the gasses which consist of the "dry" air, contains also vapor water (humidity). The vapor water arises from the evaporation of natural water reservoirs on the Earth's surface (oceans, lakes, rivers), the sublimation of ice and from the evapotranspiration of plants. The quantity of water vapor in the atmosphere is variable in space and time and the volume ratio can range between 0 and 4%. The quantity of water vapor is reduced by height and the rate of change can be described as:

$$p_z = p_o \times 10^{-z/c} \tag{1.4}$$

Table 1.1 Chemical constituents of the atmospheric air

Name of gas	Symbol	Composition		Density (g/m^3)	Molecular weight	Critical temperature	Mean molecular velocity at 0°C
		Volume	Mass				
Nitrogen	N_2	78.08	75.51	1,250	28.016	−147.2	454 m/s
Oxygen	O_2	20.95	23.14	1,429	32.000	−118.9	425 m/s
Argon	Ar	0.93	1.3	1,786	39.944	−122.0	380 m/s
Carbon dioxide	CO_2	0.03	~0.5	1,977	44.010	31.0	362 m/s
Neon	Ne	18.18×10^{-4}	120×10^{-5}	900	20.183	−228.0	–
Helium	He	5.24×10^{-4}	8.10^{-5}	178	4.003	−258.0	1,202 m/s
Methane	CH_4	$~2.2 \times 10^{-4}$	–	717	16,04	–	–
Crypton	Kr	1.14×10^{-4}	29.10^{-5}	3,736	83,7	−63.0	–
Nitrous oxide	N_2O	$(0.5 \pm 0.1) \times 10^{-4}$	–	1,978	44.016	–	–
Hydrogen	H_2	$~0.5 \times 10^{-4}$	0.35×10^{-5}	90	2.016	−239.0	1,700 m/s
Xenon	Xe	0.087×10^{-4}	3.6×10^{-5}	5,891	131,3	16.6	–
Ozone	O_3	$(0–0.07) \times 10^{-4}$ έως $(1–3) \times 10^{-4}$	$~0.17 \times 10^{-5}$	2,140	48,0	5.0	–

where, p_z and p_o are the water vapor pressure at height z (in meters) and at the Earth's surface respectively with the constant c to be equal to 5,000 m. From Eq. (1.4) it can be concluded that at a height of 10 km the quantity of water vapor is very small.

The presence of water vapor in the atmosphere is very important through its condensation for the formation of clouds. In addition, the clouds participate in the water cycle in the planet, which is essential for life existence through water precipitation. Another important property of the water vapor is that it acts as a source of thermal energy for the atmosphere. This, in relation to the intensive heat transfer between the Earth's surface and the atmosphere, had as a result the initiation of weather disturbances. The 25% from the Sun's energy which is received on the Earth's surface is consumed for the evaporation of water and this is energy is released again into the atmosphere from the condensed water vapor.

1.3.3   Atmospheric Aerosols

The atmospheric aerosols consist of solid/liquid particles which are present in air. The term aerosols includes also the air in which the particles are contained. However, in the literature many times the term aerosols refers only to the particles inside the atmosphere. The aerosols arise both from natural sources (e.g. dust from the surface, volcanoes, sea salt) and anthropogenic sources (e.g. vehicular emissions, combustion emissions from industry and houses). The aerosols which are emitted directly in particulate form are characterized as primary aerosols, whereas, aerosols which are formed from vapor molecules in air are called secondary aerosols.

The main mechanisms of aerosol removal from the atmosphere are the dry and wet deposition processes. Dry deposition occurs when a particle is deposited through gravitational settling to the Earth's surface. Wet deposition occurs with coalesce of particles with water droplets during precipitation. The residence time of aerosols in the atmosphere is dependent on particle emission characteristics and weather conditions and ranges between a few days to weeks.

After the introduction of particles into the atmosphere there are several physicochemical processes (e.g. nucleation, condensation, coagulation) which further lead to a change of their size and shape. The size of aerosols ranges from a few nanometers (nm) to several tens of micrometers (μm). The shape of aerosols is also very variable and for determination of a particle's diameter we use the term equivalent diameter, which is the particle diameter of a spherical particle which has the same drag as the particle under study.

The aerosols are separated based on their size in two categories. The fine particles are aerosols with equivalent diameter smaller than 2.5 μm, whereas the coarse particles are aerosols with equivalent diameter larger than 2.5 μm. The fine aerosols can be further divided to the nucleation mode (diameter between 0.005 and 0.1 μm) and the accumulation mode (diameter between 0.1 and 2.5 μm). The nucleation mode originates from nucleation processes or condensation of vapors and particles from the nucleation mode grow through coagulation and further condensation. The accumulation mode includes particles which result from the nucleation mode and particles in this mode have high residence times in the atmosphere due to low deposition velocity. Coarse particles are emitted mainly as primary particles in the atmosphere and have small residence time.

Aerosols have several important impacts on the environment such as:

- Act as condensation nuclei for cloud formation.
- Participate in chemical reactions in the atmosphere and affect the ecosystems (e.g. ozone hole formation in the stratosphere, eutrophication, acidification).
- Affect human health.
- Modify the Sun's radiation intensity through light scattering and absorption.
- Decrease visibility at high ambient particle concentrations.

1.4 Vertical Division of the Atmosphere – Temperature Change

The atmosphere can be divided into several layers vertically based on its temperature profile. Figure 1.7 shows the vertical profile of the temperature and pressure up to the height of 140 km in the atmosphere. Based on the temperature profile the homosphere can be divided into three distinct areas; the troposphere, the stratosphere and the mesosphere. The heterosphere can be also divided into the thermosphere and the exosphere. The zones of division between these areas are called tropopause, stratopause and mesopause.

In the following subsections the different layers in the atmosphere are examined in more detail.

1.4.1 Troposphere

The troposphere is the lowest layer in the atmosphere and generally there is a continuous drop of the temperature versus height. The troposphere is also called lower atmosphere. It extends from the surface to the height of 12 ± 4 km based on the season and the latitude of the place. In particular the troposphere's height above the polar region extends to the height of 7–8 km, whereas, above the equatorial region the height is 16–17 km.

The troposphere contains 80% of the total mass of the atmosphere and almost the whole quantity of water vapor. Therefore, the troposphere is the most important layer for meteorology since inside this layer occur all the atmospheric processes and the weather dynamics.

The energy source for the heating of the troposphere is the Earth's radiation and due to this there is a reduction of the temperature versus height. The main transport process is the vertical transport (convection). In particular, the reduction of the temperature with height in the atmosphere is due to the gradual decrease of heating from the Earth's surface (through radiation, molecular conductivity, turbulent

1.4 Vertical Division of the Atmosphere – Temperature Change

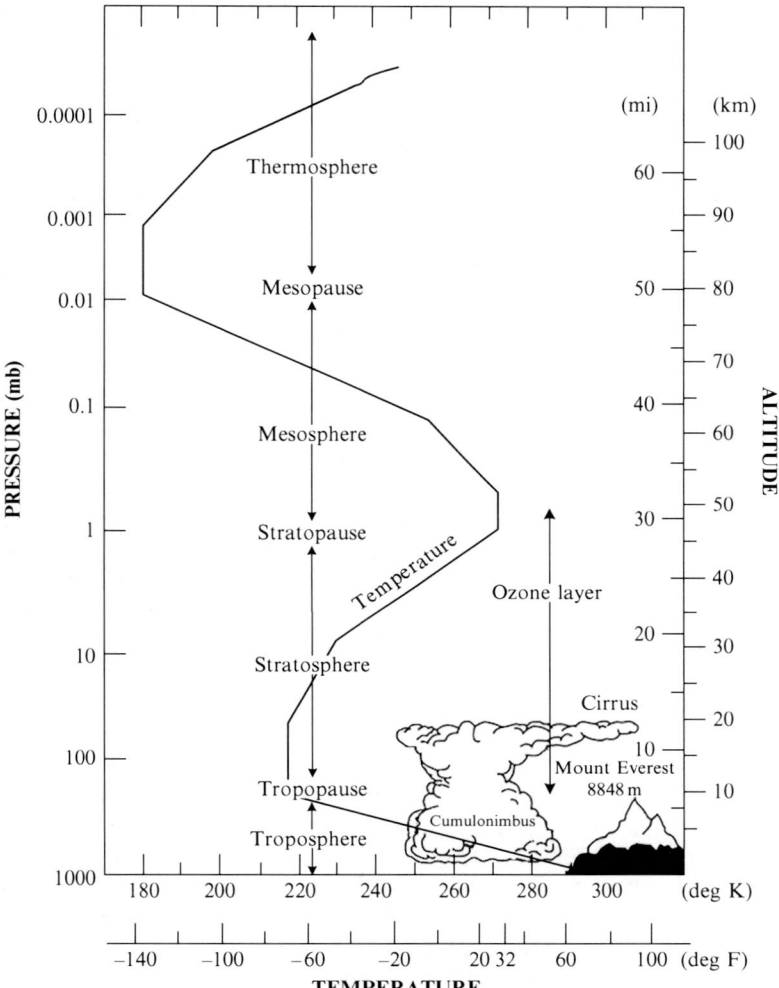

Fig. 1.7 Atmosphere's layers (Adapted from Brausseur et al. 1999)

movement and transport of latent heat from water vapor) and also due to the decrease of air density and water content and due to the temperature decrease of the ascending air arising as it expands.

The main characteristics of the troposphere are:

(a) Continuous and uniform decrease of the temperature with height. The rate of temperature decrease is close to 6.5°C/km.
(b) The wind velocity increases with height since at the lower heights there is the effect of the Earth's surface friction. The maximum velocity is observed at the upper layer of the troposphere.

(c) Almost the whole quantity of water in all three phases (solid, liquid and vapor) occurs in the troposphere with the maximum concentration occurring at the lower layers.
(d) The weather phenomena occur inside the troposphere.

1.4.1.1 Tropopause

The upper region of the troposphere is called the tropopause and it is the zone between the troposphere and the stratosphere. The height of the tropopause changes in relation with the latitude. Therefore, the tropopause can be divided into the tropical and the polar one. The tropical tropopause occurs at low latitudes and reaches the areas with latitude 35–40°, whereas the polar tropopause occurs at higher latitudes.

Quite often the tropical tropopause extends to the latitude 45° and covers the polar tropopause for a length 5–10°. It is therefore possible at these regions for two tropopauses to occur, one above the other, and their vertical distance to fluctuate between 2.5 and 5 km, based on the season and the prevailing weather conditions. It is observed that at these latitudes and at the region between the two tropopauses exists a narrow stream of air with very large velocities known as the Jet-stream.

The mean height of the tropopause at different regions is:

16–17 km at the equatorial regions,
11–12 km at the temperate regions,
7–8 km at the polar regions.

The tropopause height in relation to its mean value is higher above regions with anticyclone systems (high pressure) and lower above regions with cyclone systems (low pressure).

The temperature at the tropopause ranges between -70 to $-80°C$ above the equatorial regions and between $-55°$ to $-60°C$ above medium latitude regions. The temperature width is small and therefore the temperature change versus height is almost invariable inside the layer of the tropopause. The above fact is also an important aspect for the determination of the height and width of the tropopause in the atmosphere.

1.4.2 Boundary Layer

The lower layer of the troposphere is affected from the Earth's surface. The height of this layer is close to 1 km but it is variable during day and is affected also by meteorological conditions. The lower layer of the troposphere, which is affected

1.4 Vertical Division of the Atmosphere – Temperature Change

directly from the Earth's surface and reacts to surface changes at time intervals no larger than 1 h, is called the boundary layer. The changes occurring inside the boundary layer are caused from turbulence, evaporation, heat transfer, topography and emissions of air pollutants. The rest of the troposphere is also affected by changes at the Earth's surface but in a larger time frame.

The daily temperature variation close to the Earth's surface is not observed at higher elevations. Therefore, the daily temperature variation is one of the main characteristics for the determination of the planetary boundary layer. The planetary boundary layer is important for human life and this can be concluded from the various phenomena which occur in the boundary layer as depicted by Stull (1997). Life originates and develops inside the boundary layer as well as the meteorological processes and changes which affect human life and the environment. Absorption of the Sun's heat and air pollution affects inside this layer the human health and ecosystems. Inside the boundary layer occur also the majority of the daily variations of atmospheric circulation which is affected by the Earth's surface topography.

Due to the surface influence the air wind is directed to areas of low pressure, whereas above the boundary layer it is directed parallel to the isobaric lines. The bending of air versus height inside the boundary layer is called an Eckman spiral. The height at which the wind ceases to turn is considered to be the upper level of the boundary layer. Above the oceans the height of the boundary layer changes slowly with respect to location and time. This is due to the slow change of the sea surface temperature which has a daily cycle. The water has large heat capacity and can absorb large quantities of heat without a considerable change of its temperature.

An air volume with different temperature than the ocean will alter its temperature until a thermal equilibrium is achieved at the sea's surface. When a temperature equilibrium is achieved, the height of the boundary layer changes only by 10% at a distance of 1,000 km. More extensive changes occur at areas where two sea streams with different temperatures meet.

The boundary layer both above sea and land tends to be thinner at areas of high pressure and the air is transported to areas of low pressure. At areas of low pressure there is an ascent of air at upper levels of the troposphere. At these conditions it is difficult to determine the upper limit of the boundary layer. The base of clouds is considered as the upper limit of the boundary layer. Therefore above a low pressure system where the clouds are close to the ground the boundary layer height is low.

Above land the height and the structure of the boundary layer are well established as resulting from its daily cycle evolution. The three main parts of the boundary layer that are observed during its daily development are the mixing layer, the residual layer and the stable boundary layer (see Fig. 1.8). Near the Earth's surface there is a thin layer of air which is called a surface layer. It comprises the lowest part of the boundary layer and it is typically 10% of the boundary layer. Finally a very thin layer close to the Earth's surface is called a micro layer (few centimeters height) where the molecular diffusion is the main mechanism for the transport of air.

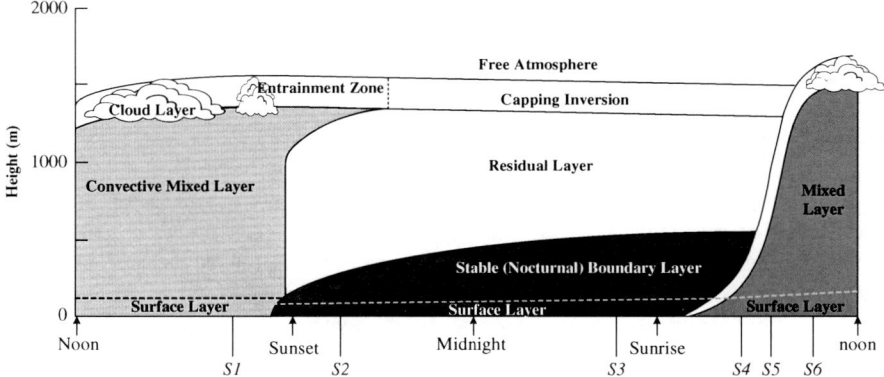

Fig. 1.8 Diurnal variation of the planetary boundary layer in fair-weather conditions. Adapted from Stull (1988)

1.4.2.1 Mixing Layer

The main characteristic of the mixing layer is the presence of turbulent flux. Turbulence in this layer is the result of heat and momentum from the Earth's surface and radiation release from the lower parts of clouds. In the first case there is an upward flow of warm air, whereas in the second case there is a downward flow of cold air.

During days without clouds the mixing layer forms, due to the Earth's surface heating from the Sun's radiation, about half an hour after sunrise. This layer is characterized by intensive mixing of air masses and inside this layer unstable conditions are observed due to vertical movement of warm air masses. The maximum height of the mixing height is observed late in the afternoon. The mixing layer can be increased with the entrainment of air from the free atmosphere. The area at which the entrainment of air occurs is called the entrainment zone.

The mixing layer is a stable layer and acts as a boundary which does not allow further ascent of the warm air masses and consequently constrains turbulence. In many occasions in this layer temperature inversion occurs and in many cases this layer is called an inversion layer. The wind velocity in the mixing layer is lower than the geostrophic wind and the wind direction has an angle in relation to the isobaric lines due to the friction effect from the Earth's surface. The vertical wind profile shows a logarithmic increase from the surface to the top of the mixing layer with velocity near to the surface close to zero. The angle between the wind and the isobaric lines increases close to the surface with an average value close to 30° and in special occasions exceeds 45°.

The major air pollution sources are located near the Earth's surface and therefore the concentration of air pollutants has higher values in the mixing layer compared to the free atmosphere. The air pollutants are transported with the warm air masses which however cannot entrain the stable atmospheric layer located above the mixing layer and are trapped. The trapping of air pollutants from an inversion

1.4 Vertical Division of the Atmosphere – Temperature Change

layer is a common phenomenon at areas of high pressure and results in elevated concentrations of pollutants which often exceed the air quality standards adopted from public organizations.

The presence of clouds results in a decrease of the ascending warm air masses. Therefore during days with the presence of dense clouds the mixing layer develops in height at a lower rate and at this layer can occur stable conditions.

1.4.2.2 Residual Layer

There is no further formation of thermal fluxes half an hour after the sunset and there is also a gradual decrease of air turbulence. Rather, a well-mixed layer of air is formed and is called a residual layer, since the air pollution levels there are the same as at the mixing layer. For example, if we suppose that the pollutant transport is not important, the inert air pollutants which will be emitted during the day will remain in the residual layer during the night. The plumes emitted from stacks under these conditions will diffuse in the same manner towards the horizontal and vertical directions forming plumes of conical shape. Inside the residual layer the temperature decreases with height.

The non-inert pollutants react chemically with other chemical compounds and form secondary pollutants. In specific cases there is the formation of new particles originating from gaseous species (nucleation). These particles will eventually be deposited on the Earth's surface through dry or wet deposition. The residual layer remains also for a short period after sunrise and therefore favors photochemical reactions inside this layer.

The residual layer is not in direct contact with the Earth's surface. During the night the nocturnal layer increases its height and therefore the base of the residual layer is changed. The remaining part of the residual layer is not affected from the transport due to turbulence and the surface's effect and therefore it is not part of the boundary layer. However, many studies include the residual layer into the boundary layer.

1.4.2.3 Nocturnal Boundary Layer

During night the lower part of the residual layer forms a stable layer of air due to influence from the Earth's surface. This layer is considered stable with some small influence from turbulence. Even though the air close to the surface has low velocity, the wind at higher elevations has velocity even higher than the geostrophic wind and this phenomenon is called a nocturnal jet.

The statistically stable layer reduces the turbulence but the presence of the nocturnal jet causes air streams which result in turbulent movements. Therefore locally intense turbulence can appear locally in this stable layer. Unlike the mixing layer during the daytime, the stable layer during the night does not have a specific upper limit. The upper limit is defined as the top of the stable layer or instead the height at which the turbulence intensity is low compared to the surface. Pollutants that are emitted inside the stable layer at night are dispersed very slowly in the vertical direction. This behavior is called fanning. The wind velocity profile is quite complex during night. Just above the Earth's surface winds are weak. At 200 m height from the surface wind velocities reach values between 10 and 30 m/s. A few 100 m higher the wind velocity is reduced and is close to the geostrophic wind value. The wind direction at the height of the nocturnal jet has intensive variations versus time, whereas at lower elevations the wind direction is parallel to the isobaric lines.

This stable layer is possible to be formed also during the day in cases that the Earth's surface is colder than the air above it. These conditions are observed during a passing of a warm air stream above a cold surface (warm front) or close to coast lines.

1.4.3 Stratosphere

The atmospheric layer which exists above the troposphere and is divided from it by the tropopause is named the stratosphere. The stratosphere extends to the height of 50–55 km (see Fig. 1.7). The air temperature from the tropopause and for the next 20 km does not show a considerable change with height. Above this elevation the temperature increases until the stratopause where it reaches a temperature close to $0°C$. The temperature increase with height is mainly due to the absorption of the ultraviolet Sun radiation (with wave lengths between 2,000 and 3,000 Å) (1 Å = 10^{-10} m) from ozone. The ozone concentration is important in the atmospheric layer which extends from a height of 15 to 40 km (ozonosphere). This process is the main mechanism for the increase of the stratosphere's temperature. The stratosphere is considered as a more stable layer compared to the troposphere since the temperature increases with height in this region. This does not mean that the stratosphere is a calm region. To the contrary, during winter and spring seasons violent temperature changes occur, mainly at high latitudes.

Often the stratosphere's region to the height of 35 km is called the "lower stratosphere", whereas the rest is called the "upper stratosphere". A common feature of the lower stratosphere is its relatively low temperature values and its high drought. This can be explained by the fact that the low temperatures (between -40 and $-50°C$) in this layer do not allow the occurrence of humidity at considerable amounts.

1.4.3.1 Stratopause

The zone between the stratosphere and the atmospheric layer above it is called the stratopause. This layer exists at a height between 50 and 55 km and is almost isothermal. In this layer is observed also the maximum temperature which occurs in the stratospheric layer.

1.4 Vertical Division of the Atmosphere – Temperature Change

The atmospheric pressure in the stratopause is close to 1 mb. Since the average pressure at sea level is 1,013.25 mb, this means that 99.9% of the atmospheric mass is concentrated in the troposphere and stratosphere.

1.4.4 Mesosphere

The atmospheric layer which follows after the stratopause and extends up to 80–85 km is called the mesosphere. A main feature of the mesosphere is an abrupt drop of the temperature versus height, which reaches values close to $-90°C$ or lower at the upper zone. The temperature decrease is due to the absence of ozone.

1.4.4.1 Mesopause

The mesopause consists of the upper limit of the mesosphere and is the coldest area of the atmosphere with temperatures as low as $-100°C$. The mesopause is also the upper limit of the homosphere.

1.4.5 Thermosphere

The thermosphere is the atmospheric layer which follows after the mesopause and extends up to the height of 400 km. Except at the isothermal base, the temperature increases monotonically with height and reaches $700°C$ or even higher depending on the Sun's activity. The parameters which are responsible for the temperature increase in the thermosphere are:

1. The large dilution of air in these heights.
2. The non-existence of molecules with three bonds.
3. The presence of Sun radiation at wave lengths smaller than 1,750 Å.
4. The energy which is released from exothermic chemical reactions.

The elevated values of the temperature confirm the absence of processes which will cool the thermosphere. The only cooling process occurs with the conduction of heating downwards.

The boundary at which the monotonic temperature increase of the thermosphere ceases is called the thermopause.

1.4.6 Exosphere

Immediately after the thermopause the atmosphere becomes isothermal and its upper layer is called the exosphere. The base of the exosphere is between 400 and 500 km and this is dependent in a large extent on the Sun's activity. The mean free path of molecules (the distance through which a molecule is transported freely until it collides with another molecule) in the exosphere is very large (average value close to 1.6 km). At these heights, with a large thermal conductivity, the neutral gaseous molecules can escape from the Earth's gravitational field.

1.4.7 Ionosphere – Magnetosphere

With respect to ions, there are two main atmospheric regions, the ionosphere and the magnetosphere. Ions are atoms and molecules that have gained or lost electrons. The ionosphere is the atmospheric region in which exists partial ionization of the atmospheric constituents from the Sun's radiation. It extends from the height of 60 km up to 300 km where the density of charged particles is maximum.

At specific heights the ions and the free electrons have higher density and form distinct layers. The most important are the regions D (60–90 km), E (90–150 km), F_1 (150–250 km) and F_2 (250–350 km). These layers play a major role in radio communications. Specifically the region D reflects AM radio waves and at the same time weakens them through their reflection.

Finally, the magnetosphere is the atmospheric region in which the movement of ions is dependent on the Earth's magnetic field and extends from 1,000 km to a height close to 10 radii of Earth's height at its illuminated part, a region which is called the magnetopause.

1.5 Change of Meteorological Parameters with Height

The vertical structure of the atmosphere is continuously changing due to several physical parameters. In meteorology, three main parameters are studied in detail:

(a) Air temperature,
(b) Air density,
(c) Air pressure.

Since atmospheric air is thin, only small quantities of heat are absorbed from the energy received from the Sun. On the contrary, the Earth's surface is heated more effectively and therefore the layers of air which are located close to the surface are heated quicker and more efficiently than the air volume which is located at an elevated height. As the air is heated it becomes lighter and is transported inside the atmosphere. Therefore heat and mass transport occur at elevated heights inside the atmosphere. At the same time, due to the heterogeneity of the Earth's surface and the unequal heating of the Earth's surface, air movements are developed vertically and horizontally which results in mixing of warm and cool air masses and, generally, the transfer of heat.

The atmospheric air can be become warmer or cooler from purely mechanical causes. These changes are called potential temperature changes. The adiabatic

1.5 Change of Meteorological Parameters with Height

changes of temperature occur continuously inside the atmosphere. When an air mass moves aloft inside the atmosphere, then it moves to an area with lower pressure and therefore is expanded and becomes cooler. The cooling of the air mass is a result of the air expansion where thermal energy is used and is transferred to kinetic energy. In this stage the air mass is not exchanging heat with the atmospheric air with which it is encircled. Rather, when the air mass is transported to an area with a higher pressure (descent of the air mass in the atmosphere) it is compressed and contracted, which leads to its heating. In this case there is also no heat exchange with the surrounding environment.

The rate of temperature change versus height is expressed with a vertical lapse rate. This is defined as the temperature decrease of the atmospheric air (∂T) per unit height (∂z) and it is expressed as:

$$\gamma = -\frac{\partial T}{\partial z}. \tag{1.5}$$

The negative sign ($-$) is used to express the temperature reduction with height inside the troposphere and γ is expressed as °C/100 m or °C/1 km. The moisture in the atmosphere affects the lapse rate and therefore its value for both dry and moist conditions must be examined.

(a) Dry vertical adiabatic lapse rate

The dry vertical lapse rate is equal to 1°C/100 m. In this theoretical condition the air mass which ascends inside the atmosphere is cooled adiabatically and does not contain any moisture. However, the air masses inside the troposphere contain moisture and it is accepted to use an average lapse rate for humid air in the troposphere equal to: $\gamma = -0.6°C/100$ m or $\gamma = -6°C/km$. At the atmospheric layers close to the Earth's surface (height of a few centimeters) the lapse rate has much higher values close to $\gamma = -1.8°C/10$ m.

(b) Moist vertical adiabatic lapse rate

The moist vertical adiabatic lapse rate takes place in an air mass which includes water vapor and during its ascent it is cooled adiabatically. It is smaller than the dry lapse rate since, during the condensation of water vapor, it releases heat (latent heat) and therefore the rate of temperature decrease with height becomes smaller. The average value of the moist lapse rate is:

$$\gamma = -0.5°C/100m. \tag{1.6}$$

The vertical temperature lapse rate is in close relation to the atmospheric stability. A more detailed treatment is given in chapter 2. If we suppose that the lapse rate keeps an almost constant value with an integration of the Eq. (1.5) it can be derived that:

$$T_{(z)} = T_{(zo)} - \gamma(z - zo) \tag{1.7}$$

where, $T_{(z)}$ and $T_{(zo)}$ are the temperature values at heights z and z_o respectively.

From Eq. (1.7) one can derive the temperature at height z when the temperature at height z_o is known. Above the height of 60 km the air temperature cannot be measured with normal thermometers. At these conditions the temperature is calculated from the corresponding value of the atmospheric pressure (p) and the molecular weight (M) of air, using the ideal law of gasses:

$$p = \rho \frac{R}{M} \cdot T \quad \text{and} \quad p = \rho R_a T \tag{1.8}$$

where, R is the universal gas constant (8.314 J mol^{-1} K^{-1}) and R_α ($\frac{R}{M}$) is the specific constant of gasses, which for dry air is equal to 287.05 J kg^{-1} K^{-1}.

As shown in Fig. 1.7 the temperature and pressure are varied with height. The temperature change is close to 1°C per 100 m, as discussed previously, and as a result the transport of a volume of air from the Earth's surface up to 1 km height has a temperature decrease from 5 to 10°C with respect to the air vapour content. Since the vapor saturation pressure is dependent drastically on the temperature, a temperature decrease with height is accompanied by an increase of the relative humidity (RH) inside the air volume. As a result the transport of air vertically by a few 100 m results in saturation (RH = 100%) and supersaturation of water vapour. The final result is cloud formation.

The inhomogeneous distribution of heating at the Earth's surface due to the Sun's radiation results in air circulation from the equatorial region to the poles. The Earth's rotation affects considerably air circulation and air transport contains both heat and mass transport. The general circulation is affected also from the local topography (e.g. mountains), the heat release due to the cloud formation as well as from local weather systems. Furthermore, the chemical compounds in the atmosphere are distributed geographically from the atmospheric circulation. Figure 1.9 shows typical vertical distributions of specific gaseous components (important for the atmospheric chemistry) in the atmosphere.

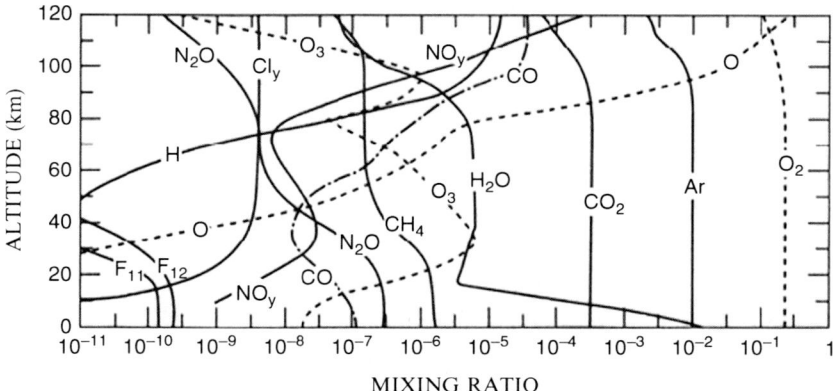

Fig. 1.9 Typical vertical concentration of chemical species in the atmosphere (F_{11} = CFCl$_3$ and F_{12} = CF$_2$Cl$_2$)

1.5.1 Temperature Inversion

There are times when the air temperature in areas in the lower troposphere increase with height rather than decrease. This phenomenon is called temperature inversion. The layer inside which this phenomenon occurs is called the inversion layer. This layer is characterized by the height of the inversion base and its own height (Fig. 1.10).

The temperature inversions can be divided according to the height at which they occur as:

Surface inversions. These kinds of inversions occur due to cooling of the Earth's surface during the night. It is known that during the night the Earth's surface emits radiation with large wave lengths (Earth's radiation) to the atmosphere and as a result it is cooled. The air molecules that are close to the Earth's surface transport heat through conduction and as a result the lower layers of air are cooled. Due to the small thermal conductivity of air, the cooling is observed at low heights with respect to the Earth's surface. Through this process an inversion temperature layer is formed with a base at the Earth's surface and top which reaches heights usually from 100 to 400 m (Fig. 1.10). Ideal weather conditions which favor the formation of surface inversion are clear weather (without clouds) during the night and the low wind speed (<3 m/s). The surface inversions are called also radiation inversions. These inversions are more intensive during the winter period since the nights have a longer duration.

Intensive Sun radiation during the day results in the release of large quantities of heat at the Earth's surface which leads also to warming of the lower layers of air.

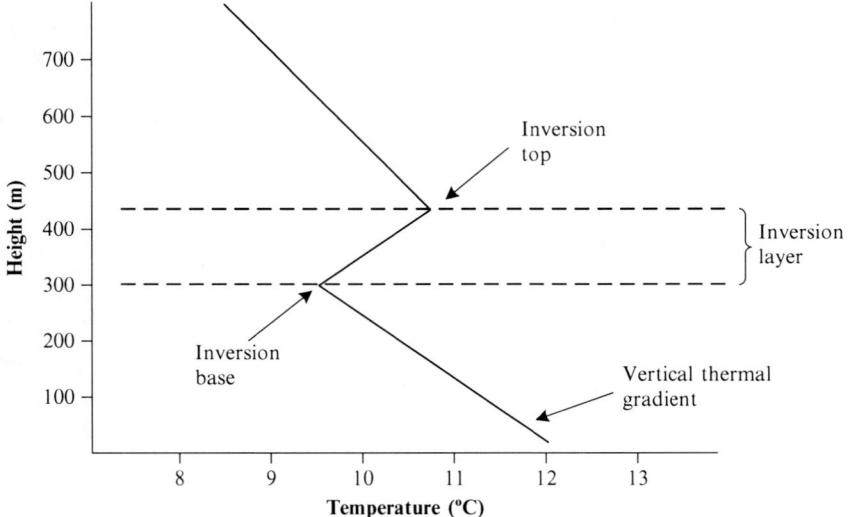

Fig. 1.10 Schematic representation of the temperature inversion

This results in a break of the inversion layer. During the presence of a high pressure field in the atmosphere, a break of the inversion layer during the day is not favoured.

It is also possible for a surface inversion to be formed during the day, especially over areas which are covered with snow. In these cases the air masses which are in contact with the snow-covered surfaces use large quantities of heat for snow melting. As a result the lower layers of air become colder than the elevated layers and thus the creation of temperature inversions is favoured.

Subsidence inversions. The subsidence inversions are caused due to the descending movement of cold air masses from the upper atmosphere to the Earth's surface. At this descending movement the air is compressed and is heated adiabatically with a consequently decreasing relative humidity. After the air mass descent, its upper layer is warmer than the lower part. With the above mechanism an elevated temperature inversion is formed which is called subsidence inversion and has similar structure with the surface inversion. Contrary to the surface inversions which last a few hours, the subsidence inversions last a few days or even longer. For this reason the subsidence inversions are related to air pollution episodes.

Frontal inversions. Frontal inversions are formed when, at a specific height in the lower troposphere, there occurs a transport of warm air masses which override a cold air layer which extends to the Earth's surface (Fig. 1.11). One of frequently occurring inversions of this kind are frontal inversions occurring at the weather fronts. Weather fronts will be examined in Chapter 2. These inversions occur due to horizontal transport of warm air above a colder air layer (warm front) and also due to transport of cold air under a warm layer (cold front). Frontal inversions have implications for air pollution in the case of warm fronts, since the warm fronts move slowly and have a small gradient of their frontal area with the Earth's surface, not allowing therefore the vertical transport of pollutants.

Thermal inversions are not very deep. The temperature inversion may occur for a few meters or a few 100 m from the Earth's surface. However, it is an important process since it is directly related to other weather phenomena such as fog and visibility reduction but also with problems related to air pollution. In general, above large urban agglomerations and industrial regions with a temperature inversion, an

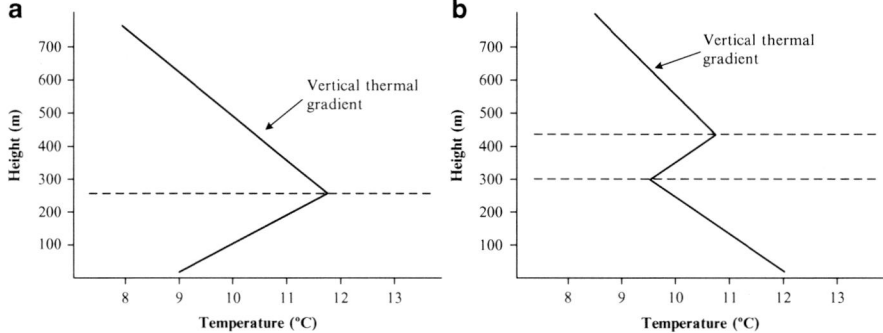

Fig. 1.11 Temperature inversions (a) surface - subsidence (b) frontal

obstacle is formed that does not allow vertical transport, as discussed previously. The atmosphere's dynamics as well as the air pollution transport in relation with temperature inversions is examined in more detail in Chapter 2.

1.5.2 Air Density Variation with Height

The equation that describes the law of ideal gasses (1.8) can be used for calculation of the density variation versus height:

$$dp = \rho R_\alpha dT + R_\alpha T dp. \tag{1.9}$$

Assuming that there is hydrostatic equilibrium in the atmosphere, the change of pressure p versus height z is described with the hydrostatic equation (see Appendix 1):

$$dp = -\rho g \, dz \tag{1.10}$$

where g is the gravitational acceleration constant.
With the help of the hydrostatic equation the Eq. (1.9) can be written as:

$$-\rho g dz = \rho R_\alpha dT + R_\alpha T d\rho \tag{1.11}$$

Dividing the above equation with the density (ρ) it results that:

$$-gdz = R_\alpha dT + R_\alpha T d\rho/\rho \Rightarrow$$

$$\frac{d\rho}{\rho} = -\frac{g}{R_a T} dz - \frac{dT}{T} \Rightarrow$$

$$d(\ell n \rho) = -\frac{g}{R_a T} dz - d(\ell n T). \tag{1.12}$$

With the hypothesis that the atmospheric layer which is examined is isothermal ($T = \bar{T}$ = constant), then $d(\ell n T) = 0$ and the Eq. (1.12) can be written as:

$$d(\ell n \rho) = -\frac{g}{R_a \bar{T}} \times dz. \tag{1.13}$$

Integrating Eq. (1.13) from the height zero to the height z with corresponding values for density ρ_0 and $\rho_{(z)}$ it can be concluded that:

$$\int_{\rho 0}^{\rho} d(\ell n \rho) = -\frac{g}{Ra \cdot \bar{T}} \int_{0}^{z} dz \Rightarrow \ell n \frac{\rho_z}{\rho_0} = -\frac{g}{Ra \cdot \bar{T}} \times z \Rightarrow$$

$$\rho_z = \rho_o \, e^{-\left(\frac{gz}{R_a \bar{T}}\right)}. \tag{1.14}$$

From the above equation, we see that, in an isothermal atmosphere, the air density decreases exponentially versus height.

In the troposphere, where the average temperature is close to 250 K ($\bar{T} = 250\,^oK$), Eq. (1.14) can be expressed as: $\rho_z = \rho_0 \times 10^{-\left(\frac{z}{17}\right)}$. This means that the air density is decreasing in the troposphere (in the same way as the pressure) with a factor of 10 every 17 km of height.

1.5.3 Change of Atmospheric Pressure with Height

The change of the atmospheric pressure with height can be described with the hydrostatic equation (see Appendix 1):

$$\frac{dp(z)}{dz} = -\rho(z)\,g \tag{1.15}$$

where, $\rho(z)$ is the air density at height z.

From the law of ideal gasses:

$$p(z)\,\frac{M_{air}}{\rho(z)} = R\,T(z) \Rightarrow \rho(z) = \frac{M_{air} p(z)}{R\,T(z)} \tag{1.16}$$

where, M_{air} is the mean molecular weight of air (28.97 g mol^{-1}), R is the gas constant (8.314 J mol^{-1} K^{-1}) and T is the temperature in Kelvin.

From Eqs. (1.15) and (1.16) we get that $\frac{dp(z)}{dz} = -\frac{M_{air}\,g\,p(z)}{R\,T(z)}$. The term $H(z) = \frac{R\,T(z)}{M_{air}\,g}$ is the characteristic length for the pressure decrease with height. Introducing the term $H(z)$, the above equation can be expressed as $\frac{1}{p(z)}\frac{dp(z)}{dz} = -\frac{1}{H(z)} \Rightarrow \frac{d\ln p(z)}{dz} = -\frac{1}{H(z)}$.

The above equation can be solved using the hypothesis that the temperature remains constant versus height. This is a reasonable approach since, if the temperature is reduced by a factor of 2, the pressure is reduced by six orders of magnitudes. If the temperature remains constant the characteristic length is independent of height (H(z) = H = constant) and a simple exponential expression results for the pressure relation versus height:

$$\int_{P_o}^{P(z)} d\ln(p(z)) = -\int_0^z \frac{dz}{H(z)} = -\frac{z}{H} \Rightarrow \ln p(z) - \ln p_o = -\frac{z}{H} \Rightarrow \frac{p(z)}{p_o} = e^{-\frac{z}{H}}.$$

$$\tag{1.17}$$

Equation (1.17) can be applied for several atmospheric gasses. It can be shown that gasses heavier than air have a smaller characteristic length (e.g. Xe and Kr). It must be noted that the above derivation for the characteristic length is valid at high elevations (above 120 km) since at lower heights the presence of turbulence modifies the results derived.

In the troposphere the average temperature is close to 250 K ($\bar{T} = 250$K) and Eqs. 1–14 can be written as:

$$P(z) = P_o \times 10^{-(\frac{z}{17})}. \tag{1.18}$$

Consequently the atmospheric pressure (also the density) inside the troposphere is lowered with a factor of 10 for every 17 km of height. Based on Eq. (1.18) one can calculate the reduction of the atmospheric pressure at Mean Sea Level:

$$P_o = P(z) \times 10^{\frac{z}{17}}. \tag{1.19}$$

Therefore the pressure is decreased by 1 mm Hg every 10–11 m.

1.6 Model of the Standard Atmosphere

Inside the atmosphere it is not possible to examine in detail the changes of all the physical parameters in relation to height, since in a very small time interval changes may occur in one parameter which affects the others. Therefore in order to solve complex meteorological problems, particularly in the construction of prognostic weather models, some simplifications and approximations are addressed in relation to the variability of atmospheric parameters versus height. In reality, inside the atmosphere not only the temperature, the pressure and the density but also the gravitational acceleration and the mean air composition are changing versus height. The applied simplifications for the changes of parameters produce several atmospheric models which are characterized by their simplicity and the application of the hydrostatic and the ideal law of gasses.

Even though the atmospheric models are continuously improving, there is no model which can describe in detail the real atmosphere. For this reason the International Civil Aviation Organization (ICAO) has adopted the model of the standard atmosphere which is an approximation of the average annual atmosphere at all latitudes. The standard atmosphere model has applications to the calibration of airplane instruments and is also an important reference for the calculation of several meteorological parameters inside the atmosphere.

For determination of the standard atmosphere the following conditions have been adopted:

1. The atmosphere is dry with stable composition and an average molecular weight equal to 28.9644.
2. The atmosphere behaves as an ideal gas.

3. The gas constant for dry air is equal to 8.314 J mol^{-1} K^{-1}.
4. The gravity acceleration at mean sea level is equal to $g_o = 9.8067$ m s^{-2}.
5. The hydrostatic equilibrium is valid.
6. The freezing point for water under pressure conditions of 1 Atm is equal to 275.16 K.
7. The air density at mean sea level (height equal to 0 m) is equal to 0.001250 g cm^{-3}.
8. The temperature at mean sea level is equal to $T_0 = 288{,}16$ K or $T_0 = 15°$C and the atmospheric pressure is equal to $P_0 = 1{,}013.25$ mbs $= 1$ Atm.
9. The vertical value of the temperature lapse rate in the troposphere (up to a height of 11 km) is equal to 0.65°C/100 m or 6.5°C/km. For heights from 11 km to 20 km the temperature remains constant and equal to $-56.5°$C. Furthermore up to the height of 32 km the temperature lapse rate is equal to 1°C/km.

As a result of the above approximations it can be concluded that between the surface (height of 0 m) and the height of 11 km, the temperature (°C) at height z is given by the relation:

$$T(z) = 15 - 0.0065 \times z \tag{1.20}$$

whereas the height z versus pressure (p) can be calculated from the following equation at heights up to 11 km:

$$z = 44.308 \left\{ 1 - \left(\frac{p}{1013.25}\right)^{0.19023} \right\} \tag{1.21}$$

and for heights above the 11 km the expression can be written as:

$$z = 11\text{km} + 6381.6 \times \ln\frac{2345}{P}. \tag{1.22}$$

1.6.1 Units of Chemical Components in the Atmosphere

The chemical components i in the atmosphere are expressed with the arithmetic density (particles per m^3) and the mass density (kg m^{-3}). In addition the term of volumetric mixing ratio is also used:

$$c_i = \frac{n_i}{n_a} \tag{1.23}$$

where n_α is the arithmetic air density. The density of the mass mixing is given by:

$$c'_i = \frac{\rho_i}{\rho_a} \tag{1.24}$$

where ρ_α is the mass density of air.

In dimensionless units the concentrations are expressed as parts per million (ppm), parts per billion (ppb) and parts per trillion (ppt) which correspond to volumetric mixing ratios of 10^{-6}, 10^{-9} and 10^{-12} respectively.

The integrated vertical arithmetic density (m^{-2}) is called vertical column abundance. For ozone, the units Dobson (DU) are used. One DU corresponds to height (in units 10^{-3} cm) which an ozone column has if the gas is under normal pressure and temperature conditions. One DU is equivalent to 2.687×10^{16} molecules cm^{-2}.

1.6.2 Unit Conversion of Concentration of Component I ($\mu g/m^3$) to Volume Concentration (Ppm)

When the concentration of a chemical compound c_i is expressed at $\mu g/m^3$, the molecular concentration (mol/m^3) is given by:

$$moles/m^3 = \frac{10^6 c_i}{M_i} \tag{1.25}$$

where M_i is the molecular weight of specie i.

The total molecular air concentration at pressure p and temperature T is equal to $c = \frac{p}{RT}$, and therefore:

$$\text{Mixing ratio of species i in ppm} = \frac{RT}{pM_i} \times m_i.$$

If the temperature T is given in Kelvins and the pressure in Pascals, the above relationship can be expressed as:

$$\text{Mixing ratio of species i in ppm} = \frac{8,314T}{pM_i} \times c_i.$$

1.7 Radiation in the Atmosphere

Energy is the basis for life and is present in different forms. All bodies in the universe that have pressure above 1 mb and temperature above absolute zero (0 K) emit energy in the form of electromagnetic waves. The emission and propagation of energy with electromagnetic waves is called radiation. This nature of radiation (particles and waves) gives the possibility to describe it as flux of particles (photons) or as an electromagnetic wave. The electromagnetic waves do not need molecules for their transfer and can be transferred even in a vacuum. On the contrary, heat needs molecules for its transfer which occurs with conduction. The basic equation which relates the wave length λ of radiation with its frequency and the velocity of light c ($c = 2.9979 \times 10^8$ m s^{-1}) is $\lambda = \frac{c}{v}$.

The energy which the Earth receives originates mainly from the Sun. The average Earth's temperature remains almost constant during a year and this implies that there is an energy balance between the receiving and the emitting energy on the Earth and its atmosphere. Due to the elliptical orbit of Earth around the Sun, the distance between Sun and Earth is not constant. The closest distance occurs in December when the Earth is at perihelion and its distance from the Sun is 147×10^6 km. In June the Earth is at aphelion which corresponds to the maximum of the elliptical orbit and its distance from the Sun is equal to 52×10^6 km. The Earth completes a rotation during a calendar year (365 dates). At the same time, due to the pitch of the Earth's rotation axis, there is a difference between the Sun's radiation that the Earth receives in the Northern hemisphere and that received in the Southern hemisphere. This difference in the Sun's radiation balance is responsible for the appearance of the four seasons on the Earth.

The radius of the Sun is equal to 695.7×10^3 km and its volume 1.412×10^{27} m³. The volume of the Sun is 1,306,000 times larger. If someone imagines the Sun as a sphere with radius of 3 m, then the Earth resembles a ping

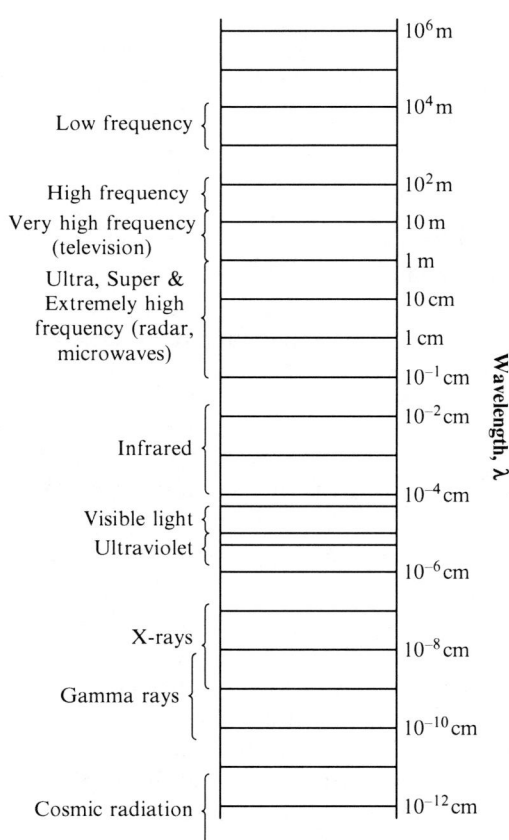

Fig. 1.12 Electromagnetic radiation spectrum

pong ball. The Sun's mass is $1{,}991 \times 10^{33}$ g and therefore 333,420 times bigger than the Earth's mass.

The Sun is composed of different layers and the most important are:

(a) Nucleus: It contains 0.9% of the total mass of the Sun and is in a vapour phase with a temperature equal to 15×10^{6}°C.
(b) Chromosphere: It extends approximately up to a height of 15,000 km.
(c) Corona: Extends above the chromosphere with a temperature that extends to 15×10^{5}°C.
(d) Photosphere: It is the visible surface of the Sun with a height of 200–300 km. The temperature at its external surface is 6,500°C. The photosphere comprises the main source of radiation and on its surface exist dark blemishes which are called sunspots and are huge (on average more than five times the Earth's diameter) and cooler regions. Sun spots have an important significance since they constitute a criterion of the total Sun activity. Several researchers have attempted to correlate the periodicity of the sun spots (close to 11 years) with the periodic change of physical variables in the Earth's atmosphere and in particular with the atmospheric temperature.

The Sun emits electromagnetic radiation and charged particles. Figure 1.12 shows the various wave lengths of radiation λ and its spectrum which is called a radiation spectrum. The Sun emits energy from the whole electromagnetic spectrum but the major part of this energy is in a narrow spectrum between 400 and 700 nm, which is called the visible part of the electromagnetic radiation. The region of the spectrum between 200 nm and 10^4 nm is the main region of the Sun's radiation.

1.7.1 Laws of Radiation

The energy which is incorporated inside an object is a result of electron vibrations inside atoms. The wave length of the radiation which is emitted from an object is dependent directly on its temperature. A higher temperature results in quicker electron vibration and shorter wave length. The radiation is produced during the electron movement to lower energy level. The energy difference during the electron jump defines the frequency of the emitted radiation and is given by the relation $\Delta E = h \times v$, where h is Planck's constant ($h = 6.626 \times 10^{34}$ J s).

The radiant ability or emission ability E of a body for a specific wave length λ and temperature T is the radiation energy emitted from the body per unit surface. This is measured in $W\ m^{-2}$ or $cal\ m^{-2}\ min^{-1}$ units. It has to be noted that the radiant ability for a specific wave length region is independent of the temperature of the environment and is dependent only on the nature and temperature of the body.

When radiation is transferred to a body, analogously to the radiation wave length and the body's structure it is possible to have the following processes:

(a) Absorption (partial) of the radiation which will result in an increase of the body's internal energy which will produce a temperature increase. The

percentage of the absorbed energy is called absorption capacity or absorption coefficient and is dependent on the wave length of the radiation and the body's properties. The absorption coefficient is represented with a symbol *a* and it is dimensionless and generally smaller than unity. Concerning the human body, there is absorption from ultraviolet radiation (UV) beneath the outer layer of the skin. Certain cells in the human body produce as a reaction a dark pigment (melanin) that absorbs part of the ultraviolet radiation. Human exposure to UV may result in malignant melanoma which is the most deadly form of skin cancer.
(b) Reflection (partial) of the radiation from the body. The percentage of the reflecting radiation is called reflection ability or reflection coefficient and has a symbol *r*.
(c) Penetration (partial) of the radiation inside the body and has a symbol *t*.

The relation which connects the above three coefficients can be written as:

$$a + r + t = 1. \tag{1.26}$$

If a body has a = 1 (r = 0 and t = 0), then the body is called a black body and absorbs the total amount of radiation for all wave lengths and temperatures. The real bodies have a $\neq$ 1 and are selective absorbers. Bodies which do not absorb radiation are called white bodies. The radiation of the black bodies can be described from the five laws of radiation which are presented below.

1.7.1.1 Kirchhoff's Law

Kirchoff's law refers to the ratio of the incoming to the emitting radiation. When a body is in thermodynamic equilibrium and receives radiation with energy E and absorption coefficient α, then the energy of absorption is equal to $E' = \alpha \times E$. Since the body is in thermodynamic equilibrium it emits the same energy E'. For a black body there is the same absorption of energy E as the emission energy. When a body has similar characteristics to those of a black body with emission coefficient $\varepsilon = E'/E$, then $\alpha = \varepsilon$, which is also the mathematical expression of Kirchoff's law. Kirchoff's law refers to the principle that, when a body absorbs energy over a specific spectrum then it emits it in the same spectrum. Therefore the law says that good absorbers are also good emitters and bad absorbers are also bad emitters at specific wave lengths. This law is valid for the whole scale of the radiation spectrum in the case of gasses.

An example is the case of snow in a forest during winter. The snow is a good absorber and emitter of energy in the infrared (in reality the snow behaves as a black body in the infrared). The tree's trunk absorbs the Sun's radiation and emits energy in the infrared which is partly absorbed by the snow and as a result the absorbed radiation is transferred to internal energy and the snow melts close to the trees.

1.7 Radiation in the Atmosphere

1.7.1.2 Planck's Law

Planck's law correlates quantitatively with the spectrum emission coefficient of the black body $\varepsilon_\lambda(T)_\mu$ with the absolute temperature T:

$$\varepsilon_\lambda(T)_\mu = \frac{c_1 \lambda^{-5}}{\exp\left(\frac{c_2}{\lambda T}\right) - 1} \tag{1.27}$$

where $c_1 = 3.7415 \times 10^{-16}$ W m^2 and $c_2 = 1.4388 \times 10^{-2}$ m K.

1.7.1.3 Wien's First Law

Wien's first law says that the product of the absolute temperature of a black body and the maximum wave length (λ_m) of the emitted radiation is constant:

$$\lambda_m T = 2897.8 \, (\text{K} \, \mu\text{m}).$$

1.7.1.4 Wien's Second Law

If in the expression of Planck's law which calculates the spectrum emission coefficient the wave length is equal to the maximum wave length (λ_m), then the expression for the maximum spectrum emission coefficient is equal to:

$$\varepsilon_\lambda(T)_\mu = 1.286 \left(\frac{T}{1000}\right)^5 \tag{1.28}$$

which comprises the mathematical expression of Wien's second law.

1.7.1.5 Law of Stefan-Boltzmann

Stefan-Boltzmann's law determines the value of the total (integral) emission spectrum coefficient of a black body $\varepsilon_\lambda(T)_\mu$ (W/m^2) with an integration of the emission spectrum coefficient from Planck's law at all wave lengths:

$$\varepsilon(T)_\mu = \int_0^\infty \frac{c_1 \lambda^{-5}}{\exp\left(\frac{c_2}{\lambda T}\right) - 1} d\lambda; \tag{1.29}$$

setting $x = c_2/(\lambda T)$ we get that $\varepsilon(T)_\mu = -\frac{c_1}{c_2^4} T^4 \int_0^\infty \frac{x^3}{e^x-1} dx = \frac{c_1 \pi^4}{15 c_2^4} T^4$.
The above expression can be written as:

$$\varepsilon(T)_\mu = \sigma T^4 \tag{1.30}$$

where $\sigma = 5.672 \times 10^{-8}$ W/m² K⁴ (Stefan-Boltzmann's constant) and it is the mathematical expression of Stefan-Boltzmann's law. In order to calculate the net quantity of radiation emission when the atmosphere has a temperature T_o, the Stefan-Boltzmann law is written as $\varepsilon(T)_\mu = \sigma \left(T^4 - T_o^4\right)$. In the case that the temperature of the environment is much lower than that of the emitting object, the mathematical expression of Stefan-Boltzmann's law is given by the equation:$\varepsilon(T)_\mu = \sigma T^4$.

The Sun and Earth, acting as black bodies, absorb almost all the received energy. However, the atmosphere has a selective absorption of the received energy from the Sun and Earth's surface and therefore cannot be characterized as a black body.

1.7.2 Sun's Radiation

The term solar radiation or solar energy refers to the radiation which the Earth receives from the Sun. The quantity of the solar energy which is emitted from the Sun's surface is 3.91×10^{26} W s⁻¹. The energy is produced from the thermonuclear fusion of four atoms of hydrogen to one atom of helium. With this continuous process 657 million tonnes of hydrogen are transformed to 652.2 million tonnes of helium per second. The difference of 4.5 million tons per second corresponds to the emitted energy from the Sun. From this energy of 3.91×10^{26} W only 1.8×10^{16} W reach the Earth's atmosphere. Mathematical calculations estimate that on a yearly basis the Earth receives energy equal to 5.78×10^{18} J, which is about ten times higher than the known natural reserves of petrol, natural gas and coal.

Besides the electromagnetic radiation the Sun emits also charged particles (mainly protons and electrons) which travel with velocities in the range of 400–3,000 km/s. Even though the intensity of this radiation is very small in relation to the electromagnetic radiation (of the order of 10^{-7} cal $\times$ cm^{-2} $\times$ min^{-1}), it plays an important role for the creation of several phenomena in the upper atmosphere.

The radiation from the Sun corresponds to the radiation from a black body with a temperature of 6,000 K. With an application of Wien's law it can be calculated that the maximum of the Sun's radiation is emitted at wave length of 0.5 μm, with 99% at wave lengths between 0.15 and 4.0 μm, and the spectrum distribution is: 7% ultraviolet, 44% visible and 49% infrared.

1.7.3 Earth's Radiation

The term "Earth's radiation" means the sum of radiation which is emitted from the Earth's surface and its atmosphere. If we treat the Earth and the lower atmosphere

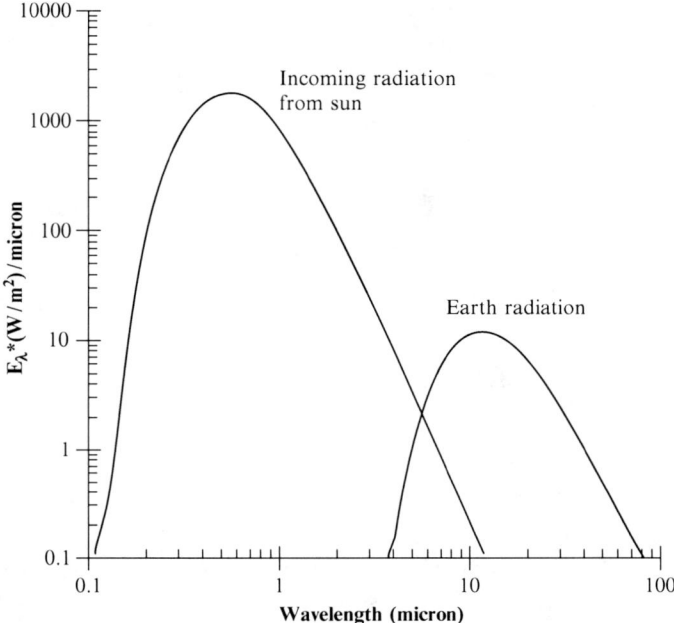

Fig. 1.13 Incoming radiation at the upper atmospheric layers and emitted Earth's after the absorption of the Sun's radiation

(up to 50 km) as a black body with temperatures on the Earth's surface lower than 80°C and for the upper atmosphere higher than −70°C, an application of Planck's law concludes that the emission spectrum of the Earth's radiation ranges between 1.5–100 μm. Specifically for the Earth's surface the mean temperature is 15° C (288 K), and its emission spectrum is between 4 and 100 μm.

Based on Wien's Law, the maximum wave length of the radiation intensity from the Earth is close to 10 μm (Fig. 1.13). The radiation emission spectrum from the Earth starts at the same wave lengths as the incoming radiation from the Sun. Therefore, the Earth emits at long wave lengths (infrared) compared to the Sun's emission spectrum.

1.7.4 Factors That Affect the Sun's Radiation Flux to Earth

Besides the astronomical factors which affect the Sun's radiation flux (distance between Sun and Earth, the height of the Sun above the horizon) there are also other factors, of which the most important are:

Table 1.2 Albedo of specific surfaces

Surface	Reflection (%)
Fresh snow	75–95
Old snow, dirty	40–60
Ice	70
Water (average value)	10
Surface without vegetation	15
Sand	18–28
Forest	5–10
Grass	15–25
Inhabited areas (towns)	14–18
Cloud (thick layer)	50–85
Cloud (thin layer)	5–50

1.7.4.1 Geographical Factors

These include:

(a) Latitude (φ).
(b) The height of the site from mean sea level.
(c) The topography of the site. It refers to the topography of the area around the site.
(d) The reflection (albedo) of the surface which receives the radiation. This depends on the surface's cover. Table 1.2 shows the albedo of several surfaces.

1.7.4.2 Geometrical Factors

These include the location of the surface in relation to the incoming radiation. These parameters are related to the change of the seasons.

1.7.4.3 Radiation Decrease

The term "radiation decrease" refers to the absorption and scattering of the radiation inside the atmosphere. Direct radiation from the Sun is the radiation quantity which reaches the Earth's surface without any scattering. Besides the direct Sun's radiation there is also a quantity of the radiation which reaches the Earth's surface after scattering (diffuse light).

1.7.5 Interaction of the Sun's Radiation in the Atmosphere

The Sun's radiation during its passing inside the atmosphere becomes weaker due to its absorption and scattering. Both these phenomena are dependent on the wave length of the incoming radiation. The reduction of the radiation I_λ with wave length λ during its transport in a homogeneous medium is given by Beer's law:

1.7 Radiation in the Atmosphere

$$\frac{dI_\lambda}{I_\lambda} = -\mu_{(\lambda)} \rho dx \tag{1.31}$$

where μ is the spectrum coefficient of radiation reduction, ρ is the density of the medium inside which the radiation is transported and dx is the transport length. Integrating the above equation for a length of transport L we conclude that:

$$I_\lambda = I_{o\lambda} \exp\left(-\int_0^L \mu_\lambda \rho dx\right) \tag{1.32}$$

where $I_{o\lambda}$ is the total incoming radiation and the integral $M = \int_0^L \rho dx$ expresses the total mass of the medium inside which the radiation transport occurs.

The reduction coefficient can be expressed as the summation of the absorption coefficient α_λ and the scattering coefficient s_λ, thus $\mu_\lambda = \alpha_\lambda + s_\lambda$. These two coefficients have units of inverse length and are dependent on the radiation wave length, the medium of transport and the radiation distance.

1.7.5.1 Absorption of the Sun's Radiation

The reduction of the Sun's radiation inside the atmosphere is due to the transformation of radiation energy to heat. Even though the absorption spectrum of electromagnetic radiation in the atmosphere is extended from X rays to small radio waves, in this chapter the focus will be on the absorption of the Sun's radiation. The Sun's radiation spectrum is between 0.1 and 20 μm and covers the areas of ultraviolet, visible and infrared radiation.

The absorption of the electromagnetic radiation of air inside the atmosphere is the main process which determines the Earth's climate and the processes responsible for the chemical reactions in the atmosphere. Figure 1.14 shows the most prominent gaseous species for the Sun's radiation which are O_2, O_3, H_2O and CO_2. The absorption spectrum of the various gasses is complicated and the absorption is dominant at specific wave lengths.

The H_2O and CO_2 are strong absorbers at infrared but weak absorbers at visible light. Other chemical species which absorb the Sun's radiation are N_2O and CH_4. Considerable absorption also occurs for the suspended particles in the atmosphere without specific spectrum preference. With the help of Fig. 1.14 the absorption spectrum of gaseous species in the atmosphere will be further analyzed.

Oxygen (O_2)

The absorption zones of radiation from oxygen are observed in the visible and infrared but are more intense in the ultraviolet from 0.1 μm up to 0.3 μm.

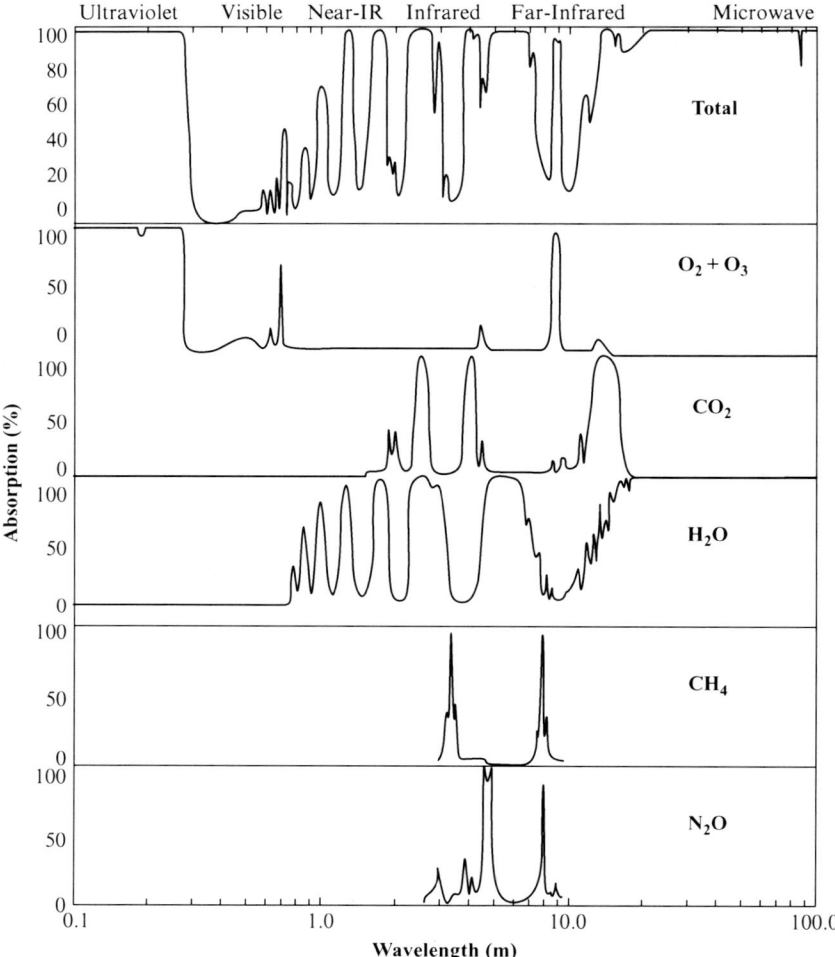

Fig. 1.14 Energy absorption from various gases in the atmosphere

Absorbtion from the oxygen radiation with wave length smaller than 0.2 μm has as a result the brake of the molecular oxygen and the production of ozone.

Ozone (O_3)

The absorption zones for ozone are similar to those for oxygen. The ozone has a large ability for absorption at the ultraviolet part of the spectrum and specifically between 0.2 and 0.3 μm with absolute maximum of absorption at 255 nm. At the visible spectrum, we observe a small zone of absorption between 0.41 and 0.8 μm. At the infrared the majority of the absorption zones of ozone are covered by the absorption zones of other gasses (carbon dioxide and water vapor) having an important narrow area of absorption between 9 and 10 μm.

The absorption of the Sun's radiation by ozone (especially in the ultraviolet region) is a basic source of heating of the upper atmospheric layers and at the same time protects the biosphere from dangerous ultraviolet radiation.

Carbon dioxide (CO_2)

The absorption zones of carbon dioxide are located only in the infrared region and specifically at the regions 1–2 μm, 2.05–2.7 μm, 4.3–12.9 μm και 12.9–17.1 μm. The strongest absorption zone is observed at the wave length of 4.3 μm, which is however of not of great importance since the Sun's and Earth's radiation in this region is quite small. Of increasing importance is the region between 12.9 and 17.1 μm because first it is a broad region and in this region occurs also the maximum thermal emission from the atmosphere.

Water Vapor (H_2O)

Water vapour has a significant importance to absorption of the Sun's radiation, not only because it is abundant in the atmosphere and variable, but also due to its wide absorption of the infrared radiation. As shown in Fig. 1.14, water vapour has a wide absorption spectrum in the infrared region.

In the atmosphere the absorption of infrared radiation increases with the increase of its water content. The expression which connects the absorption increase with the corresponding increase of the water vapour content is given by the empirical relation of Mügge-Moler:

$$\Delta I = 0.171 \times (\Delta h \times P_w)^{0.303} \text{cal} \times \text{cm}^{-2} \times \text{min}^{-1}$$

where, ΔI is the absorption of the radiation, Δh the width of the atmospheric mass inside which the radiation is passed and P_w the estimated water content in rain water (cm).

Therefore:

- For $\lambda < 0.3$ μm (ultraviolet), the Sun's energy is filtrated almost totally from the ozone and oxygen and therefore the percentage of this radiation energy which reaches the Earth's surface is very small. This ozone layer is therefore the protecting shield for the development and conservation of life on the planet.
- In the region of the visible spectrum (0.3–0.7 μm) only small absorption is observed, mainly from ozone and oxygen, allowing therefore large quantities of energy in this spectrum region to reach the Earth's surface. It has to be noted that the different colour tints in the atmosphere are related to the absorption spectrum of the visible light.
- In the infrared region ($\lambda > 0.7$ μm) a gradual increase is observed of the absorption of the Sun's radiation which is due to water vapour, oxygen and

Table 1.3 Absorption spectrum from the Earth's radiation

Spectrum Region	Ability of radiation absorption
$\lambda < 5$ μm	Insignificant absorption
5 μm $\leq \lambda \geq 8$ μm	Strong absorption from H_2O
8 μm $\leq \lambda \geq 12$ μm	Weak absorption from H_2O and O_3
12 μm $\leq \lambda \geq 17$ μm	Strong absorption from CO_2
17 μm $\leq \lambda \geq 19$ μm	Weak absorption from H_2O
$\lambda \geq 19$ μm	Strong absorption from H_2O

other gasses with important absorption zones such as carbon dioxide (CO_2), nitrous oxide (N_2O) and methane (CH_4).

Furthermore, the Earth's radiation has an absorption peak from atmospheric chemical components with high molecular weight such as water vapour (H_2O), ozone (O_3), carbon dioxide (CO_2), methane (CH_4) and nitrous oxide (N_2O). The absorption from these compounds presents the most important source of heating for the atmosphere. Table 1.3 presents the absorption of the Earth's radiation from some atmospheric chemical components. Therefore, the atmosphere's temperature is dependent significantly on the quantities of water vapour and carbon dioxide which are contained inside it. Changes of the concentration of these components will affect the thermal conditions in the atmosphere.

From Fig. 1.14 it can be seen that radiation in the range 8–12 μm has a relatively small absorption from the constituents of the atmospheric air. This fact, in relation with the fact that at this spectrum range occurs the maximum of the Earth's radiation (based on Wien's law), supports the fact that a considerable amount (close to 6%) from the Earth's radiation escapes to space.

A result of the continuous loss of energy from the Earth's surface is an observed temperature drop at the atmospheric layers close to the Earth's surface during night. The spectrum region between 8–12 μm is known as an atmospheric window. The remaining quantity of the Earth's radiation (94%) is absorbed from the atmosphere and is the main source of its heating.

Clouds are very effective emitters of infrared radiation both from their top and base, emitting respectively radiation to space and the Earth's surface. The Earth's surface emits back to clouds part of the infrared radiation. For this reason, nights with the presence of clouds are warmer in comparison with nights with clear sky. In the case that the clouds remain during the next days, they do not allow the Sun's light to reach the Earth's surface and reflect the radiation back to space. Under this condition, the Earth's surface does not receive adequate energy and as a result the next days are colder in comparison with days with clear skies.

1.7.5.2 Scattering of the Sun's Radiation

Atmospheric scattering is due to the interaction of photons with air molecules and particles in the atmosphere. During this process, particles which are found at the trajectory of an electromagnetic wave absorb part of its energy which is re-emitted at a solid angle 4π with centre the particle.

The secondary emission of electromagnetic radiation due to scattering originates from the electrons of the particles, which undergo vibrations due to the incoming radiation, and as a result the particles (molecules or aerosols) are sources of electromagnetic radiation. In the atmosphere the radiation is scattered from particles which range from the molecular size (10^{-4} μm) up to the size of coarse aerosols (10 μm) and the size of rain droplets (10^2–10^4 μm).

There are two main kinds of scattering in the atmosphere which are connected with the dimensions of particles and wave length of radiation. When the particles responsible for the scattering have dimensions much smaller than the wave length of the incoming Sun's radiation, then the scattering is called Rayleigh scattering. A typical example of Rayleigh scattering is the action of molecules of atmospheric gasses. In this case the particles have a diameter between 10^{-2} and 10^2, the wave length of the Sun's radiation, then the scattering is called Mie scattering. The Mie scattering occurs due to the existence of aerosols in the atmosphere.

A theoretical study of scattering examines radiation scattering as a sum of all the secondary emissions of the scattering centers. If the scattering centers are identical molecules and positioned at equal distances, then the result of the summation of the secondary emissions was to be zero due to the confluence of all directions, except from the direction of the incoming radiation, where the radiation is transported with velocity:

$$u = \frac{c_o}{n} \quad (1.33)$$

where, c_o is the velocity of light in vacuum and η the diffraction coefficient of the medium.

However, the molecules in the atmosphere due to their thermal motion do not have equal distance between them and the summation of their scattering emissions is not zero.

Rayleigh Scattering

Rayleigh scattering is due to gaseous molecules and aerosols with dimensions much smaller than the wave length of the incoming radiation. In this case the radiation quantity which is scattered is inversely proportional to the fourth power of the radiation's wave length. The volume spectrum scattering coefficient, which is expressed in units of inverse length, for Rayleigh scattering is given by the expression:

$$s_\lambda = \frac{32\pi^3}{3\lambda^4} \frac{(n-1)^2}{N} \quad (1.34)$$

where, n is the diffraction coefficient of the medium and N is the arithmetic density of the scattering particles. The scattering is more intense for smaller wave lengths.

The sky's blue colour can be explained by the more intense scattering of the blue colour ($\lambda \approx 425$ μm) in comparison with the red ($\lambda \approx 650$ μm). Rayleigh scattering is not homogeneous in all directions but has a maximum in the direction of radiation transport and a minimum from the directions perpendicular to this direction.

Mie Scattering

In case the scattering centers have a radius larger than 0.1 λ of the incoming radiation, then the mathematical expressions for Rayleigh scattering are no longer valid. In the atmosphere there exist aerosols which have diameters larger than the radiation wave length and result in intense scattering, especially at the lower atmospheric layers where their concentration is important. The haze and cloud droplets produce important scattering since they have dimensions in the range between 0.01 and 100 μm.

The coefficient for the volume Mie scattering is given by the equation:

$$s_\lambda = \pi r^2 (K \rho, n) \tag{1.35}$$

where N is the arithmetic density of the particles which scatter the radiation and K (ρ,n) a function which is dependent on the diffraction index of the medium and the dimensionless quantity $\rho = 2\pi r/\lambda$. Figure 1.15 presents a graphical representation of the function K(ρ,n) versus the parameter ρ. It is observed that the function K(ρ,n) obtains maximum value when the parameter ρ has values between 4 and 6. To the contrary, for large particle sizes its value remains almost constant and equal to 2.

The intensity of the scattered radiation from large particles is non-homogeneous in different directions and has a maximum in the direction of the radiation transport. For small particle dimensions, Mie scattering is equal to Rayleigh scattering.

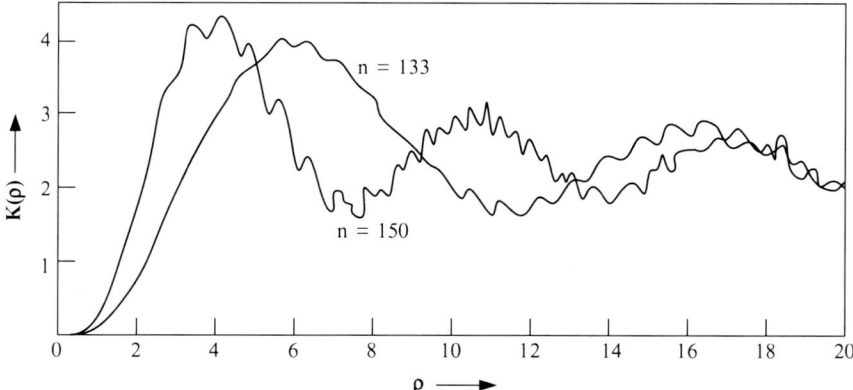

Fig. 1.15 Variability of the function K(ρ,n) versus the parameter $\rho = 2 \pi r/\lambda$ for diffraction index values 1.33 and 1.50

1.7.6 Greenhouse Effect

Based on the previous discussion it is evident that if the Earth did not have an atmosphere, then a significant part of the radiation would be reflected back (directly or after its absorption in the Earth's surface to be emitted back to space as infrared radiation) to space from the Earth's surface. Then the Earth would have a temperature equal to its effective temperature which is $-18°C$ and as a result the Earth's surface would become frozen. However, due to the existence of its atmosphere and the absorption capacities of their gaseous species, the average temperature is close to $15°C$.

During 1827 the French scientist Jean Baptist Fourier referred for the first time to a phenomenon that is related to the Earth's temperature and resembles a greenhouse. Several years later, during 1896, the Swedish scientist Svarnte Arrenius supported the hypothesis that the Earth's temperature is affected significantly from the gaseous species in the atmosphere which have the capacity to retain heat.

The atmosphere's constituents allow incoming visible and infrared radiation to reach the Earth's atmosphere (see Fig. 1.14) whereas stratospheric ozone and oxygen absorb the major portion of the ultraviolet part of it. Furthermore the greenhouse gasses absorb a considerable part of the radiation which is emitted from the Earth's surface and does not allow it to escape to space. This behavior of the atmosphere is called the "Greenhouse Effect" since it resembles, at least theoretically, greenhouses in which the glass cover allows the Sun's radiation to enter but blocks the long wave length of the radiation from exiting, resulting in a temperature increase inside the greenhouse. In reality the temperature increase inside greenhouses is due to the absence of vertical movements of heat.

Consequently the atmospheric phenomenon of absorption and re-emission of long wave length radiation from the Earth's surface is responsible for the existence of life on the planet. For the existence of equilibrium in the ecosystems it is required that the concentration of the greenhouse gasses not have considerable variations. To the contrary, large variations of the greenhouse gasses concentration may result in temperature changes on Earth and consequently climatic changes. Geological studies reveal that there were periods in the Earth's history when its surface was colder or warmer than today, during which time considerable climatic changes have occurred.

Figure 1.16 shows the temperature change in the Earth from 1880 until today as measured from a network of meteorological stations. It is evident that the temperature rose during this period. A series of similar data have led scientists to the study of the greenhouse phenomenon.

Figure 1.17 shows also another piece of evidence for the importance of greenhouse gasses to the Earth's atmosphere temperature rise. The increase of carbon dioxide (CO_2) concentrations is due to anthropogenic reasons and leads indirectly to the temperature rise in the atmosphere.

In addition, Fig. 1.18 presents some other evidence for the importance of the Greenhouse effect to the Earth's temperature change. Changes of the CO_2 concentration in the Earth's atmosphere the last 160,000 years are followed by similar temperature changes in the atmosphere.

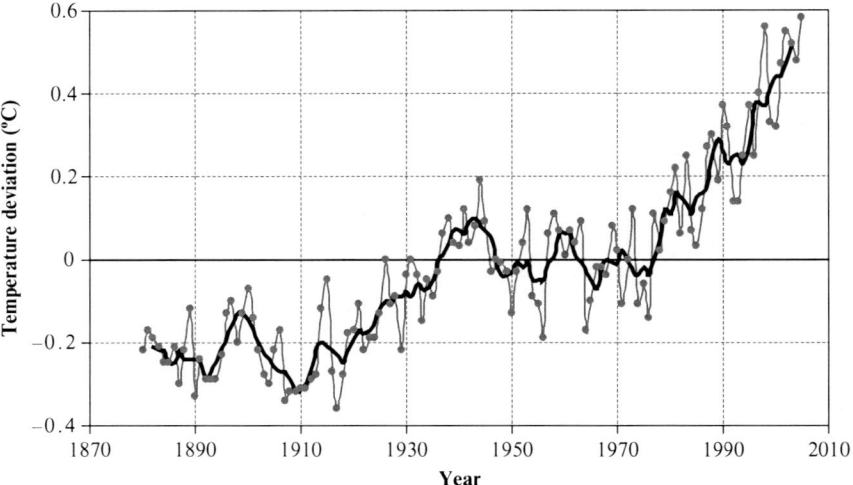

Fig. 1.16 Global average annual surface temperature as measured from a meteorological network from 1880 until today. The grey line shows the annual value and the black line the average 5 years value

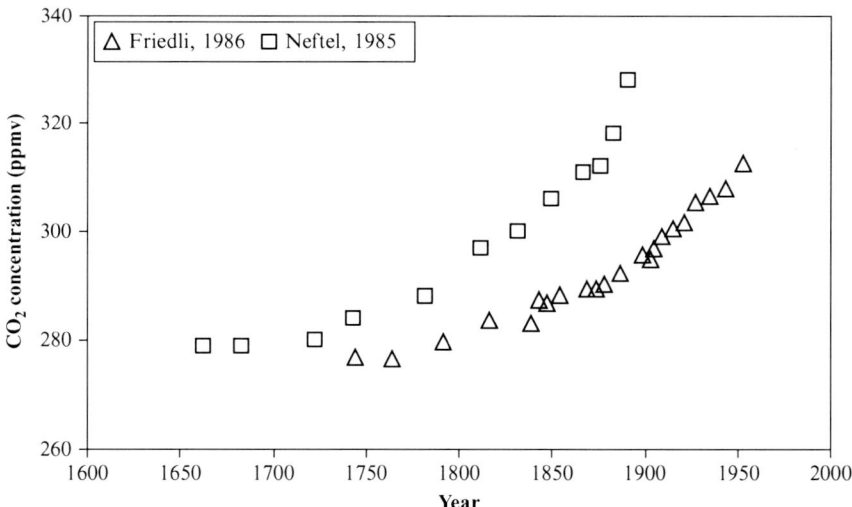

Fig. 1.17 Atmospheric concentration of CO_2 as measured at a glacier from the station Siple (Adapted from Neftel et al., 1985). The above data showed that the atmospheric concentration of CO_2 during 1750 was 280 ± 5 ppmv (parts per million by volume) and is increased by 22.5% to 345 ppmv during 1984. The data given adapted from Friedli et al. (1986) give also atmospheric concentration of CO_2 before 1,800 close to 280 ppmv

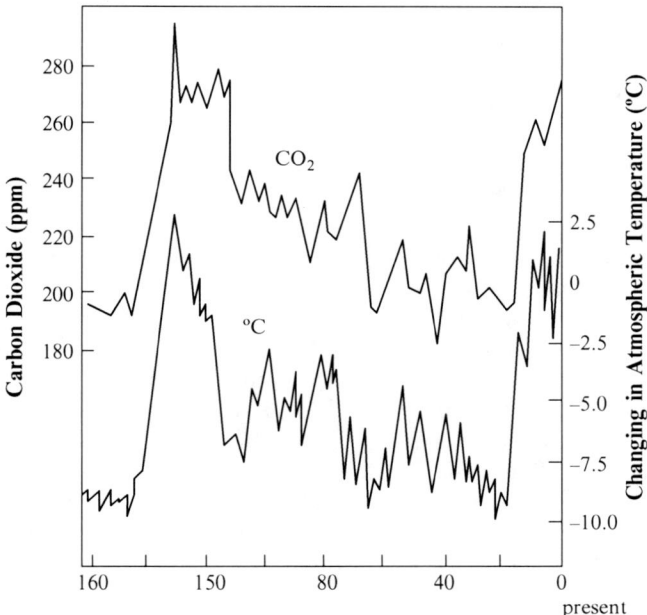

Fig. 1.18 Variation of the CO_2 concentration and the Earth's temperature the last 160,000 years (Adapted from Philander (1998))

1.7.7 Energy Balance of Earth and its Atmosphere

Even though the average temperature at different parts of Earth is variable versus time, the average changes in the thermal balance of the Earth are minimal. This fact shows that the Earth and its atmosphere as a unit emits to space the same quantity of energy as it absorbs from the Sun. The same energy balance occurs for thermal energy transfer between the Earth's surface and its atmosphere. On a yearly basis the Earth's surface has to emit back to the atmosphere the same quantity of energy as it absorbs. If there is no equilibrium, then the average value of the Earth's temperature is not constant.

The Earth's climate is determined from the quantity of the Sun's radiation that reaches the Earth's atmosphere. If the Sun emits energy with rate F from its surface, then the energy emitted in all directions is equal to the energy which reaches the mean distance Sun – Earth. Therefore, if R_s is the Sun's radius and R is the mean distance Sun – Earth then:

$$4\pi R_s^2 F = 4\pi R^2 S_o \quad (1.36)$$

where S_o is defined as the Sun's constant which corresponds to the Sun's energy which reaches the Earth's atmosphere. Calculating the energy emission rate from the Sun's surface with the law of Stefan-Boltzmann and replacing the values of R_s and R, we get the result that $S_o = 1{,}370\ Wm^{-2}$.

The Earth reflects part of the Sun's radiation back to space with a surface equal to πR^2 where R is the Earth's radius. The Earth's surface receives radiation from the Sun at its surface which is equal to $4\pi R^2$. Therefore the fraction of the Sun's constant which is received per unit surface is equal to $(\pi R^2/4\pi R^2) = 1/4$ of the Sun's constant close to $343\ Wm^{-2}$.

A fraction of the radiation received by the Sun is reflected back to space. The fraction of the Sun's radiation which is reflected in relation to incoming radiation is called reflective ability or reflectivity (R_P). The reflectivity is dependent upon the presence of clouds, reflection from air molecules and aerosols and the reflectivity of the Earth's surface which is defined as R_s. The quantity $(1 - R_P)$ corresponds to the fraction of the short wave length of the Sun's radiation which is absorbed from the system atmosphere – earth. For reflectivity $R_P = 0.3$, the above quantity is equal to $240\ Wm^{-2}$. This radiation quantity is balanced on an annual basis from the long wave length infrared radiation which is emitted to space from the system atmosphere – Earth. The flux density of the infrared radiation is close to $340\ Wm^{-2}$ and is much larger than the outward flux of $240\ Wm^{-2}$ at the limit of the Earth's atmosphere. This difference is due to intensive absorption and emission of the Sun's radiation from clouds, water vapour and atmospheric gasses such as carbon dioxide and methane.

The effective temperature of a planet is the temperature by which a black body emits the same constant flux F, which is equal to the radiation flux of the planet emitted from the upper layer of its atmosphere to space. The effective Earth's temperature (T_e) can be calculated when there is a balance between the incoming and outgoing energy of the Earth. The incoming Sun's energy (F_s) at the Earth's surface can be written as

$$F_S = \frac{S_O}{4}(1 - R_P) \quad (1.37)$$

Assuming that the Earth is a black body with temperature T_e, the emitted energy from the whole planet is equal to:

$$F_L = \sigma T_e^4 \quad (1.38)$$

Putting F_s equal to F_L results in the following expression for the effective temperature T_e:

$$T_e = \left(\frac{(1 - R_P)S_O}{4\sigma}\right)^{\frac{1}{4}}. \quad (1.39)$$

For reflection coefficient $R_P = 0.3$ the above equation gives an effective temperature $T_e = 255$ K. If the clouds from the Earth's atmosphere were totally removed, the reflection coefficient would have a value close to $R_P = 0.15$ with an effective temperature equal to 268 K. From the equation which gives an effective temperature

T_e we get that changes of T_e by 0.5 K results in changes of 10 W m^{-2} (0.7%) of the Sun's constant or equivalently changes of the reflection coefficient ($R_P = 0.3$) of the order $\Delta R_P = 0.005$.

The net radiative energy input, $P_{net} = P_S - P_L$ is equal to zero in conditions of thermal equilibrium. If there are thermal variations, then the variations of the incoming energy are connected to changes in the Sun's radiation or the long wave length radiation which is emitted from the Earth's surface. Therefore:

$$\Delta F_{net} = \Delta F_S - \Delta F_L. \tag{1.40}$$

For the existence of an equilibrium it is necessary for there to be a temperature difference ΔT_e, which can be correlated with the variation ΔF_{net} with a parameter λ_o,

$$\Delta T_e = \lambda_O \Delta F_{net} \tag{1.41}$$

where the parameter λ_o has units K(Wm^{-2})$^{-1}$ and is called the coefficient of climate sensitivity.

1.7.8 Distribution of Sun's Radiation at the System Atmosphere-Surface

The energy quantity which reaches the upper layer of the atmosphere is partially absorbed and reflected inside the atmosphere and consequently only a part of it reaches the Earth's surface. Furthermore, the surface in relation to albedo reflects part of the energy back to the atmosphere. Consequently, if Q_S is the quantity of the Sun's radiation which reaches the atmosphere, Q the quantity of the direct radiation, q the quantity of diffusional radiation, $R_s(Q + q)$ the quantity which is reflected from the surface, $(1 - R_s)(Q + q)$ the quantity which is absorbed from the surface, C the quantity which is absorbed from the atmosphere and ç the quantity which is reflected from the atmosphere, the radiation balance can be expressed as:

$$Q_S = (1 - R_s)(Q + q) + C + R_s(Q + q) + A. \tag{1.42}$$

The reflectivity of clouds, atmospheric gasses and aerosols has a total value of 0.3 which means that 30% of the Sun's radiation is reflected back to space. The clouds alone reflect 24% of the Sun's radiation at the global level. This relatively high percentage of cloud reflectivity may reach levels even higher than 50% in special cases (e.g. on the top of summer cumulonimbus with reflectivity higher than 90%). Measurements revealed that the radiation reflectivity from clouds in the southern hemisphere is higher than that of the northern hemisphere. This fact is because the cloud cover in the southern hemisphere is higher than the northern hemisphere.

The radiation percentage which is absorbed from the air's constituents is relatively small and is close to 15% of the Sun's radiation. The clouds absorb a smaller percentage which does not exceed 4%. The atmosphere absorbs greater quantities in the absence of clouds. The main absorption is observed in the ultraviolet region of the spectrum (under 300 nm) (Fig. 1.14) and is due to the stratospheric ozone.

Therefore the Earth's surface receives 51% of the Sun's radiation. The direct radiation of the Sun is responsible for 30% of the total radiation whereas the reflected part is 21%. The radiation percentage which reaches the surface at the northern hemisphere is 4.5% higher than at the southern hemisphere. This is a result of the fact that the atmosphere in the northern hemisphere is drier and the atmosphere in the southern hemisphere has higher cloud cover.

Close to 7% of the radiation which reaches the surface is lost due to thermal conductivity and vertical transport, whereas 23% is used for evaporation of the surface waters. The remaining 21% is emitted back to the atmosphere as infrared radiation.

The radiation reflected from the surface exceeds the absorbed radiation only in snow covered areas of the poles. At the southern pole about 60% of the radiation which reaches the Earth's surface is reflected, almost totally, from the snow-covered surface and returns back to space. The percentage which finally is absorbed in these regions does not exceed 10%. On the contrary, regions which are included between latitude 35°N and 35°S absorb about 50% of the radiation which reaches the atmosphere and the reflected part does not exceed 6%.

1.7.9 The Earth's Climate

The concept of weather refers to the meteorological conditions of the atmosphere in a specific region. As an example are local changes of temperature and rain conditions. In contrast, the concept of climate refers to the mean weather conditions over an extended period of time for a specific geographical region. The World Meteorological Organization (WMO) defines a period of 30 years as the minimum time period which must be examined in order to refer to climatic changes.

The mean temperature of the planet is the most appropriate and stable climatic parameter that can be examined in relation to the climate conditions and changes. This is because the Earth's mean temperature is changing very slowly versus time, since the atmosphere, the surface and the oceans have the capacity to absorb enormous quantities of heat. Therefore changes of energy flux from the Sun or other variations in the heat balance in the atmosphere are not expected to lead to considerable variations of the mean global temperature.

The climatic study of Earth's atmosphere is therefore focused on global energy balance. Parameters that change the energy flux from the Sun to the Earth, such as the absorption of energy and its reflection back to space, can be studied as climatic parameters. However, the climatic system is a quite complex physical, chemical and biological machine and needs great attention to the analysis of climatic data in order to avoid wrong conclusions. Table 1.4 presents data for the climatic

1.7 Radiation in the Atmosphere

Table 1.4 Climatic conditions at planets of our Solar system

Planet	Solar distance (AU)[a]	Albedo	Surface's temperature (K)
Mercury	0.39	0.30	460
Venus	0.72	0.70	700
Earth	1.0	0.33	290
Mars	1.5	0.20	220
Uranus	19	0.80	65

[a] astronomical units: 1 AU corresponds to 150 million kilometers which is the mean distance of the Earth from the sun

Table 1.5 Energy reservoirs on Earth (Adapted from Turco (2002))

Reservoir	Volume (Km^3)	Mass (Gt)	Temperature (K)	Heat capacity (J/g K)	Energy content (J)
Atmosphere	5.0×10^9	5.0×10^6	260	1.0	1.3×10^{24}
Land surface[a]	5.0×10^3	1.0×10^4	290	4.0	1.2×10^{22}
Land subsurface[b]	1.0×10^5	2.0×10^5	280	4.0	2.2×10^{23}
Surface oceans[c]	3.8×10^7	3.8×10^7	280	4.0	4.3×10^{25}
Deep oceans[d]	1.4×10^9	1.4×10^9	275	4.0	1.5×10^{27}

[a] The land surface depth is taken to be 4 cm thick (diurnal response)
[b] The deep soil layer depth is taken to be 80 cm (seasonal variation)
[c] The surface oceans depth is taken to be 100 m
[d] The deep oceans depth is taken to be 4 km

characteristics of specific planets of the Solar system, as well as, for the Earth. The surface temperature refers to the temperature of a black body. It is evident that the more distant a planet is from the Sun the colder is surface temperature of the planet, the only exception being the climate on Venus.

Planets that contain large quantities of carbon dioxide and water vapour in their atmospheres, such as Earth and Venus, have surface temperatures higher than the corresponding temperatures of a black body. Therefore the presence of carbon dioxide and water vapour in the atmosphere of planets leads to an increase of their temperature. Venus's atmosphere contains huge quantities of carbon dioxide which leads to its temperature increase by several 100°. The greenhouse phenomenon acts as a curtain which retains the incoming Sun's heat.

As discussed previously, the climatic system of Earth may be considered as a closed system in which there is energy influx from the Sun and energy flux back to space. The Earth's climatic system maintains a characteristic temperature and contains energy in the form of heat. The heat quantity is determined by the size of the reservoirs in the planet. The energy is the sum of separate energy quantities in the different reservoirs (see Table 1.5).

The energy deposits in the different reservoirs, as well as the energy flux balance in the entire climatic Earth's system, is shown in Fig. 1.19. Energy is calculated in units of 1×10^{16} J and energy flux in units of 1×10^{16} J/s. The energy flux from the Sun to the Earth's surface can be distinguished between direct and scattered.

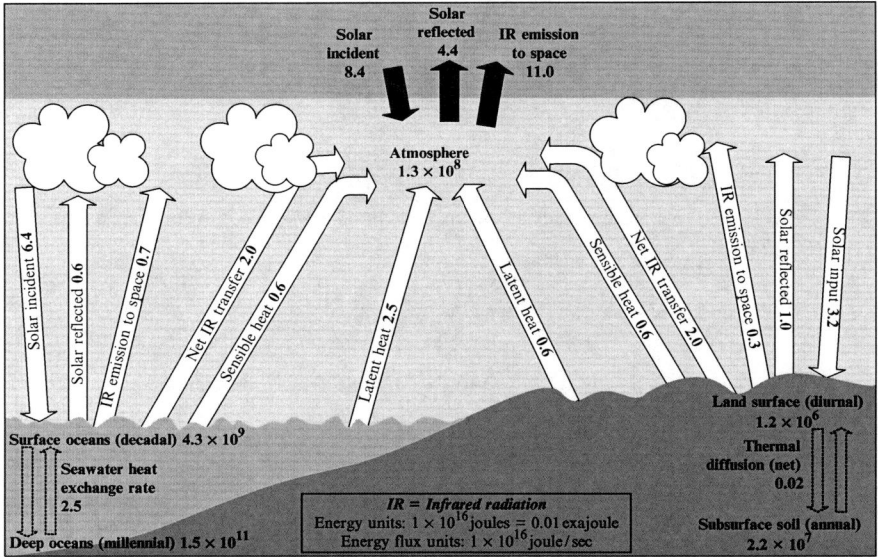

Fig. 1.19 Energy balance in Earth in relation to the climate

The atmosphere keeps energy through absorption and emits through scattering. The total energy from the Sun is 18 energy units which correspond to 1,400 W/m² (Solar constant) multiplied by the Earth's surface which is close to 1.3×10^{16} m².

The energy reservoirs are dynamic systems. A very important energy reservoir which is often overlooked is glacial ice. The most important area of glacial ice is located in Antarctica which has the size of the United States of America with a depth of 2 km and total volume larger than 20 million cubic kilometers. Permanent glacial ice exists also in Greenland and on the top of several mountains. The Antarctic ice sheets correspond to the volume of world oceans at a depth of 80 m. In the past they covered large areas of America, Europe and Asia. The Antarctic area contains glaciers which remained for a period of 200,000 years. The pole areas have an indirect effect on the Earth's climate since their high albedo means high reflectivity on their surface. The total energy which is contained in the Earth's glaciers is equal to 2.6×10^{25} J which is equivalent to 2.6×10^{9} energy units in Fig. 1.19. Therefore the energy which is incorporated in the ice sheets is comparable to the energy which is contained in the oceans surface.

1.7.9.1 Climate Change – Reasoning

The climatic system is complex and contains energy reservoirs which are connected among themselves. As discussed in previous sections, the Earth's climate is constant since there is a balance between the energy which is received from the Sun, the energy contained in the reservoirs and the emitted energy. The average Earth's temperature has been changed only by one or two degrees during a period of thousands of years.

Changes in the Earth's climate occur from changes in the Sun's energy, albedo variations and changes in the composition of the atmosphere (greenhouse effect). Important changes in the Earth's climate have been observed in the past such as during the glacier periods. We currently live in a period of relatively warm climate (Holocene) which has existed for the last 10,000 years. Geological studies have shown the existence of four glacier periods at 100,000, 200,000, 450,000 and 600,000 years ago. During the Pleiostocene period 2.5 million years ago there existed several glacier periods with variable duration. During the glacier periods the average temperature of the planet was lower between 5°C and 10°C. Furthermore a small glacier period occurred during 1450–1890 with a decrease in the average temperature of 0.5°C.

The reasons which caused the ice ages are not explained fully and there is no scientific agreement on what they are. One explanation is related to the Earth's orbit around the Sun. The first ideas that connect the ice ages with changes in the Earth's orbit were proposed by James Croll in the mid-1800s. More detailed studies were performed in the 1940s by the astronomer Milutin Milankovitch. The basis for this

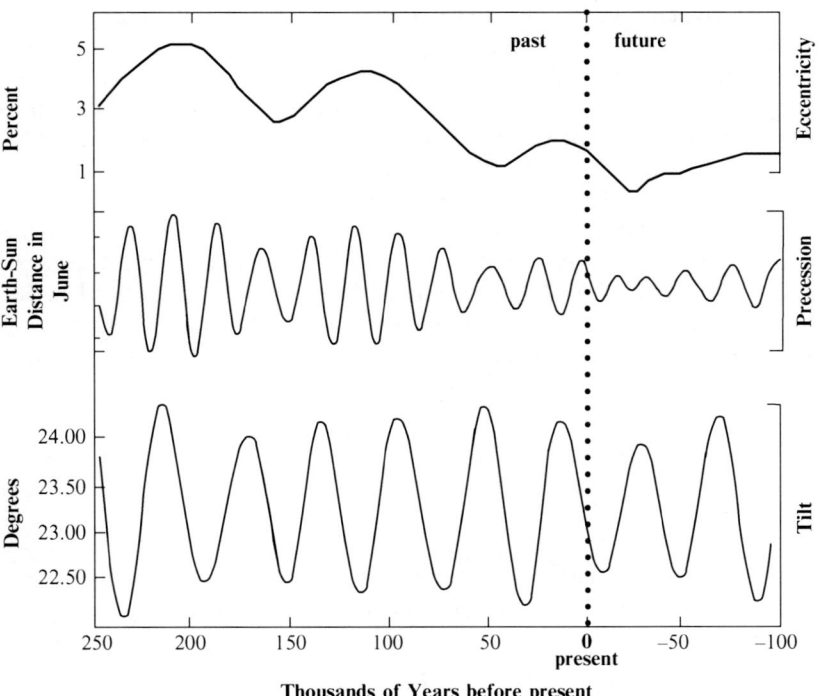

Fig. 1.20 Cycle changes of the Earth's eccentricity, the Earth-Sun distance and the rotation tilt of Earth (Adapted from Imbrie and Imbrie (1979))

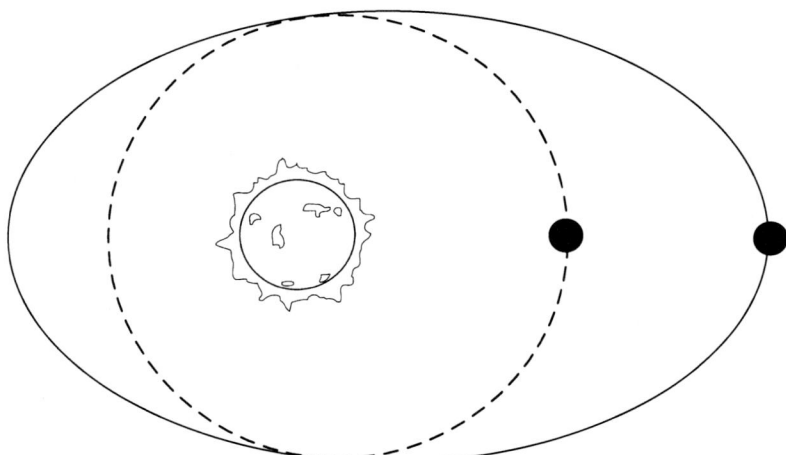

Fig. 1.21 Changes in the shape of the Earth's trajectories around the Sun at different time periods (time interval close to 100,000 years for the trajectory shift)

theory is that, during the Earth's movement in space, there are three different periodic movements which result in changes to the total quantity of Sun's radiation which reaches the Earth's surface. Changes in the gravitational attraction, which the Earth experiences mainly from the moon and Jupiter, results in changes on the distance between the Earth and Sun, the Earth's orbit and its rotation axis. The time intervals for changes in the eccentricity, tilt and Earth – Sun distance are 100,000, 41,000 and 22,000 years respectively as shown in Fig. 1.20.

Figure 1.21 shows that the Earth's orbit changes from elliptic to almost circular with a period of 100,000 years. Larger eccentricity of the orbit corresponds to larger changes of the solar energy which reaches the Earth's atmosphere. At present we are living in a period of small eccentricity with the Earth being closer to the Sun in January and more distant in June. During the 1970s, scientific findings from studies of ocean surfaces confirmed to a certain degree that the climatic changes in the past are related directly with the Milankovitch cycles.

1.8 Examples

Example 1. *At sea level the pressure and density of air are $p_0 = 1,013$ mb και $\rho_0 = 1.23$ kg/m^3 respectively. Calculate the pressure of air at a height of 100 m.*

From the hydrostatic equation it is known that $\frac{dp(z)}{dz} = -\rho(z)\,g$ ($p(z)$ is the atmospheric pressure at height z) and therefore $p_{100} - p_0 = \Delta z\,(-\rho(z)\,g)$. $\Rightarrow p_{100} = 1013\,mb - 1.23\frac{Kg}{m^3}\left(9.81\frac{m}{s^2}\right)100\,m$. It is known also that $1\,N = \frac{Kg\,m}{s^2} = 10^{-2}\,mb\,m^2$ and the final result is: $p_{100} = 1001$ mb.

1.8 Examples

Example 2. *Calculate the concentration of O_3 ($\mu g\ m^{-3}$) which corresponds to the mixing ratio of 120 ppb at $T = 298\ K$.*

Having the concentration m_i ($\mu g\ m^{-3}$) then the molecular concentration at mol m^{-3} can be expressed as $c_i = \frac{10^{-6} m_i}{MB_i}$, where MB_i is the molecular weight of the chemical component i. Using the law of ideal gasses ($c = p/(RT)$) the mixing layer expressed at ppm can be expressed as

$$ppm = \frac{RT}{p M_i} \times m_i \Rightarrow m_i = \frac{(1.0133 \times 10^5) \times (48)}{8.314 \times (298)} \times 0.12 = 235.6\ \mu g\ m^{-3}$$

Example 3. *What is the air mass weight (kg) inside a room with dimensions $5\ m \times 8\ m \times 2.5\ m$ at sea level? What is the air density at height of 2 km in the atmosphere at a temperature of $15°C$?*

The air density (ρ) versus height (z) can be written as $\rho = \rho_o e^{-(a/T)z}$ where $\rho_o = 1.225\ kg/m^3$ and $\alpha = 0.0342\ K/m$.
The air mass can be expressed as

$$m = \rho V = (1.225\ Kg\ m^{-3}) \times (100\ m^3) = 122.5\ Kg.$$

Finally the density can be written as

$$\rho = (1.225\ Kg/m^3)\ \exp[(-0.0342\ K/m)(2000\ m)/288K].$$

Therefore $\rho = 0.966\ kg/m^3$.

Example 4. *In a homogeneous atmosphere the pressure and temperature at sea level are equal to $p_o = 1,000\ mb$ and temperature $T_o = 298\ K$ respectively. Calculate the height at which the pressure equals (α) zero and (β) $p_o/2$.*

Using the hydrostatic equation we have:

$$\frac{\partial p}{\partial z} = -g\rho \Rightarrow z = \frac{p_o - p}{\rho g}.$$

When $p = 0 \Rightarrow z = \frac{p_o}{\rho g}$.
From the equation of ideal gasses we have $\frac{p_o}{\rho} = RT_o$ and as a result $z = \frac{RT_o}{g} = \frac{287\ J\ Kg^{-1} K^{-1} 298K}{9.8\ m s^{-2}} = 8728\ m$.
When $p = p_o/2$, then $z = 4,364\ m$.

Example 5. *The understanding of the potential temperature θ can be achieved using adiabatic conditions. Changes of the entropy can be correlated with changes of temperature and pressure with the equation:*

$$dS = \left(\frac{\partial S}{\partial T}\right)_p dT + \left(\frac{\partial S}{\partial p}\right)_T dp.$$

(a) Show that $dS = \frac{\hat{c}_p}{T} dT - \frac{R}{M_a} \frac{dp}{p}$.
(b) From the above equation show that for an adiabatic process ($dS = 0$) then $d\theta = 0$ and $\frac{dT}{T} = \frac{\gamma-1}{\gamma} \frac{dp}{p}$.

The thermal conductivity under constant pressure per unit mass of gas is given by the expression: $\hat{c}_p = \left(\frac{dq}{dT}\right)_p$ where

$$dq = T\, dS.$$

From the above equations can be concluded that:

$$\hat{c}_p = T\left(\frac{\partial S}{\partial T}\right)_p \Rightarrow \left(\frac{\partial S}{\partial T}\right)_p = \frac{\hat{c}_p}{T}.$$

From the Maxwell expression: $\left(\frac{\partial S}{\partial p}\right)_T = \left(\frac{\partial V}{\partial T}\right)_p$ and also $pV = \frac{1}{M_a}RT$ where $\frac{1}{M_a}$ is the number of moles per unit mass. Therefore $\left(\frac{\partial V}{\partial T}\right)_p = \frac{R}{M_a}\frac{1}{p}$, with a result:

$$dS = \hat{c}_p \frac{dT}{T} - \frac{R}{M_a}\frac{dp}{p}.$$

For adiabatic processes we have $dS = 0$ and as a result:

$$\hat{c}_p \frac{dT}{T} = \frac{R}{M_a}\frac{dp}{p}.$$

Therefore:

$$\frac{R}{M_a \hat{c}_p} = \frac{\hat{c}_p - \hat{c}_v}{\hat{c}_p} = \left(\frac{\hat{c}_p}{\hat{c}_v} - 1\right)/\left(\frac{\hat{c}_p}{\hat{c}_v}\right) = \frac{\gamma-1}{\gamma}$$

and it can be concluded that:

$$\frac{dT}{T} = \frac{\gamma-1}{\gamma}\frac{dp}{p}.$$

Example 6. *Calculate the N_2O concentration ($\mu g\, m^{-3}$) when the mixing ratio is equal to 311 ppb at pressure $p = 1$ atm and temperature $T = 298$ K.*

It is known that:

$$[N_2O]\,(\mu g/m^3) = \frac{p\, M_{N_2O}}{8.314\, T}\, \xi_{N_2O}$$

where ξ_{N_2O} is the mixing ratio.

Therefore:

$$[N_2O] = \frac{(1.0133 \times 10^5) \times 44}{8.314 \times 298} \times 0.311 = 559.66\,\mu g/m^3.$$

Example 7. *What is the electromagnetic flux which is emitted from the Earth as black body at temperature 255 K.*

Using the mathematical expressions of the Stefan-Boltzmann law the emitted energy is equal to:

$$(5.67 \times 10^{-8}\,W\,m^{-2}K^{-4}) \times (255\,K)^4 = 240\,W\,m^{-2}.$$

The emitted energy flux is not very large and is analogous to that emitted from a lamp with a 240 W front of a parabolic mirror which reflects the light with diameter 1.13 m (see Fig. 1.22). The Earth's surface is $5.1 \times 10^{14}\,m^2$ and, multiplying with the emitted energy flux, the total emission is equal to 1.22×10^{17} W.

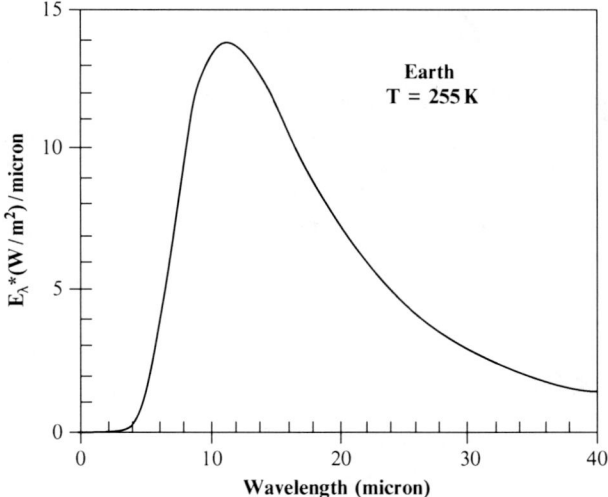

Fig. 1.22 Earth's radiation from the Earth as black body at temperature 255 K

Example 8. *Assuming that the Sun is a black body, what is the error in the estimation of its temperature when the error in the estimation of the mean distance of Earth – Sun is 2%.*

It is known that the total radiation which the Earth receives from the Sun is equal to

$$E_{o\lambda} = 4\pi R^2 I_o$$

where I_o is the mean value for the Solar constant at a specific time and R is the distance Earth – Sun.
Therefore:

$$E_{o\lambda} = 4\pi r_H^2 \sigma T_H^4 = 4\pi R^2 I_o,$$

$$T_H^4 = \frac{I_o}{\sigma r_H^2} R^2 \Rightarrow 4 T_H^3 \Delta T_H = \frac{I_o}{\sigma r_H^2} 2R \Delta R,$$

where r_H is the Sun's radius and T_H is the temperature at the Sun's surface. Dividing the above equations it results that: $\frac{\Delta T_H}{T_H} = \frac{1}{2}\frac{\Delta R}{R} = 1\%$.

1.9 Ambient Air Quality Standards

Ambient air quality standards have been introduced at national and international levels with the aim to protect human health and the environment. The threshold pollutant concentrations described in the legislation are based on detailed reviews of the scientific information related to their effects on human health. Table 1.6 presents the air quality guidelines for particulate matter, ozone, nitrogen dioxide and sulfur dioxide proposed by the World Health Organization (WHO 2006). The proposed annual mean concentration together with the 24-h, 8-h, 1-h and 10 min mean values are also presented.

Legislative rules for the PM_{10} concentration have also been proposed by the European Union (EU). The roles, goals and methods of air quality management are

Table 1.6 Air quality guidelines for particulate matter, ozone, nitrogen dioxide and sulphur dioxide proposed by the World Health Organization (Adapted from WHO 2006)

Compound	Annual mean ($\mu g/m^3$)	24 h mean ($\mu g/m^3$)	8 h mean ($\mu g/m^3$)	1 h mean ($\mu g/m^3$)	10 min mean ($\mu g/m^3$)
PM_{10}	20	50			
$PM_{2.5}$	10	25			
O_3			100		
NO_2	40			200	
SO_2		20			500

1.10 Appendixes

Table 1.7 Limit values for particulate matter (PM_{10})

	Averaging period	Limit value	Margin of tolerance	Date by which limit value is to be met
24 h limit value for the protection of human health	24 h	50 μg/m³ PM_{10}, not to be exceeded more than 35 times a calendar year	50% on the entry into force of this Directive, reducing on 1 January 2001 and every 12 months there after by equal annual percentages to reach 0% by 1 January 2005	1 January 2005
Annual limit value for the protection of human health	Calendar year	40 μg/m³ PM_{10}	20% on the entry into force of this Directive, reducing on 1 January 2001 and every 12 months there after by equal annual percentages to reach 0% by 1 January 2005	1 January 2005

determined by EU directive 96/62/EC and the later daughter directives describe the objectives for air protection policy and standards addressed to EU countries, as well as to EU candidate countries, as the platform for air quality assessment. For PM_{10} the obligatory standards have been established at levels as shown at Table 1.7.

In addition, in Table 1.8 the proposed ambient concentration levels for ozone and particulate matter proposed by the US Environmental Protection Agency (EPA) are shown.

Furthermore, ambient concentration levels of exposure to specific air pollutants proposed by the World Health Organization (WHO) are shown in Tables 1.9–1.10 (WHO 2006).

1.10 Appendixes

1.10.1 Appendix 1: The Hydrostatic Equation

Figure 1.23 presents a vertical column of air of mass M (with height z_2-z_1) in hydrostatic equilibrium. Since the air pressure is the same at all directions at specific height, the pressure which is applied vertically at the upper level of the column is equal to p_2 and in the lower level of the column is equal to p_1. The

Table 1.8 National Ambient Air Quality Standards for "criteria" pollutants (USEPA 2004)

Pollutant	Primary standard[a]	Averaging times	Secondary standard[b]
Carbon monoxide	9 ppm (10 mg/m^3)	8 h[c]	None
	35 ppm (40 mg/m^3)	1 h[c]	None
Lead	1.5 µg/m^3	Quarterly average	Same as primary
Nitrogen dioxide	0.053 ppm (100 µg/m^3)	Annual (Arithmetic mean)	Same as primary
Particulate Matter (PM$_{10}$)	Revoked[d]	Annual[d] (Arithmetic mean)	
	150 µg/m^3	24 h[e]	
Particulate Matter (PM$_{2.5}$)	15.0 µg/m^3	Annual[f] (Arithmetic mean)	Same as primary
	35 µg/m^3	24 h[g]	
Ozone	0.08 ppm	8 h[h]	Same as primary
	0.12 ppm	1 h[i] (Applies only in limited areas)	Same as primary
Sulfur oxides	0.03 ppm	Annual (Arithmetic mean)	–
	0.14 ppm	24 h[c]	–
	–	3 h[c]	0.5 ppm (1,300 µg/m^3)

[a] Primary standards set limits to protect public health, including the health of "sensitive" populations such as asthmatics, children, and the elderly
[b] Secondary standards set limits to protect public welfare, including protection against decreased visibility, damage to animals, crops, vegetation, and buildings
[c] Not to be exceeded more than once per year
[d] Due to a lack of evidence linking health problems to long-term exposure to coarse particle pollution, the agency revoked the annual PM$_{10}$ standard in 2006 (effective December 17, 2006)
[e] Not to be exceeded more than once per year on average over 3 years
[f] To attain this standard, the 3 year average of the weighted annual mean PM$_{2.5}$ concentrations from single or multiple community-oriented monitors must not exceed 15.0 µg/m^3
[g] To attain this standard, the 3 year average of the 98th percentile of 24 h concentrations at each population-oriented monitor within an area must not exceed 35 µg/m^3 (effective December 17, 2006)
[h] To attain this standard, the 3 year average of the fourth-highest daily maximum 8 h average ozone concentrations measured at each monitor within an area over each year must not exceed 0.08 ppm
[i] (a) The standard is attained when the expected number of days per calendar year with maximum hourly average concentrations above 0.12 ppm is <1. (b) As of June 15, 2005 EPA revoked the 1 h ozone standard in all areas except the fourteen 8 h ozone non attainment Early Action Compact (EAC) areas

pressure difference is equal to $p_1 - p_2$ and is in equilibrium from the gravitational force µ g:

$$p_1 - p_2 = Mg.$$

If we suppose that the surface of the air column is a unit surface, the volume of the column is equal to $(z_2 - z_1)$. The mass of the column equals $M = \rho(z_2 - z_1)$ and therefore:

$$p_1 - p_2 = \rho g(z_2 - z_1).$$

1.10 Appendixes

Table 1.9 Ambient annual average concentrations and guideline values for carbon monoxide, lead, nitrogen dioxide, ozone and sulfur dioxide (WHO 2006)

Compound	Annual average concentration ($\mu g/m^3$)	Health effect	Critical level ($\mu g/m^3$)	Guideline value ($\mu g/m^3$)	Averaging time
Carbon monoxide	500–7,000	CO critical blood level < 2.5%	Nς	100,000	15 min
				60,000	30 min
				30,000	1 h
				10,000	8 h
Lead	0.01–2	Pb critical blood level < 100–150 μg Pb/l	Nς	0.5	1 year
Nitrogen dioxide	10–150	Small change in lung function among asthmatics	365–565	200	1 h
				40	1 year
Ozone and other		photochemical oxidants	10–100	Change in lung	
function	Nς	120		8 h	
Sulfur dioxide	5–400	Change in lung function among asthmatics	1,000	500	10 min
		Increase in symptom exacerbations among adults or asthmatics	250	125	24 h
			100	50	1 year

Table 1.10 Guideline values for pollutants based on effects other than cancer or odor/annoyance (WHO 2006)

Pollutants	Time-weighted average	Averaging time	Lifetime risk (unit concentration)
Organic Pollutants			
Acrylonitrile	–	–	2×10^{-5} ($\mu g/m^3$)
Benzene	–	–	6×10^{-6} ($\mu g/m^3$)
Butadiene			
Carbon Disulfide [a]	100 $\mu g/m^3$	24 h	
Carbon Monoxide	100 mg/m^3 [b]	15 min	
	60 mg/m^3 [b]	30 min	
	30 mg/m^3 [b]	1 h	
	10 mg/m^3	8 h	
1,2-Dichloroethane [c]	0.7 mg/m^3	24 h	
Dichloromethane	3 mg/m^3	24 h	
	0.45 mg/m^3	1 week	
Formaldehyde	0.1 mg/m^3	30 min	
Polycyclic Aromatic Hydrocarbons (PAHs as BaP)	–	–	8.7×10^{-5} (ng/m^3)
Polychlorinated Biphenyls (PCBs [c])	–	–	
Polychlorinated dibenzodioxins and dibenzofurans (PCDDs/PCDFs) [d]	–	–	
Styrene	0.26 mg/m^3	1 week	
Tetrachloroethylene	0.25 mg/m^3	Annual	
Toluene	0.25 mg/m^3	1 week	

(continued)

Table 1.10 (continued)

Pollutants	Time-weighted average	Averaging time	Lifetime risk (unit concentration)
Trichloroethylene			
Vinyl Chloride			4.3×10^{-7} (μg/m^3)
Inorganic Pollutants			
Arsenic			1.5×10^{-3} (μg/m^3)
Asbestos			
Cadmium [e]	5 ng/m^3 [e]	Annual	
Chromium (Cr$^{(VI)}$)	–	–	4×10^{-2} (μg/m^3)
Fluoride [f]	–	–	
Hydrogen sulfide [b]	15 μg/m^3	24 h	
Lead	0.5 μg/m^3	Annual	
Manganese	0.15 μg/m^3	Annual	
Mercury	1 μg/m^3	Annual	
Nickel	–	–	3.8×10^{-4} (μg/m^3)
Platinum [g]	–	–	
Vanadium [a]	1 μg/m^3	24 h	
Classical Pollutants [h]			
Nitrogen Dioxide	200 μg/m^3	1 h	
	40 μg/m^3	Annual	
Ozone and other photochemical oxidants	120 (100) μg/m^3	8 h	
Particulate matter [i]	Dose Response		
PM$_{10}$	(20/50) μg/m^3	(Annual/ 24 h)	
PM$_{2.5}$	(10/25) μg/m^3	(Annual/ 24 h)	
Sulfur dioxide	500 μg/m^3	10 min	
	125 (20) μg/m^3	24 h	
	50 μg/m^3	Annual	

[a] Not re-evaluated for the second edition of the guidelines

[b] Exposure at these conditions should be for no longer than the indicated times and should not be repeated within 8 h

[c] No guideline values have been recommended for PCBs because inhalation constitutes only a small proportion (about 1–2%) of the daily intake from food

[d] No guideline values have been recommended for PCDDs/PCDFs because inhalation constitutes only a small proportion (generally less than 5%) of the daily intake from food

[e] The guideline value is based on the prevention of a further increase of cadmium in agricultural soils, which is likely to increase the dietary intake

[f] Because there is no evidence that atmospheric deposition of fluorides results in significant exposure through other routes than air, it was recognized that levels below 1 μg/m^3, which is needed to protect plants and livestock, will also sufficiently protect human health

[g] It is unlikely that the general population, exposed to platinum concentrations in ambient air at least three orders of magnitude below occupational levels where effects were seen, may develop similar effects. No specific guideline value has therefore been recommended

[h] Values in brackets are the new air quality guideline values (WHO 2006)

[i] The available information for short- and long-term exposure to PM$_{10}$ and PM$_{2.5}$ does not allow a judgement to be made regarding concentrations below which no effects would be expected. For this reason no guideline values have been recommended, but instead risk estimates have been provided

1.10 Appendixes

Fig. 1.23 The forces which are applied to a vertical column of air in hydrostatic equilibrium

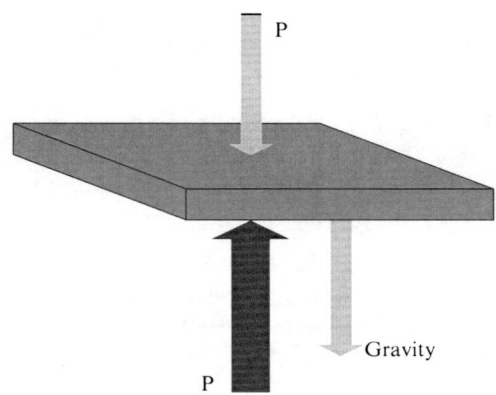

Having a column of very small width it can written that

$$\frac{p_2 - p_1}{z_2 - z_1} \cong \frac{\partial p}{\partial z}.$$

From the above equations it can be concluded that

$$\frac{\partial p}{\partial z} = -g\,\rho.$$

This is the expression of the hydrostatic equation. The minus sign expresses the fact that the pressure is decreasing with an increase of height.

Problems

1.1 If the two horizontal constituents of the air velocity are $u = -5$ m/s and $v = +5$ m/s, calculate the value of the total horizontal velocity and its direction.

1.2 The vertical column of sulphur dioxide inside the plume which originated from the mountain Pinatubo (integrated from the Earth's surface to the top of the stratosphere) was 3×10^{16} molecules/cm^2. The sulphur dioxide is transformed into sulphate aerosols. It is supposed that the whole quantity of sulphur dioxide is transformed. The aerosols have a mean diameter of 0.1 μm and are composed from sulphuric acid (75% per weight) and water (25% per weight). Calculate the surface area of aerosols integrated per unit surface area (μm^2 cm^{-2}). For the units conversion it is supposed that the aerosol concentration exists in a homogeneous layer with a height of 5 km.

1.3 The air temperature above the poles is $T_\pi = 5°C$ and above the equator is $T_I = 25°C$. Examine if the tropopause above the equator is colder than above the poles. The vertical temperature lapse rate is the same in both areas and equal to 6.5°C/km. The tropopause height above the poles is equal to $z_\pi = 8$ km and above the equator equal to $z_I = 18$ km.

1.4 A typical concentration of hydroxyl radicals [OH] is close to 10^6 molecules cm^{-3}. What is the volume ratio, which corresponds to the above concentration, at sea level and temperature 298 K.

1.5 Calculate the characteristic height (H) and the pressure value at height $z = 200$ m, when the air is dry and the pressure at height 100 m is equal to $p_d = 990$ mb and the mean temperature between the heights $z = 100$ m and $z = 200$ m is equal to $T = 284$ K.

1.6 Calculate the solar constant (the solar energy received on the Earth's atmosphere surface perpendicular to the Sun's rays). The Sun's radius is $R_{sun} = 6.96 \times 10^5$ km, the temperature at the Sun's surface is equal to $T_{sun} = 5,780$ K and the distance Earth – Sun equals 1.495×10^8 km.

1.7 What is the quantity (volume) of water vapour that must condense on the surface of an aluminum tin that holds a soda in order to heat the soda from $1°C$ to $16°C$.? The density of water and soda are equal to $\rho = 1,025$ kg m^{-3}, the thermal capacity of water and soda are $4,200$ J Kg^{-1}, the volume of the can is 354 ml and the latent heat of water is equal to $2.5\ 10^6$ J kg^{-1}.

1.8 Calculate the characteristic height (H(z)) for the pressure drop versus height in the atmosphere at height $z = 200$ m from the Earth's surface. The air is dry and the pressure at height $z = 100$ m is $p = 990$ mb, and the average temperature between 100 and 200 m height is $T = 284$ K.

1.9 Calculate the characteristic height (H(z)) for the pressure drop versus height in the atmosphere and the pressure at height $z = 10$ km from the Earth's surface. At this height the pressure is equal to $p = 250$ mb and the temperature equal to $T = 218$ K (base of the troposphere). What is the atmospheric pressure at height $z = 10.5$ km ?

1.10 (a) Calculate the arithmetic concentration of air molecules in the atmosphere at sea level pressure and temperature $T = 288$ K. The gas constant is equal to 0.083145 m^3 mb mole^{-1} K^{-1}. (b) What is the arithmetic concentration at atmospheric pressure 1 mb? Discuss the results.

1.11 (a) Calculate the wave length that corresponds to the maximum of emitted radiation from the Earth's and Sun's surfaces. The effective temperature at the Sun's surface is $T = 5,785$ K and the average surface temperature of Earth equal to $T = 288$ K. (b) Increasing by a factor of two the temperature of a black body, what will be the result to its radiation emission?

References

Ahrens, C. D. (1994). *Meteorology today – An introduction to weather, climate and the environment* (5th ed.). New York: West Publishing Company.

References

Brausseur, G. P., Orlando, J. J., & Tyndall, G. S. (1999). *Atmospheric chemistry and global change*. USA: Oxford University Press.

Erisman, J. W., Otjes, R., Hensen, A., Jongejan, P., van den Bulk, P., Khlystov, A., et al. (2001). Instrument development and application in studies and monitoring of ambient ammonia. *Atmospheric Environment, 35*(11), 1913–1922.

Fenger, J. (2009). Air pollution in the last 50 years – From local to global. *Atmospheric Environment, 43*, 13–22.

Friedli, H., Lotscher, H., Oeschger, H., Siegenthaler, U., & Stauffer, B. (1986). Ice core record of 13C/12C ratio of atmospheric CO_2 in the past two centuries. *Nature, 324*, 237–238.

Grennfelt, P., & Schjoldager, J. (1984). Photochemical oxidants in the troposphere – A mounting menace. *Ambio, 13*(2), 61–67.

Imbrie, J., & Imbrie, K. P. (1979). *The ice ages: Solving the mystery*. Cambridge: Harvard University Press.

Intergovernmental Panel on Climate Change (IPCC), et al. (2007). Fourth assessment report. In S. Solomon, D. Qin, & M. Manning (Eds.), *The physical science basis*. New York: Cambridge Univ. Press.

McNeill, J. R. (2001). *Something new under the sun. An environmental history of the twentieth century world*. New York – London: W.W. Norton & Company.

Morawska, L., Ristovski, Z., Jayaratne, E. R., Keogh, D. U., & Ling, X. (2008). Ambient nano and ultrafine particles from motor vehicle emissions: characteristics, ambient processing and implications on human exposure. *Atmospheric Environment, 42*, 8113–8138.

Philander, S. G. (1998). *Is it temperature rising? The uncertain science of global warming*. Princeton University Press: Princeton, NJ.

Stull, R. B. (1997). *An Introduction to boundary layer meteorology*. Dordrecht: Kluwer Academic Publishers.

Sutton, M. A., Asman, W. A. H., Ellermann, T., Van Jaarsveld, J. A., Acker, K., Aneja, V., et al. (2003). Establishing the link between ammonia emission control and measurements of reduced nitrogen concentrations and deposition. *Environmental Monitoring and Assessment, 82*(2), 149–185.

Turco, R. P. (2002). *Earth under siege. From air pollution to global change*. New York: Oxford University Press.

U.S. Environmental Protection Agency (USEPA) (2004). Air quality criteria for particulate matter, US Environmental Protection Agency, Research Triangle Park, NC, USA (Report EPA/600/P-99/002aF and bF).

World Health Organization (WHO). (2006). *WHO Air quality guidelines for particulate matter, ozone, nitrogen dioxide and sulfur dioxide - Global update 2005 - Summary of risk assessment*. Geneva, Switzerland: WHO Press. http://www.euro.who.int/Document/E90038.pdf.

Zimmerman, M. (2004). Paleopathology and study of ancient remains. In *Encyclopedia of medical anthropology* (Vol. I). US: Springer.

Chapter 2
First Principles of Meteorology

Contents

2.1	General Aspects of Meteorology	67
2.2	Vertical Structure of the Temperature and Conditions of Atmospheric Stability	70
	2.2.1 Dry Vertical Temperature Lapse Rate	71
	2.2.2 Wet Vertical Temperature Lapse Rate	74
	2.2.3 Temperature Inversion	75
2.3	Atmospheric Variability – Air Masses – Fronts	77
	2.3.1 Air Masses	77
	2.3.2 Classification of Air Masses	78
	2.3.3 Fronts	79
	2.3.4 Wave Cyclone	85
2.4	Turbulence – Equations for the Mean Values	86
2.5	Statistical Properties of Turbulence	87
2.6	Atmospheric Temperature	92
	2.6.1 Temperature Season Variability	92
	2.6.2 Temperature Daily Variability	96
	2.6.3 Heating of the Earth's Surface and Heat Conduction	98
	2.6.4 Distribution of Temperature in the Air	100
2.7	Humidity in the Atmosphere	101
	2.7.1 Mathematical Expressions of Humidity in the Atmosphere	102
	2.7.2 Dew Point	104
	2.7.3 Clouds in the Atmosphere	106
	2.7.4 Precipitation	107
	2.7.5 Study of Precipitation Scavenging	109
2.8	Applications and Examples	112
References		117

Abstract The second chapter examines general aspects of meteorology including conditions of atmospheric stability in conjunction with the vertical temperature lapse rate. It also studies large-scale weather changes which are related to changes in pressure systems. Air masses have different thermodynamic characteristics based on their origin and the morphology of the surfaces above which they move. A classification of the air masses is created, together with a mathematical description of the statistical properties of the air masses' transport. Furthermore, the temperature and humidity variability in the atmosphere is studied. Finally the clouds in the atmosphere and general aspects of precipitation are also studied.

2.1 General Aspects of Meteorology

The study of atmosphere dynamics is subsumed under the science of meteorology. The atmosphere contains thousands of chemical species in trace quantities (ppm to ppt levels) (Finlayson Pitts and Pitts 1986; 2000). Therefore the troposphere can be viewed as a huge container that includes gaseous and particulate matter pollutants. The atmosphere is a dynamic system with continuous exchange of its gaseous components between the atmosphere and the earth's surface, including the vegetation and the oceans. Emissions of pollutants are transported into the atmosphere at long distances from their sources. The dynamic of the atmosphere and the chemical reactivity of the pollutants, as well as the size of particulate matter, determine their residence time and their effects on humans and ecosystems (Seinfeld and Pandis 2006). Table 2.1 presents the different spatial scales of pollutant transport in the atmosphere and related physico-chemical processes.

Figure 2.1 presents the different time and length scales related to atmospheric processes ranging from molecular diffusion to climatic impacts. In the atmosphere the chemical composition of atmospheric species can be divided into four main groups, namely sulfur, nitrogen, carbon and halogen containing compounds (Finlayson-Pitts and Pitts 1986; Seinfeld and Pandis 2006). Of course there are chemical compounds in the above groups that include atoms from other groups such as compounds that include both sulphur and carbon atoms. The chemical compounds which are emitted into the atmosphere eventually are removed and there exists a cycle for these compounds. This cycle is called biogeochemical cycle of the compound. The term "air pollution" is used when chemical compounds emitted from mainly anthropogenic activities are at concentrations above their normal ambient levels and have measurable effects on humans and ecosystems. In addition, Fig. 2.1 shows the time and spatial scales related to air pollution, turbulence, clouds, weather and climate.

Understanding of the complex sequence of events starting from the emissions of air pollutants to the atmosphere with human health effects as a final event is necessary for the prognosis of potential risk to humans from specific chemical compounds and mixtures of them (see Fig. 2.2). Furthermore, understanding of the chemical composition/size distribution characteristics of particulate matter (PM) and the chemical reactivity of gaseous pollutants together with their indoor-outdoor

Table 2.1 Spatial scales of pollutant transport in the atmosphere and related phenomena

Scale	Dimension	Examples of physical and chemical processes in the atmosphere
Molecular scale	$\ll 2$ mm	Molecular diffusion
Microscale	2 mm–2 km	Industry emissions, clouds
Mesoscale	2–2,000 km	Cloud coalesce, storms, air pollution at urban centers
Synoptic scale	500–10,000 km	Low and high pressure systems, ozone hole in the Antarctica
Global scale	$>$ 10,000 km	Decrease of stratospheric ozone, planetary wind systems

2.1 General Aspects of Meteorology

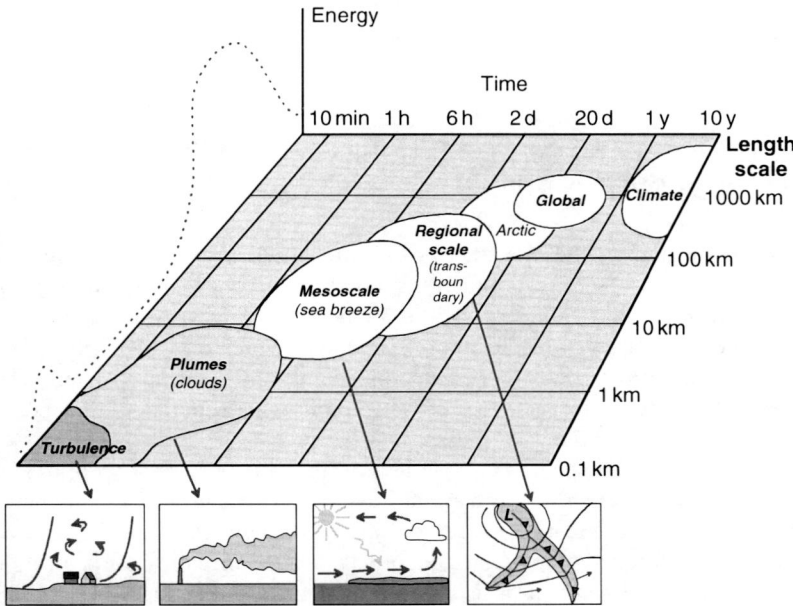

Fig. 2.1 Time (s), energy and length (m) scales in the atmosphere and related phenomena

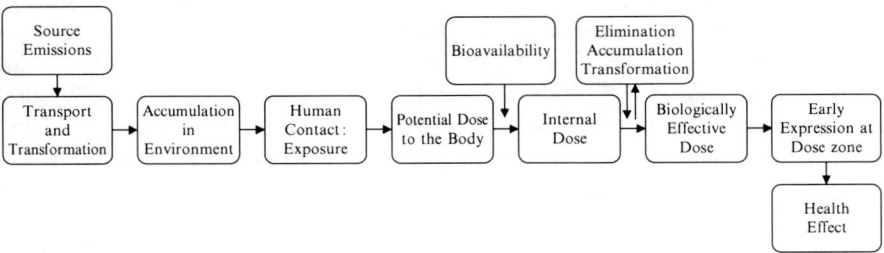

Fig. 2.2 Schematic representation of the complex sequence from emissions of air pollutants to health effects (Adapted from Lioy 1990)

characteristics and their relation to human exposure and internal dose are necessary steps for the quantification of human exposure to air pollutants.

Many problems of air pollution occur on several scales, such as the acidification problem which extends from the mesoscale to a regional scale. The majority of air pollution episodes occur in the lower part of the atmosphere, which is called the planetary boundary layer as discussed in Chapter 1. This layer is defined as the lowest part of the troposphere which is affected from the surface forces in time scales of 1 h or lower. The structure of the boundary layer is not static but dynamic and contains usually the first 1,000 m from the Earth's surface.

2.2 Vertical Structure of the Temperature and Conditions of Atmospheric Stability

Meteorology examines the movement of air molecules in the atmosphere. Atmospheric stability is examined under the conditions in which there is a small displacement of an air volume with the application of an external force. In the case that the external force results in bringing the air volume to its original position prior to its displacement, then there are stable conditions in the atmosphere. On the contrary, if the external force brings the air volume away from the original position, there are unstable conditions.

Stability in the atmosphere is dependent on the vertical profile of temperature and humidity of ambient air. Warm air has lower density than cold air and therefore it is lighter. A similar situation occurs for humid air which has lower density than dry air and therefore is lighter. Consequently, a warmer or more humid air volume than the surrounding ambient air is characterized as unstable and will ascend into the atmosphere. On the contrary, an air volume that is colder or drier than the surrounding ambient air is characterized as stable and will descend into the atmosphere until it reaches equilibrium.

The stability conditions in the atmosphere are related to the atmosphere's ability to mix and spread out pollutants. These conditions determine also the turbulent conditions in the atmosphere and the cloud formation.

Atmospheric air absorbs less heat than the Earth's surface due its lower heat capacity. Therefore the layer of the atmosphere which is closer to the Earth's surface receives more energy, and consequently more heat, than the above layers. Due to the heating these layers of air become lighter than the above layers and are lofted above with consequent expansion and cooling. Since this air volume expansion occurs at small time intervals without significant heat exchange with the surrounding air environment, it is an adiabatic process. Of course the process is not pure adiabatic since the air masses are not thermally insulated. However, since the expansion of the air masses happens quickly and the heat exchange with conduction and radiation is slow, the process of adiabatic expansion in the atmosphere is assumed. Therefore, when an air mass is moved to an area with lower pressure it is adiabatically expanded and cooled. On the contrary, when an air mass is moved to an area with a higher atmospheric pressure it is contracted adiabatically and heated. Indeed, observations have shown the importance of adiabatic processes in the atmosphere in relation to weather.

In the International Standard Atmosphere (ISA) the temperature is decreased with height at a rate of $0.65°C/100$ m. This rate is called the temperature lapse rate. The determination of the temperature lapse rate is presented in the following subsections.

2.2 Vertical Structure of the Temperature and Conditions of Atmospheric Stability

2.2.1 Dry Vertical Temperature Lapse Rate

The calculation of the temperature lapse rate for a volume of dry air is examined here assuming adiabatic expansion as discussed earlier. The first law of thermodynamics can be expressed as:

$$dU = dQ + dW, \qquad (2.1)$$

where, U, Q and W are the internal energy of the air volume, its heat and produced work respectively. The internal energy is given from the expression

$$dU = C_v \, dT, \qquad (2.2)$$

where, C_v is the thermal capacity of the system at constant volume. The expression which gives the work is given by

$$dW = -p \, dV. \qquad (2.3)$$

The law of ideal gasses can be also expressed as

$$pV = mRT/M_{air} \Rightarrow d(pV) = mR \, dT/M_{air} = p \, dV + V \, dp \qquad (2.4)$$

where m is the air mass.

Since adiabatic conditions occur: $dQ = 0$.

Combining equations (2.1) to (2.4) it can be found that

$$C_v \, dT = V \, dp - \frac{mR \, dT}{M_{air}} = \frac{mRT \, dp}{M_{air} \, p} - \frac{mR \, dT}{M_{air}} \Rightarrow \frac{dT}{dp} = \frac{mRT/M_{air} \, p}{C_v + mR/M_{air}}. \qquad (2.5)$$

In Chapter 1 the change of pressure versus height was described with the hydrostatic equation: $\frac{dp}{dz} = -\frac{M_{air} \, g \, p}{RT}$. With the combination of the hydrostatic equation and equation (2.5) it can be concluded that:

$$\frac{dT}{dz} = -\frac{mg}{C_v + mR/M_{air}} = -\frac{g}{\hat{c}_v + R/M_{air}}, \qquad (2.6)$$

where $\hat{c}_v$ is the thermal capacity at constant volume per unit mass. Finally, it can be written that $\frac{dT}{dz} = -\frac{g}{\hat{c}_p}$ since $\hat{c}_p - \hat{c}_v = \frac{R}{M_{air}}$.

The ratio $\frac{g}{\hat{c}_p}$ for dry air is equal to 0.976°C/100 m, has the symbol Γ and is called dry lapse rate.

If Λ is the dominant lapse rate in the atmosphere, then the following cases of stability exist for the volume of air:

$$\Lambda = \Gamma, \text{ neutral}$$
$$\Lambda > \Gamma, \text{ unstable}$$
$$\Lambda < \Gamma, \text{ stable}$$

Usually, unstable conditions exist in the first 100 m from the Earth's surface on a sunny day. Neutral conditions exist at day or night with clouds and wind and stable conditions near the Earth's surface at night.

Figure 2.3 shows examples of different stability conditions in the atmosphere. The solid and semi-continuous lines refer to the lapse rates of the air volume under study and the atmosphere respectively. The circle denotes the air volume under study and its position in the atmosphere. At this position the air volume temperature is equal to the atmosphere's temperature. Unstable conditions occur when the temperature of the rising air volume is higher than the temperature of the surrounding air. In this case the air volume will accelerate upwards. Stable conditions occur when the temperature of the rising air volume is lower than the temperature of the surrounding air. In this case the air volume tends to return to its initial position of equilibrium. Neutral conditions prevail when the temperature of the rising air volume is the same as the temperature of the surrounding air. In this case the air volume follows the movement of the surrounding atmosphere.

At the upper part of the Fig. 2.3 is shown a relative gravitational example for the stability conditions where a ball is located at the top of a hill (unstable condition), at

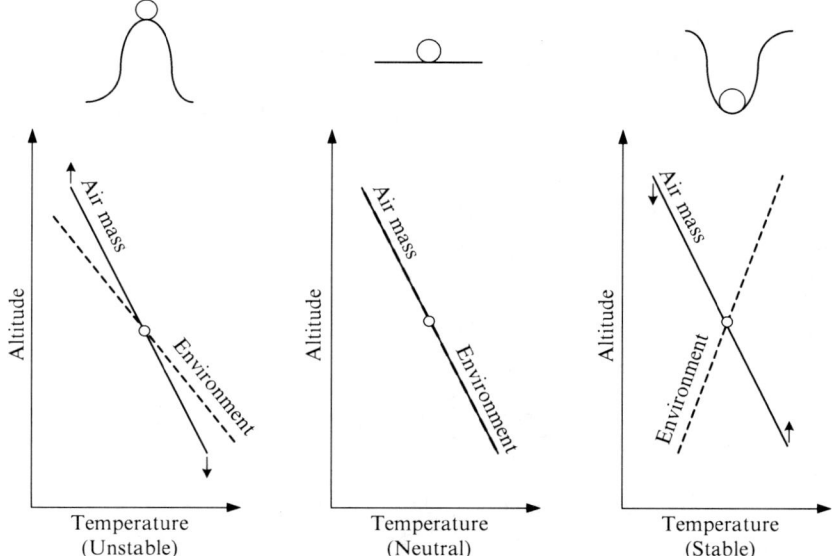

Fig. 2.3 Graphical representation of the relation between the temperature and the height for an air volume for unstable, neutral and stable conditions in relation to the surrounding atmosphere (environment) (Adapted from Hanna et al. 1981)

2.2 Vertical Structure of the Temperature and Conditions of Atmospheric Stability

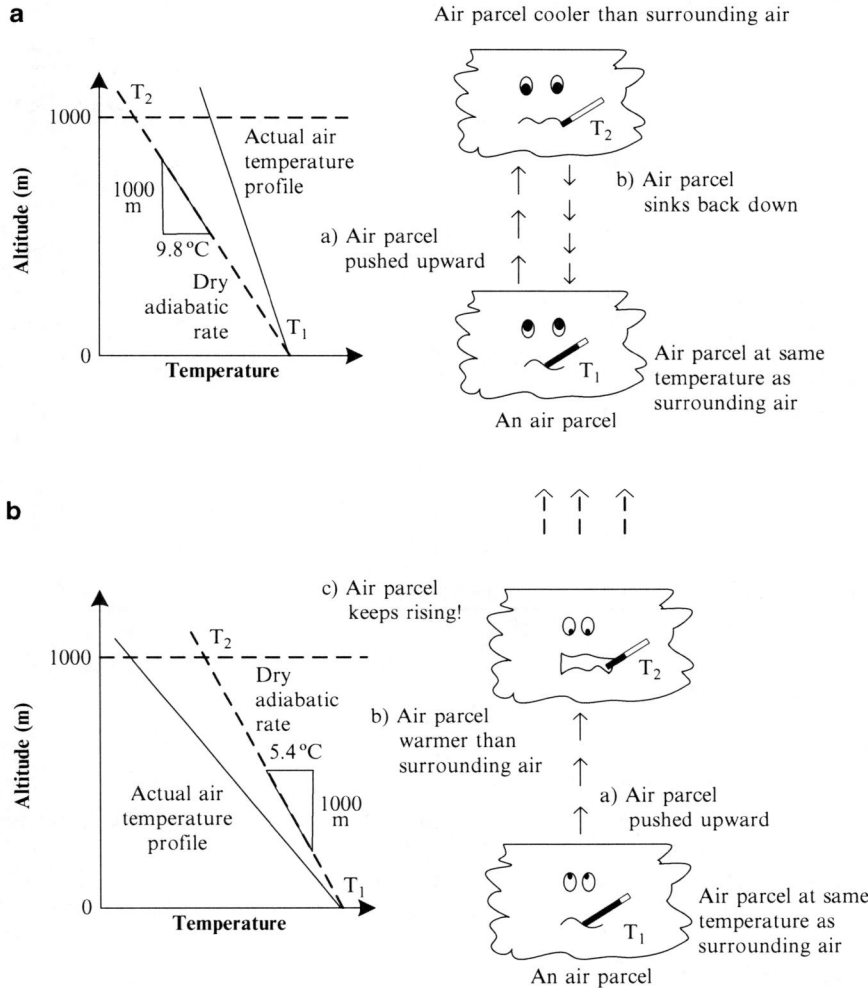

Fig. 2.4 The relation of the temperature versus height for an air volume for (**a**) stable and (**b**) unstable conditions (Adapted from Hemond and Fechner-Levy 2000)

a flat surface (neutral condition) and at the base of a valley (stable condition). Another representation of the unstable and stable conditions in the atmosphere is shown in Fig. 2.4.

A more widely accepted methodology for the calculation of atmospheric stability has been introduced by Pasquill. The methodology is based on measurements of the wind speed at a height of 10 m and intensity of the Sun's radiation during the day and cloud cover during the night (Table 2.2). However, a more practical approach is the use of a radiosonde for determination of the vertical profile of several meteorological parameters such as temperature and pressure.

Table 2.2 Stability conditions in the atmosphere using the Pasquill methodology

Wind velocity at height of 10 m (m/sec)	Day			Night	
	Incoming sun's radiation intensity			Cloud cover	
	High	Medium	Low	>4/8	<3/8
<2	A	A-B	B	-	-
2–3	A-B	B	C	E	F
3–5	B	B-C	C	D	E
5–6	C	C-D	D	D	D
>6	C	D	D	D	D

A Very unstable
B Moderate unstable
C Slightly unstable
D Neutral (it is applied for conditions of total cloud cover both day and night)
E Slightly stable
F Moderate stable

Table 2.3 Equivalence of stability classes with the methodologies of the vertical temperature change and the Pasquill methodology

Stability class	Vertical temperature change dT (°C/100 m)	Pasquill class
Unstable	$dT < -1$	A + B + C
Neutral	$-1 \leq dT < 0$	D
Slightly Stable	$0 \leq dT < 1$	E
Stable	$dT \leq 1$	F

Another methodology for determination of stability classes is based on examination of the vertical temperature profile. This methodology is used widely for the application of Gaussian models and the equivalence of this stability class methodology with the Pasquill methodology is presented in Table 2.3.

2.2.2 Wet Vertical Temperature Lapse Rate

When the air contains water vapor, the thermal capacity $\hat{c}_p$ of air has to be corrected. If w_v is the ratio of the mass of water vapor to the mass of dry air in a specific air volume, then the new thermal capacity coefficient $\hat{c}_p'$ is given by the expression:

$$\hat{c}_p' = (1 - w_v)\hat{c}_{pa} + w_v \hat{c}_{pv} \qquad (2.7)$$

where, $\hat{c}_{pv} > \hat{c}_{pa}$ and therefore $\hat{c}_p' > \hat{c}_{pa}$. The symbol *a* refers to air and *v* to water vapor.

Therefore the rate of cooling of rising humid air inside a cloud is smaller than that of dry air. The humid air volume will continue to rise until the partial pressure of the water vapor becomes equal to the equilibrium water vapor pressure.

2.2 Vertical Structure of the Temperature and Conditions of Atmospheric Stability

This condition will lead to the condensation of water vapor. If ΔH_v is the heat of sublimation, then the release of heating due to the condensation of water vapor is given by the expression

$$dQ = -\Delta H_v \, m \, dw_v. \tag{2.8}$$

Following the same derivation as in the case of dry air, using the first law of thermodynamics, it can be concluded that

$$C_v \, dT = -\Delta H_v \, m \, dw_v + \frac{m \, R \, T}{M_{air}} \frac{dp}{p} - \frac{m \, R \, dT}{M_{air}}$$

$$\Rightarrow \left(C_v + \frac{mR}{M_{air}} \right) dT = -\Delta H_v \, m \, dw_v + \frac{m \, R \, T}{M_{air}} \frac{dp}{p}$$

with a final result being

$$\frac{dT}{dz} = -\frac{g}{\hat{c}_p} - \frac{\Delta H_v}{\hat{c}_p} \frac{dw_v}{dz}. \tag{2.9}$$

The term $\frac{dw_v}{dz}$ has negative value for a rising volume of air where water vapor condensation takes place inside Therefore the cooling rate of a humid air inside clouds is lower than that of dry air. The reason is that with the increase of height the percentage of water vapor is decreasing due to its condensation. The term $\frac{dw_v}{dz}$ is dependent on temperature since the equilibrium vapor pressure of water vapor is increasing considerably with the temperature.

2.2.3 Temperature Inversion

As we have examined in Chapter 1, on many occasions the temperature of air instead of decreasing with height increases at some places in the atmosphere. This is called temperature inversion. The atmosphere is in stable equilibrium inside the inversion layer and therefore it does not favor vertical movements of air. The temperature inversion may result in increased concentration of air pollutants at the lower layers of the atmosphere.

There are three main factors which may result in the occurrence of thermal inversions:

- The cooling of lower atmospheric layers (radiation or surface inversion).
- The adiabatic warming of descending air (subsidence inversion).
- Horizontal transport of warm or cold air.

A typical profile of a plume during a subsidence inversion is shown in Fig. 2.5 and in the case of radiation inversion in Fig. 2.6. The layer which extends from the

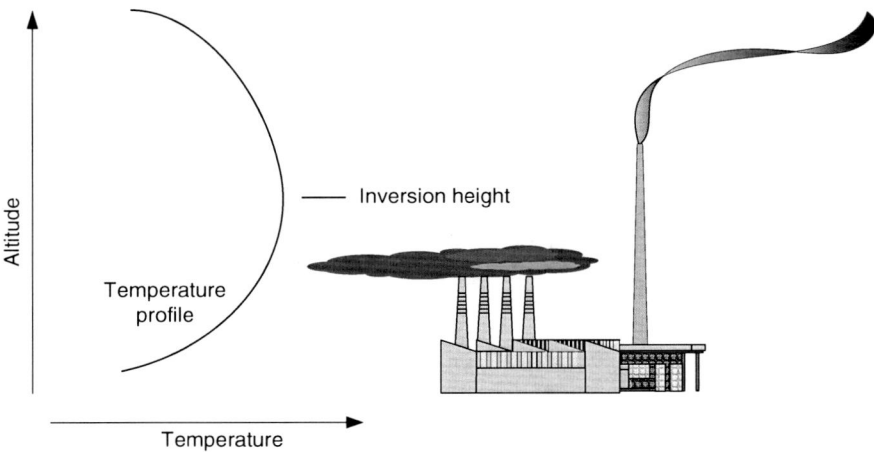

Fig. 2.5 The plume from the lower chimneys is trapped inside the inversion, whereas, the plume from the tall chimney which is located above the inversion is rising, mixing and transported

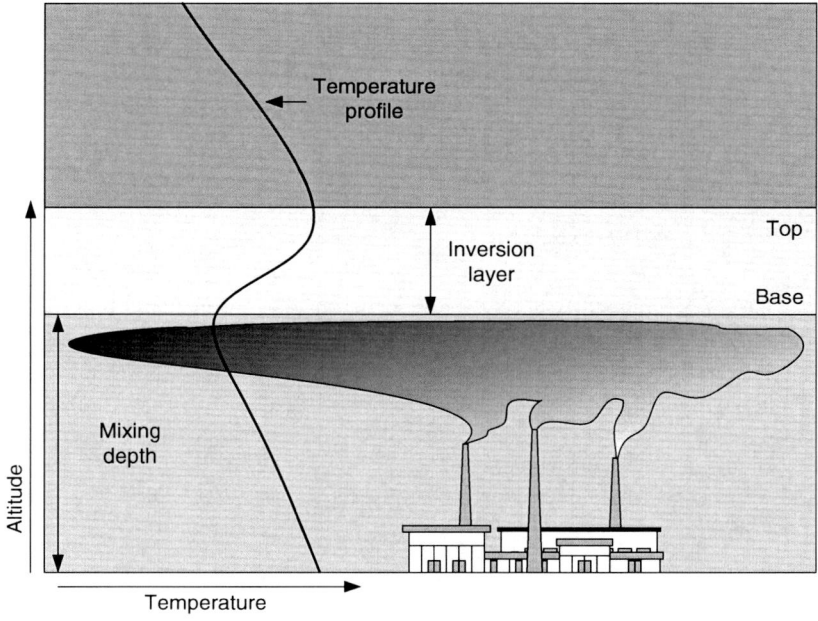

Fig. 2.6 The inversion height is not allowing the escape of gaseous pollutants under it. If the inversion height is becoming lower, then the mixing height is reduced and the pollutants are trapped in a smaller air volume

Earth's surface up to the base of the inversion layer is called the mixing layer. The height of this layer is called the mixing height.

It can be noted that the relatively unstable air below the inversion allows vertical mixing of air pollutants up to the base of the inversion layer. The stable air in the inversion layer does not allow vertical mixing inside the inversion layer and acts as an obstacle to the pollutant entrainment upwards. If the inversion moves upwards, then the height of the mixing layer is increasing and the pollutants will mix in a bigger air volume and, on the other hand, if the inversion moves downwards then the mixing height will be lower and the pollutants will be concentrated in a smaller air volume. In the latter case the concentration of pollutants is increasing and in urban areas may lead to concentrations above the health limits. Since the atmosphere tends to be usually more unstable during the afternoon and more stable in the morning, then there is higher mixing height at the afternoon and lower early in the morning.

2.3 Atmospheric Variability – Air Masses – Fronts

Weather changes of large scale are related to changes of the pressure systems. The first step for weather prognosis is given by the Scandinavian School of meteorology which realized the importance of the formation and transport of High and Low pressure systems. The next step was the discovery of the characteristics of the air masses and front zones, the areas in which air masses are met with different thermodynamic characteristics. This led to a more detailed study of the air masses which form the basis for the weather phenomena.

Generally the meteorological conditions which are persistent in an area at specific time are dependent on the characteristics of the air masses which pass above this region or from the interactions of two different air masses which meet there. Therefore the weather above a region is quite uniform with small differences which are due to local morphological characteristics of the region such as the topography, the vegetation and distribution of land and sea.

2.3.1 Air Masses

Air masses are large bodies of atmospheric air which have similar properties of temperature and humidity distribution. It is possible for the diameter of an air mass to be larger than 1,500 km, covering large continental and ocean regions. Its height may reach the tropopause. With the air masses is accomplished the general atmospheric circulation and the transport of large quantities of heat from the equator to the poles. Inside an air mass, important factors are its humidity content, its temperature and especially the temperature change with height. The quantity of humidity influences the cloud type and consequently the rain quantity. The vertical temperature distribution affects the stability of the air mass.

In order for an air mass to attain similar temperature and humidity characteristics, it is necessary to remain for several days stagnant above a region which also has similar characteristics. This region is called the source of the air mass. At which degree the air mass attains the characteristics of the source region is dependent on its residence time and the temperature difference between the initial air temperature and the temperature of the region's surface. The air mass during its residence above the region becomes at a certain degree homogeneous. Under the influence of wind flow the air mass starts to move and retains during its transport at high degree the characteristics of its regional source.

From a thermodynamic point of view the air masses are classified in two categories, the warm and cold ones. The warm air masses are warmer than the surface above which they are transported and for this reason are becoming colder at their base and are also stable. The freezing of air masses is more effective on the layers which are contiguous to the surface and extends slowly above, mainly due to turbulent movements and not due to heat transfer. This is also the reason for the formation of temperature inversions. When the winds are weak there are often formations of mist and dew, whereas, with stronger winds there is the formation of low clouds (Stratus), below the upper limit of the temperature inversion.

Cold air masses are colder than the Earth's surface and for this reason receive heat from below. The heating of those layers which are located close to the surface results in a sudden drop of their temperature versus height. This temperature drop results also in an increased temperature lapse rate and therefore unstable conditions and enhanced ascending air movements. If these cold air masses contain sufficient quantities of humidity or if they are supplied with humidity from the warmer layers below, then there is a formation of clouds of vertical development which also results in intense rainfall.

2.3.2 Classification of Air Masses

The classification of air masses is based on the following criteria:

- The source of the air mass. Air transported above oceans absorbs humidity and tends to be saturated in the lower layers. On the contrary, continental air masses remain dry since there is not enough water quantity for evaporation on the surface.
- The trajectory which is followed above the Earth's surface. The polar air which is transported at lower latitudes is receiving heat from below and becomes unstable. On the contrary, the tropical air which is transported at higher latitudes becomes more stable since it becomes colder at its base.
- If the air is characterized as diverging or converging. An air mass which is affected from the divergence of air at a high pressure system at the Earth's surface will move downwards at a low rate and converted to a warmer, drier and

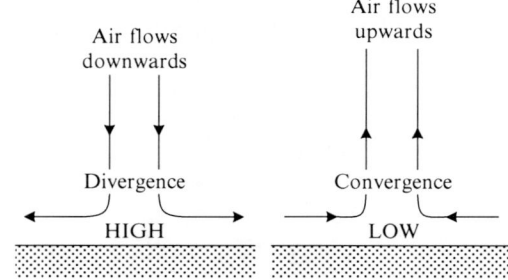

Fig. 2.7 The descending air, which is resulted from divergence at the surface is stable, whereas the ascending air which is resulted from convergence at the surface is unstable

more stable air mass. On the contrary, an air mass which is affected from the convergence of air at a low pressure system at the Earth's surface will move upwards and will be converted to a colder and more unstable air mass (Fig. 2.7).

Based on the above characteristics one can distinguish two main categories of air:

- Polar air masses at high geographical latitudes and
- Tropical air masses at low geographical latitudes.

These two basic categories are further divided into continental and ocean masses based on whether the air mass moves above a continent or a sea as follows:

- Polar Maritime air (Pm)
- Polar Continental air (Pc)
- Tropical Maritime air (Tm)
- Tropical Continental air (Tc)

In addition to the above basic air masses, there are also the Arctic masses. Sources of the arctic masses are the arctic regions close to the poles. These are cold air masses through all seasons and especially during winter. The behavior and alteration of the arctic air masses that intrude at lower latitudes is dependent on the part of the Earth's surface above which they are transported. In general if they are transported above oceans, then the lower layers become warmer and receive large quantities of water vapour. This results in the formation of unstable clouds of vertical development and extension of the weather characteristics in larger areas. Transport above cold continental areas results in stable air masses with mild weather characteristics, the main feature being intense cold.

2.3.3 Fronts

We pointed out previously that air masses have different thermodynamic characteristics based on their source of origin and the morphology of the surfaces above which they move. When two air masses with different characteristics come in contact, they mix very slowly and form an inhomogeneous surface whose vertical profile is called a frontal surface. The cross section of this surface with the Earth's surface is called a front.

The frontal surfaces have small depth and are steep in relation to the horizon, with the warmer mass always being above the colder one. The most important fronts are the following:

2.3.3.1 Polar Front

At the mid-latitudes, polar cold air masses collide with tropical warm ones. The separation zone between these tropical and polar air masses is called a polar front. The position of the polar front is variable. The strong polar air moves the warm tropical air at some places whereas, in some other places the polar front declines at northern positions due to pressure from the tropical air. As a result the polar front has a corrugated form as shown in Fig. 2.8.

However, the polar front seldom is found as a continuous zone encompassing the whole hemisphere as shown in Fig. 2.8. There are locations around the hemisphere where the transition between the polar and tropical air masses is so smooth that the dividing curve does not appear. Therefore the polar front is not continuous,

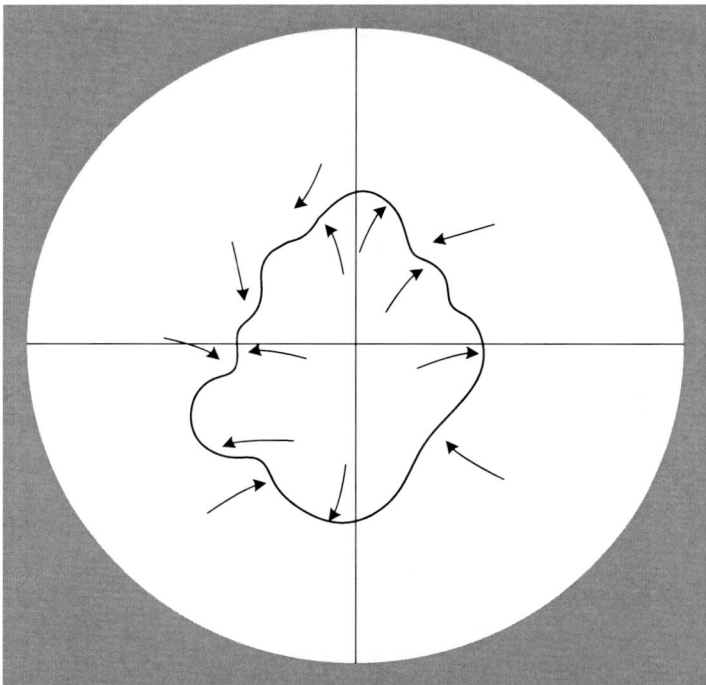

Fig. 2.8 Semi-hemispheric view of the polar front

2.3 Atmospheric Variability – Air Masses – Fronts

especially during the summer period in the northern hemisphere where the front is pressed northerly of the 60° parallel. During winter the polar front that is located usually at medium latitudes moves south and invades the tropical zones.

2.3.3.2 Cold Front

Even though the polar front represents a main zone of discontinuity in each hemisphere, fronts can be formed at any location on Earth if there is a coalescence of air masses with different thermodynamic characteristics. During the meeting of the air masses, two types of fronts are formed, the warm and cold fronts.

When two air masses (a cold and a warm one) come in contact and move so that the cold mass displaces the warm mass, then the surface that divides them is called a cold frontal surface. A side view of a cold front is shown in Fig. 2.9.

In a cold front, the warm mass which is located at the lower layers moves slower than the cold air. Therefore the cold air which moves quicker and penetrates through the lower levels of the warm air, causes a violent vertical upward movement. It has to be noted that the warm air is ascending at both warm and cold fronts and is responsible for the weather changes. It is furthermore interesting to refer to the characteristics of the warm and cold fronts, since their dynamics determine the extent of cloud coverage, the intensity and duration of rain and finally the direction and velocity of the wind.

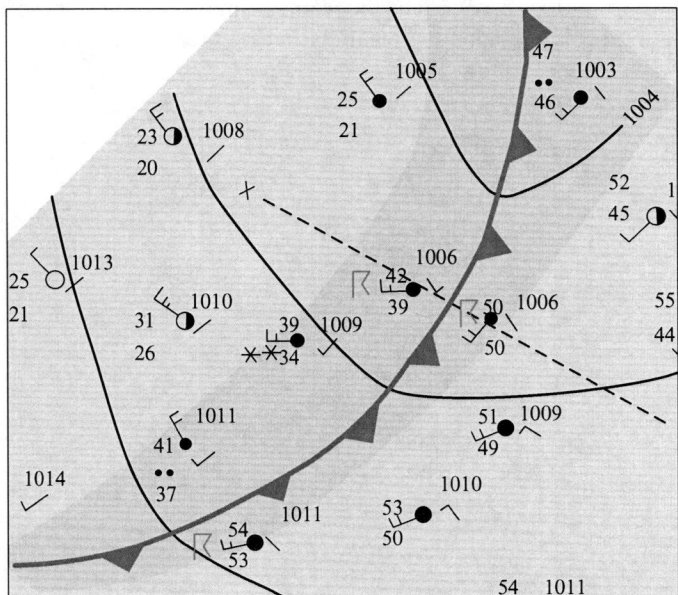

Fig. 2.9 Surface weather associated with a cold front. The dark band denotes the region of the weather phenomena

The peak of a cold frontal surface in relation to the Earth's surface is close to 1/50 and the frontal weather covers a narrow region ranging from 30 to 50 miles. The physical process which occurs for the formation of the cold front is related with the movement of warm air which is transported faster than the warm air and, because it is heavier, is located under it. Therefore the warm air which is lighter has an upward movement along the frontal surface. During the upward movement process it is cooled adiabatically due to expansion and reaches saturation. Furthermore there is a condensation process and cloud formation. Since the ascending warm mass is transported quickly, there is formation not only of stratocumulus clouds but also of clouds of vertical development (storms). The weather phenomena which are related with a cold front are intense (intense rain, storms, hail) and of small time scale.

The distribution of clouds as well rain is dependent on the atmosphere's stability and the humidity percentage of the rising warm air. The limit between the two air masses on the Earth's surface is depicted in meteorological maps as a line with a blue shading and solid triangles which point towards the moving front. The cold front moves quite quickly.

2.3.3.3 Warm Front

When two air masses (warm and cold) are in contact and moving so that the warm mass forces the cold mass to move, then the surface which divides them is called a warm frontal surface. The intersection of this surface with the Earth's surface is called a warm front (Fig. 2.10). Since the warm air mass is lighter, it ascends above the cold air and cools adiabatically resulting in extended condensation.

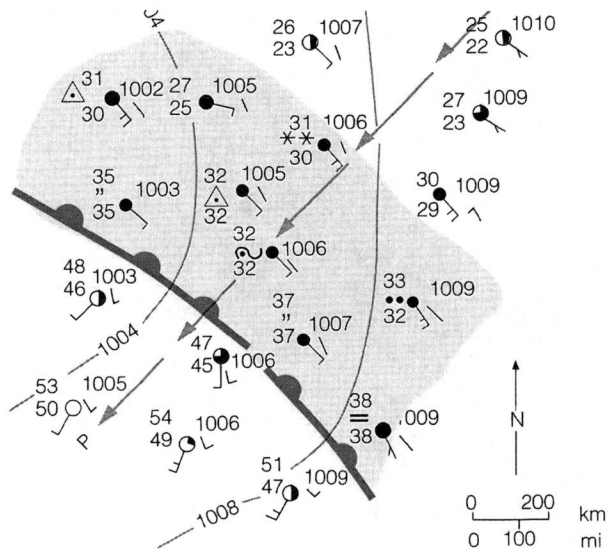

Fig. 2.10 Surface weather associated with a warm front

2.3 Atmospheric Variability – Air Masses – Fronts

In a warm front the presence of warm air at the upper atmospheric layers prevents the mixing of air masses vertically. The presence of warm air masses that are rich in humidity and that move above cold air masses has as a result a volume increase in air mass and the formation of clouds and rain with gradual cooling of the air. At the warm fronts there is the formation of extended cloud layers and storms with the scavenging of chemical compounds which are inside them. Transport from the lower layer is not easy due to the absence of clouds.

The gradient of the warm front in relation to the surface is close to 1/100. The warm air, since it is lighter, is moving faster than the cold air and is ascending above it along the warm frontal surface. During its ascent the warm air is getting colder adiabatically and after the saturation point starts the condensation process with the formation of extended cloud layers. The width of these cloud layers extends between 100 and 300 miles and its length may extend 1,000 miles. Since the gradient of the warm front is small, then the upward movement of the warm air masses occurs slowly. The warm air masses are characterized by stable conditions due to the uniformity of the temperature and relative humidity characteristics inside them, and the clouds which are formed are stratocumulus. These clouds are very extensive geographically and give rain over a range up to 2,000 miles which may be light or medium in intensity but has long duration. In special cases the warm air can be unstable so there is a possibility of formation of clouds with vertical development (storms).

The limit on the Earth's surface between the two air masses is depicted in the meteorological maps as a red line with semi-circles which show the movement direction of the warm front. In Table 2.4 are presented the main characteristics of

Table 2.4 Characteristics of warm and cold fronts

Characteristics	Cold front	Warm front
Distribution of air masses	The cold air mass press the warm air mass	The warm air mass ascends above the cold air mass and pushes it away
Gradient of the frontal surface (in relation to the surface)	Large gradient (1/50)	Small gradient (1/100)
Weather phenomena	Intense weather phenomena (storms, strong winds, strong rain etc.)	Moderate weather phenomena (weak to moderate rain, moderate winds)
Band of weather phenomena	Close to 150 km (100 km front of the cold front and 50 km behind it)	400 km (or more) The weather phenomena occur before the warm front
Duration of the weather phenomena	Small duration	Large duration
Symbol	Solid blue line with triangles along the front showing its direction of movement	Solid red line with half circles along the front showing its direction of movement

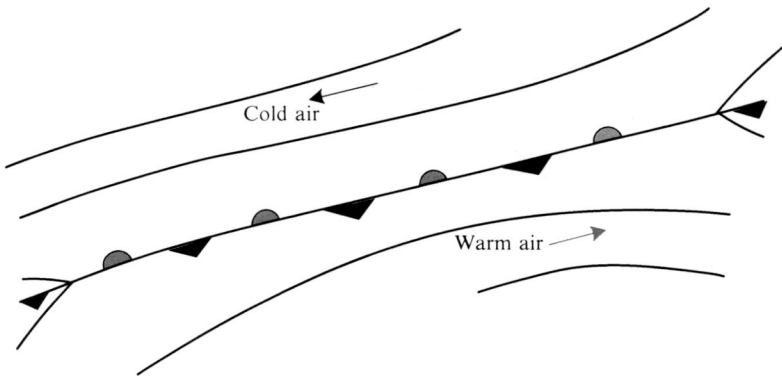

Fig. 2.11 Stationary front at a weather surface map

the warm and cold fronts. These characteristics can be modified to a larger extent since they are dependent on many parameters, such as the season, the thermodynamic characteristics of the air masses, the humidity content of the warm air mass and others.

2.3.3.4 Stationary Fronts

When warm and cold air masses are in contact but are not moving and therefore neither of them is tending to displace the other, then the intersection of their dividing surface with the Earth's surface is called a stationary front (see Fig. 2.11).

The wind is moving parallel to the stationary front and the weather conditions at these fronts are similar to the warm fronts. However, they have weaker intensity and can cover a larger area. The stationary fronts can remain for several days or weeks and their exact weather conditions are difficult to be forecast. In surface maps they are depicted with a solid line which has red arrows and blue semi-circles.

2.3.3.5 Occluded Fronts

Cold fronts move with higher velocity in relation to warm fronts and as a result the warm region (the area between the cold and warm fronts) is reduced and finally the cold front catches up and overtakes a warm front with the formation of an occluded front. This can occur at the final stages of a wave cyclone. In the formation of an occluded front, three air masses are involved and their position vertically is dependent on their temperature difference (Fig. 2.12)

The occluded front is depicted with a line on which triangles (blue) and hemi-circles (red) denote the direction of the front movement.

The clouds that accompany an occluded front are dependent on the clouds which were present at the warm and cold fronts. Severe weather conditions may occur at

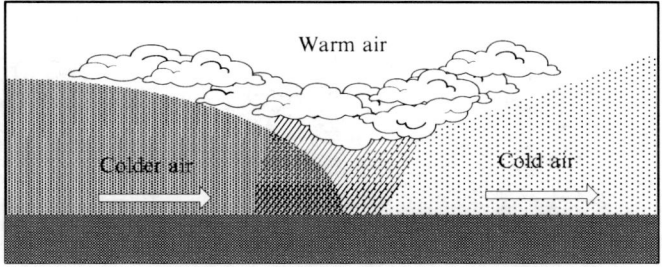

Fig. 2.12 Vertical structure of an occluded front

the first stages of the formation of an occluded front due to the unstable air mass which is forced to move upward. However, this stage lasts only for a short time interval.

2.3.4 Wave Cyclone

The low barometric conditions at the middle latitudes are developed at regions where there is a formation of fronts. The fronts are formed due to the general atmospheric circulation close to the Earth's surface.

The process starts with a stationary front, where the cold air is located north and the warm south. A front has the properties of a wave. Due to the front tilt there is a formation of an initial oscillation at the intersection between the front and the surface. Thus the warm air forms a cavity inside the cold air. The pressure starts to drop at the top of the cavity and starts the formation of a cyclonic movement of air. The low pressure system which is formed and is accompanied with frontal movements (warm and cold front) is called a wave cyclone. When the wave cyclone is formed it is moving in an easterly direction, passing successively through stages as depicted in Fig. 2.13 until its dissolution (life cycle of a wave cyclone).

The first stationary front is divided into a warm and a cold front and the low system which accompanies it starts to deepen. The sector between the warm and cold fronts is called a warm sector. The whole system moves eastward and the pressure at the centre of the low system continues to decrease. The transport velocity of a wave cyclone is equal to the velocity of the geostrophic wind in the warm region.

A continuing pressure drop causes convergence inside the low system and consequently an upward flow. This upward flow forms due to adiabatic cooling of water condensation and other weather phenomena. The weather conditions are discussed in the sections related to warm and cold fronts. As the air of the warm sector ascends in conjunction with the quick arrival of the cold front, a narrow area of the warm sector develops. Then the cold front occupies the warm section and the two fronts are combined. The combination starts at the Earth's surface. The front which is formed from the above process is the occluded front. Finally, the barometric low gradually disappears.

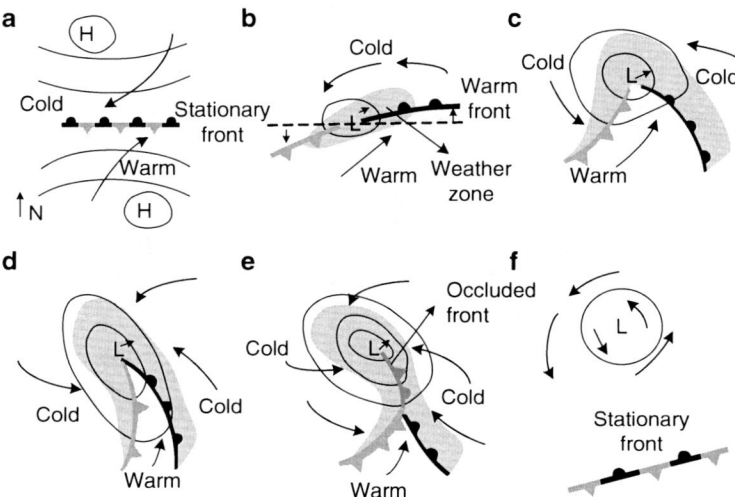

Fig. 2.13 A life cycle of a wave cyclone

2.4 Turbulence – Equations for the Mean Values

Turbulence characterizes the atmospheric boundary layer; but, due to the complex structure and variability of the layer, a deterministic description of its turbulence is difficult. A description of turbulence can, however, be constructed through its statistical properties. A suitable methodology is to divide the flux into turbulent and non- turbulent terms. An example is the calculation of the changes in small spatial and temporal changes, where the equations are expanded to average and instant values. This methodology is known in the literature as Reynolds analysis since it was developed by Osborne Reynolds.

As an example, the gaseous number concentration can be divided as the sum of the average and instant values:

$$N = \overline{N} + N' \tag{2.10}$$

where, $\overline{N}$ is the average gaseous concentration (molecules/(m^3 s)) and N' is the instant value. The average concentration is calculated with integration of time and volume:

$$\overline{N} = \frac{1}{\Delta t \, \Delta x \, \Delta y \, \Delta z} \int_t^{t+\Delta t} \left\{ \int_x^{x+\Delta x} \left[\int_y^{y+\Delta y} \left(\int_z^{z+\Delta z} N \, dz \right) dy \right] dx \right\} dt. \tag{2.11}$$

The atmospheric flux is turbulent. Turbulent flux does not have specific forms but exhibits a random behavior in relation to time. These random changes of velocity give changes to the rates of change of momentum, heat and mass which are higher

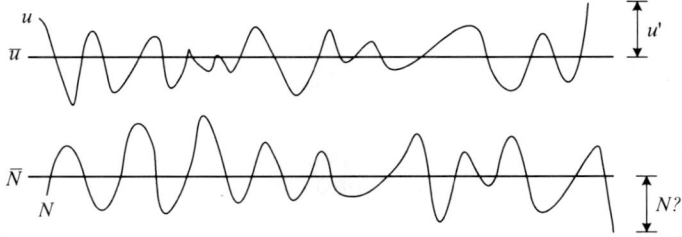

Fig. 2.14 Actual, average and instant values for the velocity and gaseous concentration. Every point in the horizontal axis depicts variations at specific time and space values

by several orders of magnitude in relation to the molecular diffusion. During turbulent flux there is a continuous conversion of the kinetic energy to internal energy. The energy source in turbulent dispersion is the shear flux in the flux field.

If $u_i(t)$ is the instant value of the velocity of a molecule, then the average velocity value versus time is given by the expression

$$\bar{u}_i = \lim_{\tau \to \infty} \frac{1}{\tau} \int_{t_o}^{t_o+\tau} u_i(t)\, dt. \qquad (2.12)$$

Since the velocity is not exactly constant versus time, the average velocity can be expressed as

$$\bar{u}_i(t) = \frac{1}{\tau} \int_{t-\tau/2}^{t+\tau/2} u_i(t')\, dt'. \qquad (2.13)$$

Figure 2.14 shows the time series of velocity u and the gaseous concentration N, depicting the difference between average and instant values.

2.5 Statistical Properties of Turbulence

Turbulence is a characteristic of the atmospheric boundary layer and due to its non - deterministic nature its description is formulated with the help of statistics. In this section statistical methods for the study of turbulence are examined.

Turbulence has a spectrum analogous to the colour spectrum of light after passing through prism. Similar analysis is performed for the description of the colour spectrum and turbulence such as the contribution of different components of flux to the total turbulent kinetic energy. Figure 2.15 shows an example of the function of the probability density for temperature in the atmospheric boundary layer. It is important to understand that variations of different components in the atmosphere make necessary their statistical evaluation.

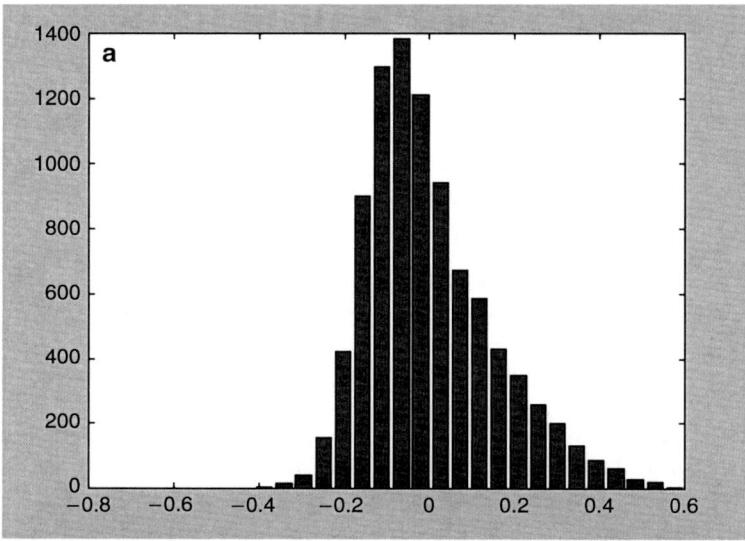

Fig. 2.15 Typical form of the function of the probability density for temperature fluctuations in the atmospheric boundary layer (x-y plane at z = 389 m). The y-axis denotes frequency of occurrence, and the units of the x-axis are degrees Kelvin. Mean temperature is 301 K (Adapted from Housiadas et al., 2004)

The function for the mean value constitutes one of the main functions which will be examined. There are several ways to examine the mean value of a function. These include mean values of time $^t(\overline{})$, space $^s(\overline{})$, and from the ensemble $^e(\overline{})$ (Stull 1997). Therefore for a variable A(t,s) that is function of time (t) and space (s), the following expressions for discontinuous and continuous conditions can be used:

$$^t\overline{A}(s) = \frac{1}{N}\sum_{i=0}^{N-1} A(i,s) \quad or \quad ^t\overline{A}(s) = \frac{1}{P}\int_{t=0}^{P} A(t,s)\,dt, \qquad (2.14)$$

$$^s\overline{A}(t) = \frac{1}{N}\sum_{j=0}^{N-1} A(t,j) \quad or \quad ^s\overline{A}(t) = \frac{1}{S}\int_{t=0}^{S} A(t,s)\,ds, \qquad (2.15)$$

$$^e\overline{A}(t,s) = \frac{1}{N}\sum_{i=0}^{N-1} A_i(t,s) \quad or \quad ^e\overline{A}(t) = \frac{1}{E}\int_{t=0}^{E} A(t,e)\,de. \qquad (2.16)$$

A statistical criterion for the variation of data around its mean value is the dispersion coefficient which is defined as

$$\sigma_A^2 = \frac{1}{N}\sum_{i=0}^{N-1}\left(A_i - \overline{A}\right)^2. \qquad (2.17)$$

2.5 Statistical Properties of Turbulence

The dispersion coefficient is a good measure of the variations inside the boundary layer. Another coefficient which is used more often for groups of measurements can be defined as

$$\sigma_A^2 = \frac{1}{N-1} \sum_{i=0}^{N-1} (A_i - \overline{A})^2, \tag{2.18}$$

where the instantaneous variations can be written as

$$a_i' = A_i - \overline{A}. \tag{2.19}$$

Finally we can write as a result,

$$\sigma_A^2 = \frac{1}{N} \sum_{i=0}^{N-1} a_i'^2 = \overline{a'^2}. \tag{2.20}$$

In addition, a coefficient can be defined for the intensity of turbulence for an average velocity value $\overline{U}$ as

$$I = \sigma_U / \overline{U}. \tag{2.21}$$

The covariance coefficient denotes the correlation between variable A and B:

$$\text{covar}(A, B) \equiv \frac{1}{N} \sum_{i=0}^{N-1} (A_i - \overline{A})(B_i - \overline{B}). \tag{2.22}$$

Using the Reynolds methodology ($\overline{(AB)} = \overline{A}\,\overline{B} + \overline{a'b'}$) for the average value, then the following expression is derived for the covariance (covar(A,B)):

$$\text{covar}(A, B) \equiv \frac{1}{N} \sum_{i=0}^{N-1} a_i' b_i'. \tag{2.23}$$

As an example for the covariance coefficient, we examine the case that the coefficient A is the temperature (T) and B is the vertical velocity component w. On a warm summer day the warmer air ascends (positive T' and w') and the colder air descends (negative T' and w'). This denotes that the product w' × T' is positive and therefore the temperature and vertical velocity components change co-instantaneously. Figure 2.16 shows the vertical fluctuations of the velocity at the surface x − y for a typical situation of the atmospheric boundary layer, whereas, Fig. 2.17 shows the vertical fluctuations of the velocity at the surface y − z.

The normalized covariance (r_{AB}) can be defined as

$$r_{AB} \equiv \frac{\overline{a'b'}}{\sigma_A \sigma_B}. \tag{2.24}$$

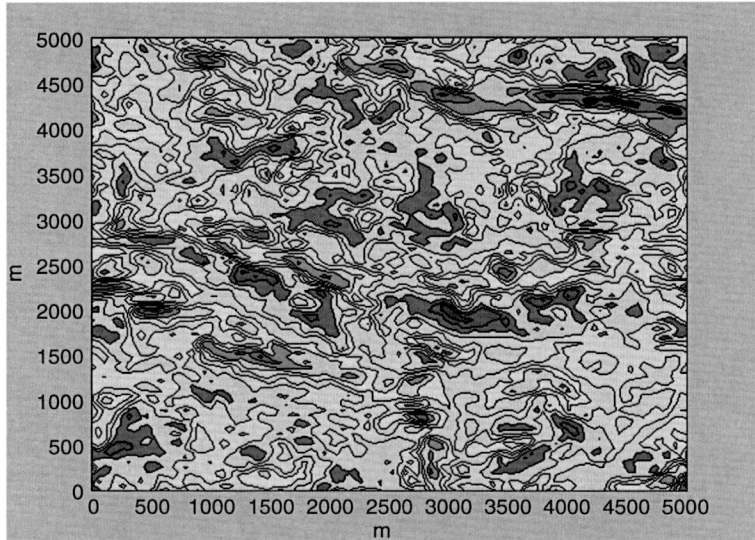

Fig. 2.16 Instantaneous w-fluctuations (m) in the x-y plane at z = 190 m (middle of domain) (Adapted from Housiadas et al., 2004)

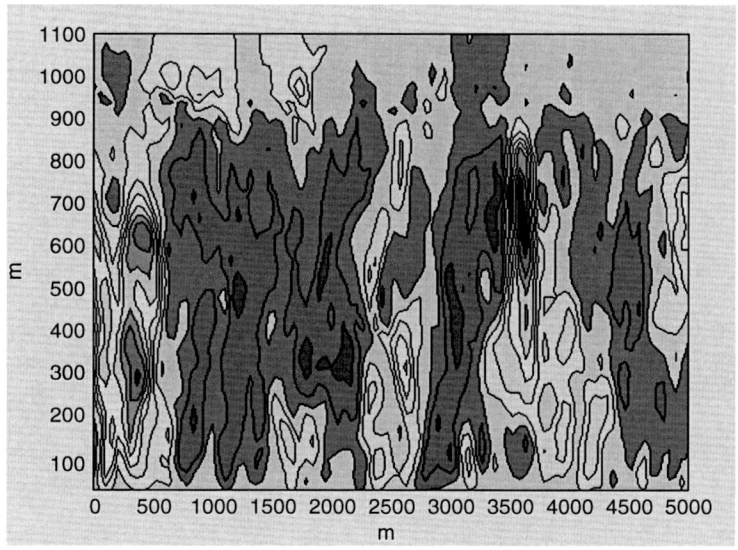

Fig. 2.17 Instantaneous w-fluctuations (m) in the y-z plane at x = 2,500 m (middle of domain). (Adapted from Housiadas et al., 2004)

2.5 Statistical Properties of Turbulence

The coefficient r_{AB} has values which range between $+1$ and -1. When two variables change in the same manner, then $r_{AB} = 1$. When the variables change in opposite ways, then $r_{AB} = -1$.

The kinetic energy of mass m is given by the expression $E_K = 1/2\, m\, v^2$, where v is the velocity. The kinetic energy per unit mass is written as $E_K/m = 1/2\, v^2$. The kinetic energy of flux can be divided into two terms, one which is related to the mean velocity and one which is related to the turbulence. Therefore the expression for the mean kinetic energy can be written as

$$\frac{MKE}{m} = \frac{1}{2}\left(\overline{U^2} + \overline{V^2} + \overline{W^2}\right), \qquad (2.25)$$

where $\overline{U}, \overline{V}, \overline{W}$ are the three components of the mean velocity.

The expression for the turbulent kinetic energy can be written as

$$\frac{TKE}{m} = \frac{1}{2}\left(\overline{u'^2} + \overline{v'^2} + \overline{w'^2}\right), \qquad (2.26)$$

where, u', v' and w' are the three components of the instantaneous velocity.

Furthermore we study the flux concept in the atmosphere, which is defined as the transport of a quantity per unit surface and unit time. In the boundary layer there is usually a study of flux of mass, heat, humidity, momentum and pollutants. As an example for the mass we define the pollutant flux ($\widetilde{M}$) which is expressed in units $\left[\frac{Kg_{air}}{m^2 s}\right]$. The kinematic flux is defined as $M = \frac{\widetilde{M}}{\rho_{air}}$ (units $\left[\frac{m}{s}\right]$). Furthermore the vertical kinematic heat flux due to turbulent flux (vertical kinematic eddy heat flux) is defined as $\overline{w'\theta'}$ where θ is the temperature. Another quantity that is important in the study of atmospheric flux is stress, which is actually a force that can produce a deformation of a body. Pressure is a kind of stress which is applied in fluids under equilibrium. The Reynolds stress for a fluid in turbulent movement is given by the expression

$$\tau_{Reynolds} = -\rho\,\overline{u'w'} \quad \text{(Reynolds stress)} \qquad (2.27)$$

or can be expressed as

$$\tau_{ij} = \mu\left(\frac{\partial U_i}{\partial x_j} + \frac{\partial U_j}{\partial x_i}\right) + \left(\mu_B - \frac{2}{3}\mu\right)\frac{\partial U_k}{\partial x_k}\delta_{ij}, \qquad (2.28)$$

where, μ_B is the viscosity coefficient and μ is the dynamic viscosity coefficient.

The tensor of the Reynolds stress is symmetric and is given as

$$\begin{bmatrix} \overline{u'u'} & \overline{u'v'} & \overline{u'w'} \\ \overline{u'v'} & \overline{v'v'} & \overline{v'w'} \\ \overline{u'w'} & \overline{v'w'} & \overline{w'w'} \end{bmatrix} \qquad (2.29)$$

where u, v, w are the air velocities in the three directions.

A typical value of the kinematic coefficient of the Reynolds stress in the atmosphere is 0.05 m²/s². The Reynolds stress is a characteristic property of the flux and not of the medium.

In addition the Reynolds stress at directions x, y and z can be expressed as

$$\tau_{xz} = -\overline{\rho}\, \overline{u'w'_s}, \quad \text{(viscous stress)} \tag{2.30}$$

$$\tau_{yz} = -\overline{\rho}\, \overline{v'w'_s}, \tag{2.31}$$

with a total Reynolds stress

$$\left|\tau_{Reynolds}\right| = \left[\tau_{xz}^2 + \tau_{yz}^2\right]^{1/2}. \tag{2.32}$$

Finally the friction velocity u∗ can be written as

$$u_*^2 \equiv \left[\overline{u'w'_s}^2 + \overline{v'w'_s}^2\right]^{1/2} = \left|\tau_{Reynolds}\right|/\overline{\rho}. \tag{2.33}$$

2.6 Atmospheric Temperature

2.6.1 Temperature Season Variability

The Sun is the largest heat source in our solar system and of course also for the Earth. Heat is an energy form that is dependent on the composition of materials and can be moved to other bodies or can be transferred to other energy forms.

The Earth spins on its own axis (with a period of one day – 24 h) and at the same time revolves with a velocity of thousands of kilometers per hour around the sun with elliptical trajectory (with a period close to 365 days) (Fig. 2.18). The direction of rotation is anticlockwise and has a velocity of several 100 kilometers per hour. This is the reason that sun, moon and stars rise in the east and decline in the west.

Since the Earth performs an elliptical orbit around the sun, the distance between Sun – Earth changes during the year. Earth is closer to the Sun in January (distance of 147 million km) than in June (distance of 152 million km). At the Northern Hemisphere the average temperature in June is higher than that of January and thus the change of seasons is determined by the quantity of the Sun's energy which reaches the earth. The quantity of radiation from the Sun which reaches the earth's surface is determined from the angle that radiation strikes the surface and the time period that the sun radiates at a specific latitude.

2.6 Atmospheric Temperature

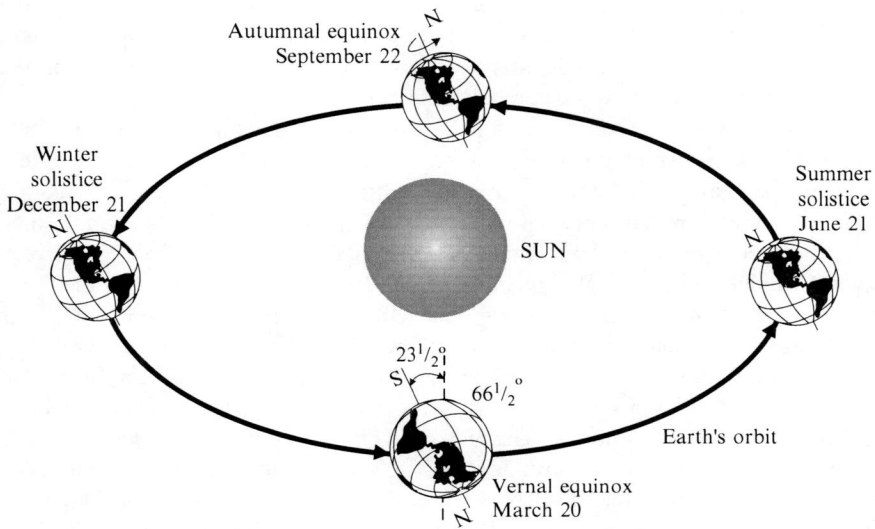

Fig. 2.18 Earth's orbit around the Sun. During this movement the Earth is rotated around an axis which has an angle of 23 ½ ° from its vertical axis The rotation axis is constant during the year. The result is that in June, when the northern hemisphere is turned towards the Sun, there is more direct light and longer duration of the day there. This results in warmer weather, compared to December when the northern hemisphere is turned away from the Sun

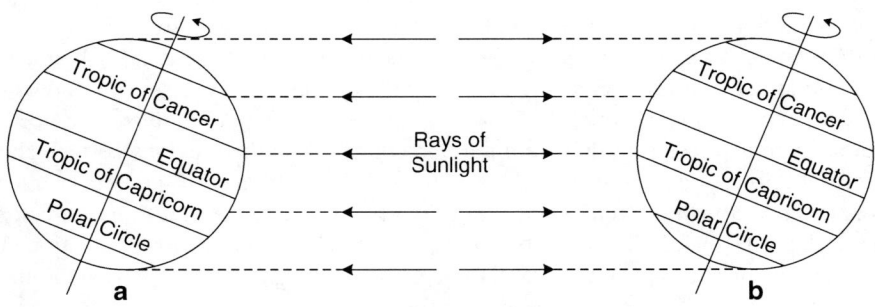

Fig. 2.19 Earth's direction (**a**) 21st June (**b**) 21st December

Figure 2.19 shows also the earth's direction in relation to the Sun's radiation during 21st June and 21st December. On 21st June the Sun's radiation strikes the earth vertically at $23^{1/2}$ ° north (N) (Tropic of Cancer). On 21st December the darkness lasts for 24 h for all areas above $66^{1/2}$ ° north (N) and $66^{1/2}$ ° south (S).

When the Sun's radiation penetrates the atmosphere, a part of it is absorbed or scattered from the atmospheric gasses, whereas another part is reflected from clouds. This means that the greater is the thickness of the atmosphere which the Sun's radiation has to penetrate, the higher is the possibility that the radiation will be absorbed or scattered. This occurs during summer and high latitudes such as in the Scandinavian countries. The Sun in these countries is never very high on the horizon and therefore the radiation must penetrate a quite thick atmospheric layer before it reaches the Earth's surface. Also due to high cloud content of the atmosphere the radiation is scattered effectively before it reaches the ground.

The seasons are determined from the quantity of the Sun's radiation which reaches the planet, which is determined from the duration of the day and the incident angle of the Sun's radiation. As a result the higher latitudes lose more radiation (from the emitted radiation at the higher wave length of the Earth's surface) than what they receive (short wave length of the Sun's radiation). On the contrary, at lower latitudes there is a higher quantity of Sun's radiation which reaches the Earth than is emitted from it. Due to the global wind circulation that results from this temperature difference there is a transport of heat from the lower to higher latitudes.

As shown in Figs. 2.18–2.19, on 21st June the Sun's radiation has a maximum at the Earth's surface at latitude 30°N. This day the Sun is located above the latitude 23 1/2 °N (Tropic of Capricorn). The reason that the maximum of incident energy is at latitude 30°N and not at latitude 23 1/2 °N is due to two reasons. First, the duration of day at the area of 30°N is larger than the area 23 1/2 °N on 21st June. Secondly the area close to the point 30°N is characterized by desert areas, clear sky and dry air, whereas at the area 23 1/2 °N the climate is more humid with clouds which reflect the Sun's radiation.

After the 21st of June the sun is lower at noon in the sky and the summer days become shorter. With the beginning of September there is the start of autumn. On September 22nd the Sun is located above the equator. On this date, except at the poles, the day and night have the same duration (autumnal equinox). At the north pole the Sun appears on the horizon for 24 h and then disappears from the horizon for 6 months. At other latitudes in the north hemisphere after the 22nd September the Sun at noon appears gradually lower in the sky and the day lasts fewer hours.

On 21st December, which is 3 months after the autumnal equinox, the north hemisphere is directed farther from the Sun compared to the previous period. The nights are long and the days are short. This is the shortest day of the year (winter solstice) and it is the astronomical start of winter. This day the Sun shines directly above latitude 23 1/2 °S (Tropic of Capricorn). It is located at its lower position at the middle of the day and its radiation passes through a large portion of the atmosphere and affects a large portion of the Earth's surface.

On 20th March there is the astronomical start of spring and it is the vernal equinox. This day the Sun shines directly above the equator, whereas at the north pole the Sun appears on the horizon after being 6 months absent. The following period the days last longer and there is warmer weather in the north hemisphere.

After 3 months, 21st June, the sun's radiation reaches a maximum. It is obvious that even though the sun's radiation is more intensive during June, the warmest period appears a few weeks later during July or August. The reason is that during June the outgoing radiation from earth is smaller than the incoming and there is no thermal equilibrium. When a thermal equilibrium is achieved, there is the highest temperature in the atmosphere and this is achieved a few weeks after the 21st June. The same situation occurs during winter when the outgoing energy from the

2.6 Atmospheric Temperature

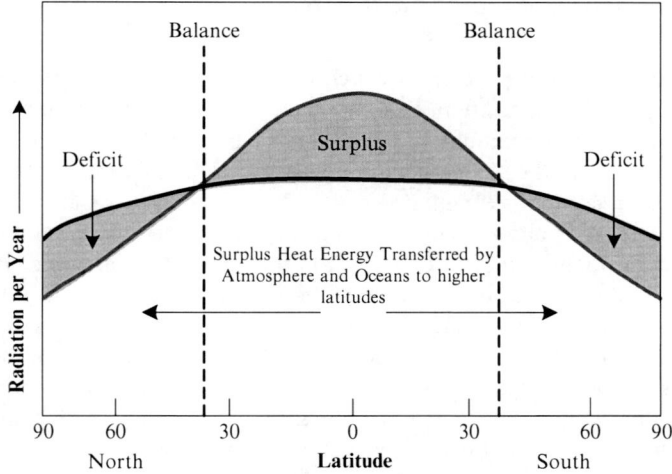

Fig. 2.20 The average annual incoming solar radiation (grey line) absorbed by the earth and the atmosphere compared with the average emitted infrared radiation from the earth and the atmosphere (Adapted from Ahrens 1994)

atmosphere is higher than the incoming and the temperature decreases. Since the outgoing radiation is higher than the incoming then for a few weeks after 21st December there is the coldest period mainly during January and February.

Whereas in the whole planet there is energy equilibrium between the absorption and emission of solar radiation, this is not valid at every latitude. Figure 2.20 shows the energy equilibrium at different latitudes.

The temperature is expressed in degrees, which are subdivisions of thermal scales. The best known and most used are the Celsius centigrade scale (°C) and the Fahrenheit scale (°F) which is used mainly in the United States and finally the Kelvin scale (K) which is used mainly for scientific purposes. Thermal energy is observed in the system Earth – atmosphere with the following forms:

- As absolute heat which can be measured with the help of thermometers and
- As latent heat which occurs during specific physical processes related to the phase changes of water (evaporation, condensation etc.).

As discussed earlier, the main source of heat for the Earth and its atmosphere is the Sun, since the energy from the Earth's interior is negligible. It has been calculated that if there was no contribution from the Earth's interior its mean temperature would be reduced by less than 0.1°C. Therefore the short length Sun's radiation affects the planet's temperature and controls, together with the long length radiation, the temperature on the Earth's surface. The processes of heat exchange (between warm and cold regions) at the lower atmospheric layers due to uneven heating of the earth's surface are the main reasons for the formation of weather phenomena inside the atmosphere.

2.6.2 Temperature Daily Variability

The intensity of the Sun's radiation which is received from the Earth's surface is dependent on the Sun's position. During a sunny day the radiation intensity has a simple variation with maximum close to noon. During a day without clouds and turbulent atmosphere the air temperature varies with the minimum a few minutes after the Sun's rise and maximum 2–3 h after the noon hour. As shown in Fig. 2.21 during a normal day without clouds the upward part of the daily variation of the temperature has higher gradient compared to the downward part, since during the night there is no great loss of thermal radiation.

During the daily temperature variation in air the upper layer of the Earth's surface absorbs and emits the Sun's radiation, which is responsible for the temperature changes. During the night there is often a temperature inversion with an increase of the temperature versus height since there is cooling of the Earth's surface with the emission of electromagnetic radiation with large wave length.

The difference between the maximum and minimum temperature during a day is called daily thermal width. This is larger above the mainland and smaller above the

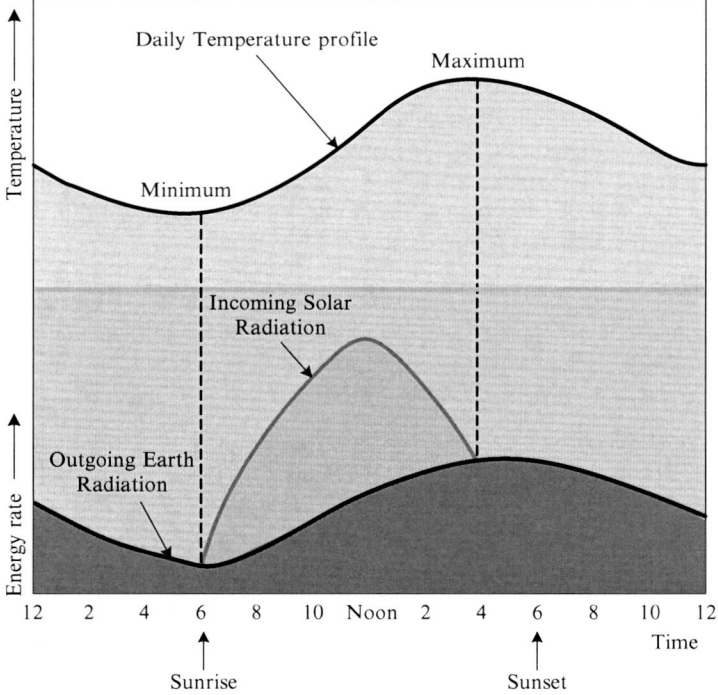

Fig. 2.21 Daily changes of the temperature and incoming and outgoing radiation. When the incoming Sun's energy is higher than the outgoing energy then the air temperature increases. Contrary when the outgoing energy is higher than the incoming then the air temperature decreases. (Adapted from Ahrens 1994)

2.6 Atmospheric Temperature

seas due to the larger heat capacity of water. Its value decreases gradually from equator to poles as the latitude increases.

In the northern hemisphere, over a year's time, the temperature of air in temperate regions shows a simple variation. The maximum occurs most of the time above mainland during July and above sea during August. The minimum is observed above mainland during February and above sea during March.

The difference between the mean temperature of the warmest and coldest month of a year is called annual thermal width. This is larger above mainland and smaller above sea and increases from the equator to the poles.

The first classification of climate is based on the annual thermal width:

- Tropical moist, when the annual width is smaller than 10°C.
- Temperate, when the annual width is between 10°C and 20°C and
- Moist mid-latitude climates with severe winters, when it is larger than 20°C.

A detailed characterization of climate is demonstrated by the Köppen classification system (Ahrens 1994).

The air temperature is the most important climatic component and the most important parameter for climate classification. Meteorologists and climatologists examine temperature values at different elevations inside the atmosphere. When someone refers to air temperature, they mean the temperature in shadow inside a special shelter (meteorological cage) and at height 1.5–2.0 m above ground.

For mainly climatological reasons the air temperature in a given location can be described with the following parameters:

- Average daily temperature ($\overline{T}_{day}$) which is defined from the expression $\overline{T}_d = \frac{1}{24} \sum_{i=1}^{24} T_{h(i)}$ where $T_{h(i)}$ is the hourly value (i = 1, 2, 3..., 24). This expression is used when the meteorological station has the capability of hourly temperature measurements. The expression which is used by the World Meteorological Organization for the calculation of T_d is: $\overline{T}_d = \frac{1}{4}(T_{06} + T_{12} + 2T_{18})$, (where 06, 12 and 18 are the time in UTC – Universal Time Coordination). The use of this expression is adopted in cases where there is need to have direct comparison between the meteorological values in an extensive geographical area. For climatological use the data have to be extended for at least a period of 30 years.
- Absolute maximum (T_{max}) and minimum (T_{min}) value of the air temperature which occurs during a whole day (24 h).
- Average monthly temperature ($\overline{T}_{mo}$) which can be calculated from the expression $\overline{T}_{mo} = \frac{1}{v} \sum_{i=1}^{v} \overline{T}_{d(i)}$, where v is the number of days during the month.
- Daily thermal width. This is defined as the difference between the maximum and minimum temperature value (M.T.V.) during 24 h: $MTV = T_{max} - T_{min}$.
- Yearly temperature width (Y.T.W.). This is defined as the difference between the average air temperature of the coldest month and the average temperature of the warmest month during the year: $Y.T.W. = \overline{T}_{mo(warm)} - \overline{T}_{mo(cold)}$

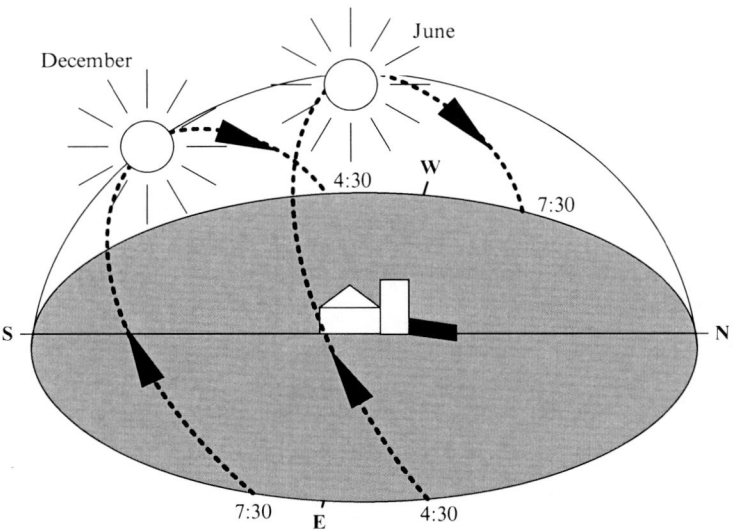

Fig. 2.22 Variability of the Sun's orbit at mean geographical widths at the northern hemisphere (Adapted from Ahrens 1994)

Figure 2.22 shows the trajectory of the Sun at middle latitudes in the northern hemisphere during a year. During winter the sunrise starts at the south-east and the sunset occurs at south-west. During summer the sunrise occurs in the northeast and the sunset in the northwest. Consequently a house with windows facing south receives more light than a house with windows facing north. The same applies more intensively for mountainous regions. Regions which are oriented to the south receive more light and as a result tend to be warmer and drier than regions which are oriented to the north. This has a direct effect on the flora of the region. Vineyards which are located at areas oriented to the south produce a better quality wine. On the contrary plants which withstand cold are located at areas oriented to the north. The architecture of houses is also affected from the position of the sun and considerable benefits for energy consumption are derived from an appropriate orientation of houses and the rational use of windows (bioclimatic architecture). Another example is the operation of ski resorts which have a north orientation.

2.6.3 Heating of the Earth's Surface and Heat Conduction

Radiation from the sun is the main heat source for the Earth's surface. The heating of the surfaces and their temperature is dependent on a number of parameters such as:

- The special heating (**heat capacity**) of a surface. By the term heat capacity is meant the quantity of heat that is needed to increase the temperature of one gram

2.6 Atmospheric Temperature

of a given material by 1°C. Since the heat capacity of water is twice that of the ground, then more heat is needed in order to increase the temperature of a water surface by 1°C compared to the ground. As a result the ground becomes heated (or cooled) more quickly compared to the sea. In comparison with the sea, the ground is warmer during the day but it is cooler during the night.

- The **absorbency** of a surface. Everybody that receives a quantity of radiation absorbs part of its energy. The percentage of absorption is dependent on the body's nature and the radiation.
- The **reflectivity** of a surface. If the total amount of the sun's radiation is reflected from a surface, there is no absorption or transformation to thermal energy. The surfaces of snow and water have high reflectivity and are not heated, such as surfaces with low reflectivity (e.g. cultivated areas, dense jungle etc.).
- The **conductivity** of a surface. The ocean streams transport large quantities of heat (inside oceans) through the water movement. With this process the sea is heated to a larger depth compared to the mainland.
- The **cloud cover** of a surface. A factor which plays a significant role in the temperature of surfaces is the cloud cover, which during the day does not allow penetration of the Sun's radiation to the Earth's surface (Fig. 2.23). This results in a reduction of the Earth's temperature. Therefore the air which is in contact with the Earth's surface will receive lower heating during the day. During the night the cloud cover results in the opposite phenomenon since it does not allow part of the thermal energy which it receives from the Earth to escape to space. The atmosphere under clouds experiences lower cooling and therefore higher temperatures occur.

The thermal energy is spread from one body to the next or re-distributed inside a body in different ways. Some of these mechanisms are:

- **The radiation**. All bodies emit energy in the form of electromagnetic radiation. Higher temperatures result in smaller wave length of the radiation. Consequently the wave length of the Sun's radiation is smaller than the re-emitted radiation from the Earth's surface which is much colder than the Sun's surface.

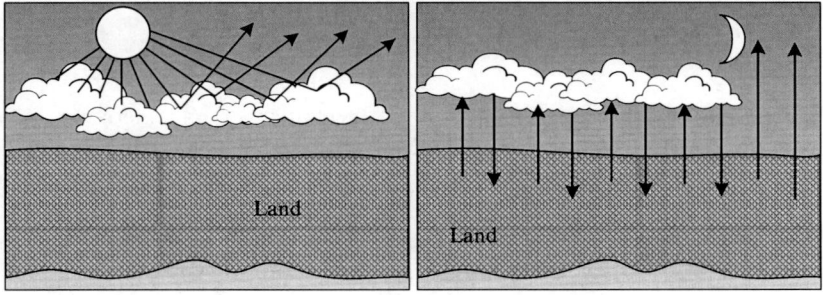

Fig. 2.23 The cloud cover reduces the warming of the Earth's surface during day and its freezing during the night

- **The conductivity.** The thermal energy can be conducted inside a body or from one body to the other with contact through conductivity. For example iron is a good conductor of heat, whereas the wood or air are poor heat conductors. An air molecule which is in contact with the Earth's surface is heated through conductivity. This is an important factor for creation of the weather phenomena.
- **The heat transfer.** A moving body of air transports also its thermal energy. The heat transfer occurs vertically and horizontally inside the atmosphere and this process is the most important for the weather phenomena.
    a. **Vertical heat transfer**: An air mass which is heated on the Earth's surface starts to expand and then becomes less dense and ascends. With its uplift it expands adiabatically and transfers its thermal energy at upper atmospheric layers.
    b. **Horizontal heat transfer**: An air mass moves horizontally to fill the gap of air which is created from the vertical movement of another air mass. This air mass which is moving horizontally transports together its thermal energy and humidity.

2.6.4 Distribution of Temperature in the Air

The temperature distribution of air above a region (small, large or even above the whole globe) is described with isothermal curves (lines with the same temperature along them).

The most important factors which control the temperature distribution in air are:

- The season and the latitude
- The distribution of land and sea
- The vegetation and general nature of the surface
- The elevation
- The slope of the Earth's surface
- The existence of snow or ice on the surface
- The cloud cover
- The prevailing winds and
- The sea streams.

Each of these parameters acts in a different manner and therefore the air temperature does not decrease smoothly from the equator to the poles. The highest temperatures are not observed at the equator but at latitudes 10°–20° south and north from it. This is due to the fact that at the equator there are extensive clouds and rainfall.

An important factor is the geographical distribution of land and sea. During summer the land is warmer than the sea. The temperature increases mainly above land. During winter the land is colder than the sea and therefore lower temperatures are observed above land at higher latitudes (e.g. North Canada, Siberia, Greenland). Winds have a great influence on temperature distribution at different locations. Therefore the west winds in northern temperate zones transport warm ocean air masses to the west regions of Europe and America and cold air masses to the east regions of America and Asia.

The constant ocean currents finally influence considerably the temperature distribution. These are moving to the poles carrying warm water to colder regions, whereas the currents moving to the equator transport cold water masses to warmer regions.

2.7 Humidity in the Atmosphere

The term humidity refers to the water vapour which is contained in the atmosphere at a specific time. The air humidity is a very important parameter since it determines the cloud formation and the rain formation.

The atmosphere and especially troposphere contains water vapour at variable quantities which come mainly from water evaporation. The quantity of water vapour that the air contains is specific and is dependent directly on the air temperature.

When the air contains the maximum quantity it is called saturated. When only a part of the maximum quantity is contained, then it is called unsaturated. The term humid air is often used when there is an elevated amount of water vapour in the atmosphere. The term dry air is used in the absence of water vapour.

The water in the atmosphere is not only in the vapour phase but it exists also in the liquid phase (cloud, rain) and in the solid (snow, drizzle). It is known that in the atmosphere there is a water cycle. The water enters the air through evaporation from all the water surfaces and especially from oceans. Therefore huge quantities of water evaporate from the Earth's surface and are moved higher in the atmosphere and are condensed forming extensive cloud layers. Further, the clouds are transported to other regions and through wet deposition the water comes back to the Earth's surface.

Taking into account that $\frac{2}{3}$ of the planet are covered by water, millions of tons of water are evaporated daily and are moved to the upper troposphere in the form of water vapour. The 84% of water vapour originates from oceans whereas the other 16% originates from lakes, rivers, wet surfaces, vegetation and the expiration of animals.

Every year the planet receives from water precipitation (e.g. rain, snow) 400 km^3 of water. An equal quantity is introduced to the air through evaporation. Of the total water precipitated on the Earth's surface, close to 100 km^3 precipitate on land and the other 300 km^3 on oceans and water surfaces. Evaporation of 400 km^3 water into the atmosphere during 1 year requires 3×10^{29} cal. This energy corresponds to 23% of the total energy which the Earth receives during a year from the sun.

The water in the atmosphere can change from one phase to another and for the water condensation the atmosphere has to be saturated (relative humidity equal to 100%). However in specific cases there is no condensation even at relative humidity above 100%. This is due to the absence of a sufficient number of available condensation nuclei in the atmosphere. Without these nuclei, a relative humidity close to 420% is required in order for condensation to occur. Existence of sufficient condensation nuclei will lower the relative humidity required for condensation to 100%. With the presence of nuclei of sodium chloride the necessary relative humidity for condensation is close to 97%, whereas for sulphur oxide or phosphate oxide nuclei the relative humidity drops to 80%. Therefore in industrial regions the dense fog is a consequence of the existence of a large number of condensation nuclei. To the contrary, the number of condensation nuclei drops at higher elevations in the atmosphere.

As the water vapour condenses onto condensation nuclei, droplets or ice crystals form and with time their size increases. The formation of droplets or ice crystals is dependent on the temperature and pressure. At higher elevations where the temperature is far below $0°C$ and the pressure is also low, the water exists in the liquid phase. This is an unstable condition which is called super critical melting. When this unstable condition of water is disturbed by the passage of an aircraft, there is an immediate icing of the water droplets above the airframe and the wings which has consequences for the flight. Water at super critical melting exists at temperatures up to $-15°C$ and, though less often, at $-40°C$. The most dangerous region is between $0°C$ and $-15°C$.

Generally the percentage of water vapour which comes from water evaporation from the soil is about 15% of the total water vapour which exists in air. The other 85% comes from the oceans. It is surprising to think that if the total amount of water in the atmosphere precipitated simultaneously onto the Earth's surface, it would cover the whole surface of the planet to a height of 2.5 cm.

Figure 2.24 shows the average spatial distribution of rainfall in Europe during the period 1940–1995. Lower values are observed in Southern Europe.

2.7.1 Mathematical Expressions of Humidity in the Atmosphere

The term humidity refers to the water content inside air. We present here some different methodologies that can describe this phenomenon.

2.7.1.1 Absolute Humidity (B)

Absolute humidity β is defined to be the ratio of a mass of water vapour to the air volume in which the mass is contained. Absolute humidity denotes the density of water inside an air volume and usually is expressed as grams of vapour inside a cubic meter of air. For example if the water vapour inside a cubic meter of air weights 25 g, then the absolute humidity β of air is 25 g/m^3.

The air volume changes with fluctuations of its elevation due to differences in air pressure. This results in changes in absolute humidity even though the quantity of

2.7 Humidity in the Atmosphere

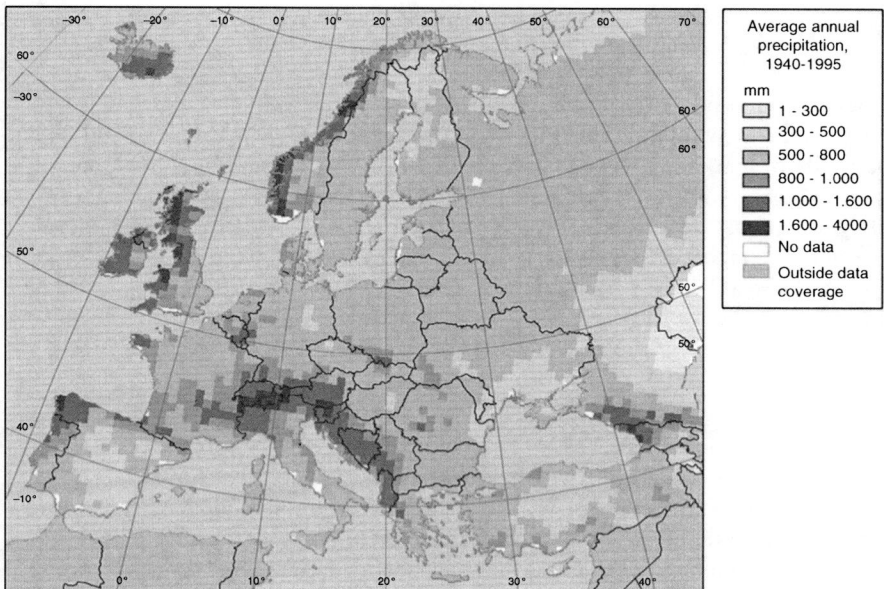

Fig. 2.24 Map of the spatial distribution of precipitation (mm) in Europe in the period 1940–1995

water vapour inside the volume remains constant. For this reason the term absolute humidity is not often used in atmospheric sciences.

2.7.1.2 Specific Humidity (Q)

Specific humidity q is the ratio of the water vapour of the mixture divided by the total air mass and is expressed as water vapour grams per gram of humid air.

2.7.1.3 Mixing Ratio (R)

Mixing ratio is the ratio of water vapour to the mass of dry air.

2.7.1.4 Relative Humidity (RH)

Relative humidity of the atmospheric air is called the ratio of the water vapour mass which is contained in a specific air volume to the mass of the water vapour in the same volume under saturation conditions at the same pressure and temperature conditions. The relative humidity is actually the ratio of water vapour pressure to its equilibrium pressure at the same temperature. Therefore the relative humidity can be expressed as

$$RH = 100 \frac{p_{H_2O}}{p^o{}_{H_2O}}, \qquad (2.34)$$

where the multiplication by 100 occurs since RH is expressed in percentage.

The relative humidity is an important parameter which is involved directly in the daily life of humans. An example is the study of the transfer of cold outdoor air to indoors. Dry arctic air incorporates small quantities of water vapour. Saturated air at temperature $-25°C$ includes only 0.5 g of water vapour per 1 kg air. When this air is introduced indoors at temperature 20°C, then its capacity to incorporate water vapour increases by 29 times to the value 14.7 g/Kg. This results in relative humidity indoors of

$$RH = 100 \times \frac{0.5\,g/Kg}{14.7\,g/Kg} = 3\% \qquad (2.35)$$

Low levels of humidity have direct results on the quality of life indoors. Plants which exist indoors become dry due to the fast evaporation of the humidity from the soil. Furthermore, human skin becomes dry, there are effects on the nose and the larynx, and bacteria can be more easily introduced into the body.

High values of relative humidity on warm days during summer result in respiratory problems for sensitive members of the population, especially older people and children. When the temperature is high, the main path for lowering of body temperature is through sweating. In the case of low air humidity, evaporation of sweat from the skin occurs quickly and the temperature feels lower than its actual value. In the case of high air humidity, sweating is difficult and the body cannot easily reduce its temperature.

2.7.2 Dew Point

The dew point denotes the temperature at which the air has to be cooled without changes of pressure and humidity in order to have saturation. The dew point is an important parameter which is used for the prognosis of ice and fog. At ground level there is no considerable variation of atmospheric pressure and as a result the dew point corresponds to the humidity quantity in air. High values of the dew point indicate a high humidity content and low values of the dew point indicate low humidity content.

The difference between the air temperature and the dew point indicates if the relative humidity is high or low. When the air temperature and the dew point are equal the relative humidity is 100%. This occurs during a snow storm. But the air in a desert area, where there is a large difference between the air temperature and the dew point, has quite low relative humidity. It is interesting to note that the desert air with a higher dew point contains more water vapour than the air in a snowstorm.

There are different phenomena, other than the dew phenomenon, which are related directly with the water vapour condensation. An example is fog formation. When water vapour condenses, visibility is reduced due to growth in size of the atmospheric particles. The size of the fog droplets further increases when the relative humidity is high and the droplets become visible. When the visibility is reduced to less than 1 km, a cloud appears close to the Earth's surface and this is called fog.

The air above urban areas is usually polluted with elevated concentrations of airborne particles. Therefore the fog above urban areas is denser than above the sea under the same atmospheric conditions. Examples include the dense fog which was formed above London in the 1950s. The fog was so dense that the Sun's rays could not penetrate the air and it was necessary to use lamps during the day. The fog can be acid when there are interactions with gaseous pollutants such as sulphur and nitrogen oxides. Furthermore, acid fog has negative consequences to the public health and especially to people with breathing problems.

Fog is formed during evaporation processes, when cold air comes into contact with warm quantities of water vapour. The fog consists of small droplets which are produced from the water vapour condensation at layers of air close to the Earth's surface. A fog is therefore a cloud which is formed in stable layers of air and its base is usually the Earth's surface.

Necessary conditions for the occurrence of fog are:

- small difference between temperature and due point (large relative humidity)
- presence of condensation nuclei
- weak surface winds
- cold Earth's surface with warm and humid air above or warm sea and cold air above.

Usually the fog is formed in coastal areas where the humidity is elevated and also in industrial areas where there is increased concentration of airborne particles which serve as condensation nuclei.

The occurrence of fog is a result of two main phenomena:

2.7.2.1 The Evaporation of Water at Cold Air

One phenomenon is called evaporation fog and occurs close to the sea surface (or lake), when the air temperature is very low and there is a large temperature difference between sea and air. Its formation is due to the quick evaporation of the sea (or lake) water, and its surface resembles the view of a large boiler that emits large quantities of vapor which are condensed quickly inside the cold air. The layer of fog is small and seldom reaches 30 m and also the visibility changes are highly variable.

2.7.2.2 The Freezing of Humid Air

The fog which is formed with freezing of humid air can be divided into different categories such as radiation fog, advection fog, mixing fog and sea smoke. Different mechanisms are responsible for these effects. For example the radiation fog is formed during the night without clouds, with a light wind, when a thin layer of humid air is located close to the Earth's surface and under a layer of dry air. The thin layer of humid air does not absorb adequate infrared radiation from the Earth's surface and freezes quickly, while and at the same time it freezes the dry air above. When the temperature equals the dew point, then the condensation of water vapor starts and consequently also the formation of droplets and the fog formation. The interested reader can consult the book by Ahrens (1994) to study the different forms of fog.

2.7.3 Clouds in the Atmosphere

We explained previously that adiabatic processes together with ascending warm air from the surface results in its expansion and cooling. The unsaturated air mass is cooled adiabatically with a rate close to 1°C/100 m. When the air is colder, it contains smaller quantities of water vapour. Therefore the air mass which ascends and becomes cooler has also an increase of its relative humidity. At the height at which the temperature of the air mass is equal to the dew point (when the relative humidity reaches 100%) the excessive water vapour starts to condense with the formation of small droplets. The formed droplets coagulate and forms larger droplets. Note that an average rain droplet consists of about 10^3 water droplets. Finally a very extensive number of water droplets form a cloud.

The cloud formation occurs with: (1) radiation of heat from an air mass to the environment, (2) with the air transport to a cooler region and (3) adiabatic rise of an air mass. Figure 2.25 shows the steps in the formation of clouds during adiabatic cooling of warm air masses from the Earth's surface.

As the saturated air mass continuous to rise it will be further cooled with increase of the quantity of water droplets and the cloud size. The temperature lapse rate is not equal to the dry lapse rate since the air mass receives the latent heat which is released from the water vapour during the condensation process. Therefore the temperature drop of the ascending saturated mass is close to 0.5°C/100 m which is known as wet lapse rate.

It is accepted that most clouds are formed with adiabatic cooling which occurs inside air masses when they ascend inside the atmosphere. The ascending movement of air masses is due to:

- The vertical air transport after intense surface heating,
- The impact of air masses on mountains,
- The convergence of air masses due to barometric systems,
- The movement of air at the warm and cold fronts.

For the formation of clouds it is not enough to have only adiabatic cooling but it is necessary that condensation nuclei exist in the atmosphere. The water vapour condensation can occur on the surface of soluble (such as NaCl) or insoluble particles (such as dust or diesel particles) and on ions in the atmosphere.

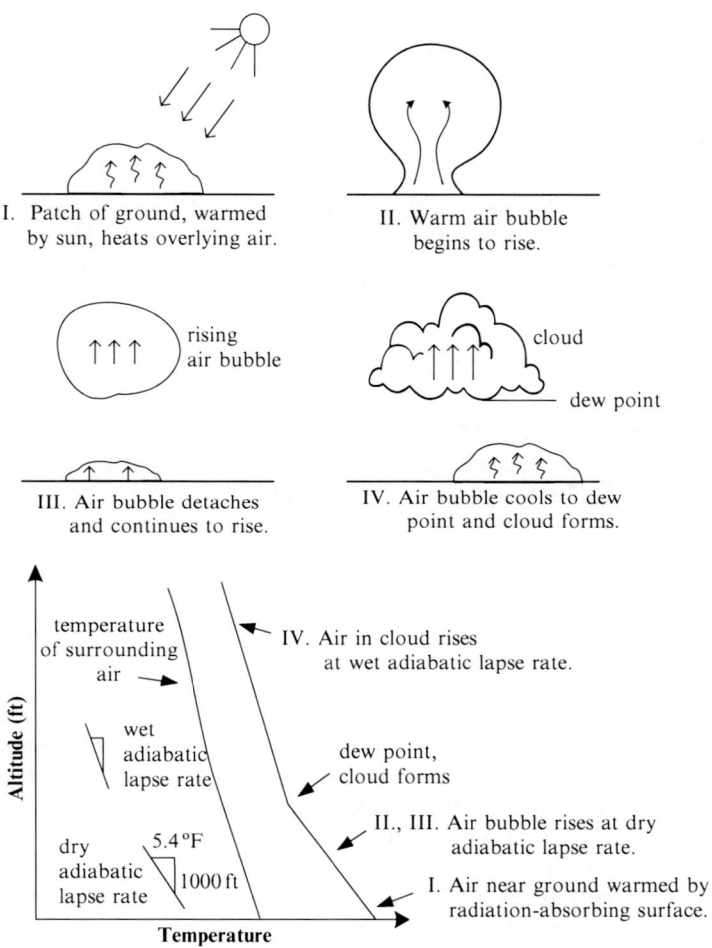

Fig. 2.25 Schematic representation of cloud formation in the atmosphere with adiabatic cooling of a warm air volume from the Earth's surface (Adapted from Hemond and Fechner 2000)

2.7.4 Precipitation

The term precipitation refers to all condensation products of water that fall from the atmosphere as rain, snow, hail and soft hail. The formation of precipitation inside clouds is determined from the air temperature and the turbulence conditions in the atmosphere. Precipitation is one of the most important meteorological and climatological parameters. During precipitation a factor which is important to be studied is the water quantity that drops on a surface and is referred to as rainwater. This quantity expresses the rainwater height (more commonly referred to as depth) on a horizontal surface and is measured with a rain-gauge. Another useful parameter in climatology is the rain intensity which expresses the resulting rainwater height per unit time. Internationally the measurement unit of rainwater height is mm or cm. For example 1 mm rainwater expresses a water quantity of 1 kg water per 1 m^2 surface.

In respect to the droplet size and precipitation conditions, rain has different names such as "shower" which is produced from clouds of vertical development and has sudden starts and stops as well as abrupt changes in intensity. Drizzle is characterized by small and many droplets which are suspended and follow the air currents.

For the study of the water precipitation in a region it is necessary to consider the following parameters:

- Average Monthly Precipitation (A.M.P.): The average total water precipitation per month (T.P.). For example the average total water precipitation in January in the period 1959–2004 is equal to the summation of the total January precipitation per year divided by the total number of observation years:

$$\text{A.M.P.(JAN)} = \frac{T.P.(JAN1959) + T.P.(JAN1960) + \ldots\ldots\ldots + T.P.(JAN2004)}{46}$$

(2.36)

- Total maximum precipitation of 24 h (T.max.P.24 h): The maximum water precipitation during 24 h. The maximum water precipitation is observed and logged during 24 h in a period of 1 month for the total number of observation years. This value is possible to be characterized as extreme and it is important to know the year that it occurred. The T.max.P.24 h has to be considered for the occurrence of floods.
- Average Precipitation Days (A.P.D.): The average number of days in a month at which there is water precipitation. The minimum water amount which is necessary to precipitate in order to be considered as water precipitation is different among countries but are generally close to 0.1 mm. For example for the calculation of the average precipitation days of January in the period 1959–2004 are equal to the summation of the precipitation days per year divided by the total number of observation years:

$$\text{A.P.D.} = \frac{P.D.(JAN1959) + P.D.(JAN1960) + \ldots\ldots\ldots + P.D.(JAN2004)}{46}$$

(2.37)

The atmosphere is a dynamic system and changes in the form of precipitation can occur during transport from its origin to the Earth's surface. Figure 2.26 shows some physical processes which can be associated with precipitation.

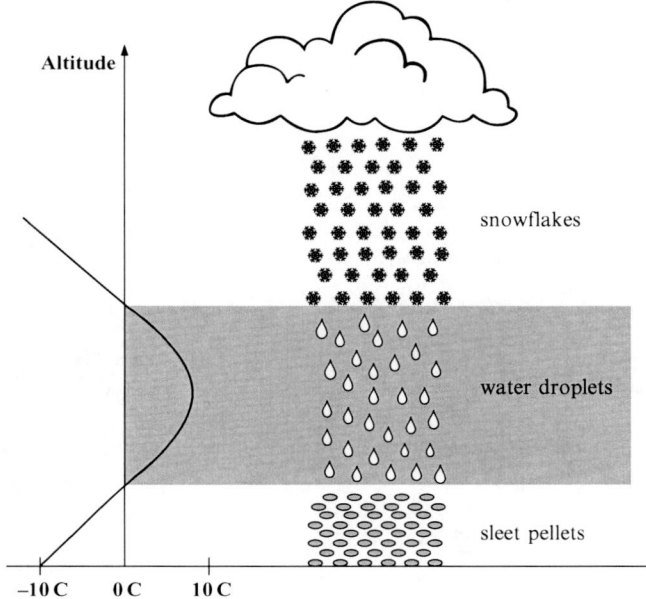

Fig. 2.26 Snowflakes originating from a cloud can melt during their downward movement when they encounter a warm layer of air and after can freeze again and form sleet pellets when they meet a colder air layer

2.7.5 Study of Precipitation Scavenging

There are several ways for precipitation to occur as discussed in the current chapter. The most common ones are through rainfall and snowfall. The flux of gasses and particles from the atmosphere to the Earth's surface through rain can be defined as (Seinfeld and Pandis 2006)

$$W^i_{gas/rain} = \Lambda_{ig}\, C_{i,gas}, \tag{2.38}$$

$$W^i_{aeros/rain} = \Lambda_{ip}\, C_{i,part}, \tag{2.39}$$

where Λ_{ig} and Λ_{ip} are the scavenging coefficients for the components i for the gaseous and particulate phase respectively. The total scavenging $F_{bc}(t)$ (Kg m^{-2} h^{-1}) under clouds, when the concentration of the pollutants exists in a horizontally homogeneous atmosphere, is $C_g(z,t)$; then it results that

$$F_{bc}(t) = \int_0^h \Lambda_g(z,t)\, C_g(z,t)\, dz, \tag{2.40}$$

where h is the height of the cloud base and Λ_g is the scavenging coefficient which is dependent on time (s^{-1}). The total scavenging is the sum of the scavenging inside the clouds (washout) and under the clouds (rainout). For a homogeneous atmosphere under the clouds it can be written that

$$F_{bc}(t) = C_g(t) \int_0^h \Lambda_g(z,t) \, dz = \overline{\Lambda}_g \, h \, C_g(t), \qquad (2.41)$$

where $\overline{\Lambda}_g$ is the average value of the scavenging coefficient.

The washout ratio can be defined as

$$w_r = \frac{C_{i,precip}(x,y,0,t)}{C_{i,air}(x,y,0,t)}, \qquad (2.42)$$

where $C_{i,pecip}(x,y,0,t)$ is the concentration of component i which is contained inside the rain at the Earth's surface and $C_{i,air}(x,y,0,t)$ is the concentration of component i inside the air at the Earth's surface. Therefore the flux F_w of the wet scavenging can be expressed as (Seinfeld and Pandis 2006)

$$F_w = C_{i,precip}(x,y,0,t) \, p_o, \qquad (2.43)$$

where p_o is the rain intensity (mm h^{-1}). For light rain the rain intensity is equal to $p_o = 0.5$ mm h^{-1}, whereas for heavy rain $p_o = 25$ mm h^{-1}. Furthermore, the velocity of wet deposition can be defined as $u_w = \frac{F_w}{C_{i,air}(x,y,0,t)}$.

In meteorology the rain is defined when the descending droplets have a diameter larger than or equal to 0.5 mm. After an intense rain there is usually better visibility since rain is scavenging a large number of particles. The rain has also the ability to absorb water soluble chemical components and remove them from the atmosphere. It is important to calculate the scavenging rate of gaseous components in the atmosphere based on knowledge of the rain's characteristics such as density and droplet size, as well as from the physico-chemical characteristics of the gaseous species.

The transfer of gasses on the droplet surface can be calculated from the expression

$$W_t(z,t) = K_c \left(C_g(z,t) - C_{eq}(z,t) \right), \qquad (2.44)$$

where K_c is the coefficient of mass transfer (cm s^{-1}), C_g is the concentration of the pollutant under study in the gaseous phase and C_{eq} its concentration on the droplet surface which is in equilibrium with the concentration in the aqueous phase.

Using Henry's law we can derive that

$$C_{eq} = \left(\frac{1}{H}\right) C_{aq}, \qquad (2.45)$$

2.7 Humidity in the Atmosphere

where, H is Henry's coefficient and C_{aq} is the pollutant concentration in the aqueous phase.

Consequently the equation (2.44) can be written as

$$W_t(z,t) = K_c \left(C_g(z,t) - \frac{C_{aq}(z,t)}{H} \right). \tag{2.46}$$

The coefficient of mass transfer from molecules in the gaseous phase to the droplet can be calculated from the expression (Seinfeld and Pandis 2006)

$$K_c = \frac{D_g}{D_p} \left[2 + 0.6 \left(\frac{\rho_{air} U_t D_p}{\mu_{air}} \right)^{1/2} \left(\frac{\mu_{air}}{\rho_{air} D_g} \right)^{1/3} \right], \tag{2.47}$$

where, D_p is the droplet diameter, D_g is the diffusivity in the gaseous phase, ρ_{air} is the air density, μ_{air} is air viscosity and U_t is the droplet velocity. In the above equation the term $Sh = K_c D_p/D_g$ is the Sherwood number, $Re = \rho_{air} U_t D_p/\mu_{air}$ is the Reynolds number and $Sc = \mu_{air}/\rho_{air} D_p$ is the Schmidt number. The concentrations C_{aq} and C_g are functions of height and time.

In addition to precipitation occurring under clouds, transport of chemical components to rain droplets can occur inside clouds. Gaseous species such as HNO_3, NH_3, and SO_2 can be absorbed inside rain droplets. If the concentration of particles inside a cloud is equal to $N(D_p)$ then the transfer rate of a gas W_{ic} which has larger concentration in air than in the liquid phase (e.g., HNO_3) is given by the expression (Seinfeld and Pandis 2006)

$$W_{ic} = C_g \int_0^\infty K_c \pi D_p^2 N(D_p) dD_p = \Lambda C_g, \tag{2.48}$$

where Λ is the scavenging rate. A typical concentration of droplets can be given by the expression $N(D_p) = a e^{-bD_p}$ where $\alpha = 2.87 \text{ cm}^{-4}$, $b = 2.65 \text{ cm}^{-1}$ at $D_p = 5\text{--}40$ μm. With a replacement in equation (2.48) it can be concluded that $\Lambda = 0.2 \text{ s}^{-1}$ for a cloud which consists of 288 droplets/cm^3. The above calculations denote that the processes which transfer chemical species at droplets are very fast, of the order of a few seconds in comparison with the transport of the chemical species inside a cloud or with changes of the ratio of condensation and evaporation. Several chemical species, such as SO_2 are not absorbed directly from rain droplets. Their absorption is dependent on other factors such as the presence of other chemical species inside droplets and the pH of the water. In addition to the absorption of gaseous species there is also absorption of particles inside clouds. A common process is the increase of particle size through absorption of water vapour and coagulation with other particles.

2.8 Applications and Examples

Example 1. *An air volume is rising adiabatically from height z_1 to z_2. Prove that the relationship between temperature and pressure at two different heights in the atmosphere are $\frac{T(z_2)}{T(z_1)} = \left[\frac{p(z_2)}{p(z_1)}\right]^{(\gamma-1)/\gamma}$.*

From the theory it is known that

$$\frac{dT}{dp} = \frac{m R T / M_{air} \, p}{C_v + mR/M_{air}},$$

where

$$\hat{c}_p - \hat{c}_v = \frac{R}{M_{air}}$$

And

$$\gamma = \frac{\hat{c}_p}{\hat{c}_v}.$$

Therefore:

$$\frac{dT}{T} = \frac{mR}{M_{air}} \bigg/ \left(C_v + \frac{mR}{M_{air}}\right) \frac{dp}{p} \Rightarrow \frac{dT}{T} = \frac{R}{M_{air}} \bigg/ \left(\hat{c}_v + \frac{R}{M_{air}}\right) \frac{dp}{p}$$

$$\Rightarrow \int_{T(z_1)}^{T(z_2)} \frac{dT}{T} = \int_{p(z_1)}^{p(z_2)} \left(\frac{\gamma-1}{\gamma}\right) \frac{dp}{p} \Rightarrow \ln \frac{T(z_2)}{T(z_1)} = \left(\frac{\gamma-1}{\gamma}\right) \ln\left(\frac{p(z_2)}{p(z_1)}\right).$$

Finally: $\frac{T(z_2)}{T(z_1)} = \left[\frac{p(z_2)}{p(z_1)}\right]^{(\gamma-1)/\gamma}$.

If z_1 is the height of the Earth's surface and air volume, which is initially at conditions T, p is rising adiabatically to pressure p_o, then the temperature θ at pressure p_o is given by the relation $\theta = T \left(\frac{p}{p_o}\right)^{-\frac{(\gamma-1)}{\gamma}}$ and is called potential temperature

Example 2. *Show that if the atmosphere is isothermal, then the temperature change of the temperature of an air volume which, ascending adiabatically, is given by the expression $T(z) = T_o \, e^{-\Gamma z/T_o'}$ where T_o and T_o' are the temperatures of the air volume and the atmospheric air at the surface respectively. $\Gamma = \frac{g}{\hat{c}_p}$.*

For the air volume (T' is the air temperature and T is the temperature of the air volume under study) we have

$$\frac{dp}{dz} = -\frac{M_{air} \, g \, p}{R T'}.$$

2.8 Applications and Examples

For the air volume which is ascending we can write that

$$\frac{dT}{dp} = \frac{m R T/M_{air} \, p}{C_u + mR/M_{air}} \Rightarrow \frac{dT}{dp} = -m\,g\, \frac{\frac{dz}{dp}\frac{T}{T'}}{c_v + m R/M_{air}}$$

which results in $\frac{dT}{dz} = -\Gamma \frac{T}{T'}$.
With the integration of the above expression it follows that

$$\int_{T_o}^{T} \frac{dT}{T} = \int_0^z -\Gamma \frac{dz}{T'} \Rightarrow \ln\frac{T}{T_o} = -\frac{\Gamma}{T'}z \Rightarrow T(z) = T_o\, e^{\left(-\frac{\Gamma z}{T_o}\right)}.$$

Example 3. *Concentrations of two gaseous species ($N_1 = 8$ και $N_2 = 4$) with velocities ($u_1 = 3$ and $u_2 = -1$) have been measured at two different points at specific times. Calculate the following variables: $\overline{N}, N'_1, N'_2, \overline{u}, u'_1, u'_2, \overline{u'N'}, \overline{u}\,\overline{N}, \overline{uN}$.*

Using the expressions for the average and instant values it can be written that

$$\overline{N} = (N_1 + N_2)/2 = 6,$$

$$N'_1 = N_1 - \overline{N} = 2,$$

$$N'_2 = N_2 - \overline{N} = -2,$$

$$\overline{u'N'} = \left(u'_1 N'_1 + u'_2 N'_2\right)/2 = 4,$$

$$\overline{uN} = \overline{u}\,\overline{N} + \overline{u'N'} = (u_1 N_1 + u_2 N_2)/2 = 10:$$

$$\overline{u} = (u_1 + u_2)/2 = 1,$$

$$u'_1 = u_1 - \overline{u} = 2,$$

$$u'_2 = u_2 - \overline{u} = -2,$$

$$\overline{u}\,\overline{N} = 6.$$

Example 4. *A meteorological station measures with anemometers the components U and W of the wind. Velocity measurements have been performed every 6 s for 1 min. The measurements are given in the following 10 pairs:*

U (m/s) 5 6 5 4 7 5 3 5 4 6,
W (m/s) 0 −1 1 0 −2 1 2 −1 1 −1.

Calculate the average value and dispersion of each component of the velocity, as well as the correlation coefficient between U and W.
Using the expressions for the average and correlation coefficient values, the following values can be written.

$$\overline{U} = 5\,m\,s^{-1}$$

$$\overline{W} = 0\,m\,s^{-1}$$

$$\sigma_U^2 = 1.20\,m^2\,s^{-2}$$

$$\sigma_U = 1.10\,m\,s^{-1}$$

$$\sigma_W^2 = 1.40\,m^2\,s^{-2}$$

$$\sigma_W = 1.18\,m\,s^{-1}$$

$$\overline{u'w'} = -1.10\,m^2\,s^{-2}$$

$$r_{UW} = -0.85$$

From the above expressions can be concluded that the turbulence component of velocity W is higher than the U component even if the average value for the velocity W is zero. From the negative value of the correlation coefficient r_{uw} it can be concluded that the variations of the components U and W are occurring mainly in opposite directions.

Example 5. *The concentration of a gas A is equal to 10 μg m^{-3} under a cloud. Suppose that the washout coefficient is constant and equal to 3.3 h^{-1}. Calculate the concentration of gas A in the atmosphere after 30 min of rain and the total flux of the wet deposition. The cloud base is 2 km (Adapted from Seinfeld and Pandis, 2006).*

The concentration variability can be expressed as

$$\frac{\partial C}{\partial t} = -W_{air/rain} + R + E,$$

where since there no emissions or chemical reactions ($R = 0, E = 0$) it can result that

2.8 Applications and Examples

$$\frac{\partial C}{\partial t} = -\Lambda C,$$

where Λ is the washout coefficient. The solution of the equation gives

$$C = C_o e^{-\Lambda t}.$$

After 30 min the resulting concentration is equal to $C = C_o \times 0.19 = 1.9 \ \mu g \ m^{-3}$. Therefore, C_o-$C = 8.1 \ \mu g \ m^{-3}$ and the total flux for a column of 2 km is equal to $8.1 \times 2{,}000 = 16.2 \ mg \ m^{-2}$.

Problems

2.1 It is proposed that the problem of air pollution in the city of Los Angeles can be solved by digging tunnels in the surrounding mountains and pumping the air outside to the surrounding areas which are mainly deserts. Calculate the energy that would be needed for the transport of air from Los Angeles. Los Angeles covers an area of 4,000 km² and the polluted air is located under the boundary layer which has an average height of 400 m. The viscosity coefficient of air which is transported above the Los Angeles area is 0.5 and the minimum energy which is required for the air flux is equal to the energy which is consumed from the surface friction. Calculate the energy which would be required for an air mass transport with velocity 7 km/h. Compare the result with the capacity of Hoover Dam (power station of energy production) which is equal to 1.25×10^6 KWh.

2.2 The value of the vertical temperature lapse rate in an area is equal to 5 °C/km and the temperature of air on the surface is equal to 20°C. If an insulated balloon full of dry air with temperature 50°C is allowed to ascend from the surface, then calculate the height which the balloon can reach (the balloon temperature lapse rate is equal to $\gamma_d = 10°C/km$).

2.3 Simultaneous measurements of the air temperature at four points A, B, Γ and Δ which are located downwind of a mountain chain are the following:

Locations	A	B	Γ	Δ
Height (m)	1,530	1,396	690	378
Temperature (T) (°C)	6.3	6.9	11.9	14.4

Calculate the value of the vertical lapse rate between the positions (a) $\Delta \ \Gamma$, (b) Γ B, (c) B A and (d) Δ A. Additionally make a comparison of the mean value of the vertical lapse rate above the mountain chain with the corresponding value of $-6.5°C/km$ and determine the difference.

2.4 A potential temperature Θ is determined under the hypothesis of adiabatic processes. Changes in entropy can be connected with changes in temperature and pressure with the expression

$$dS = \left(\frac{\partial S}{\partial T}\right)_P dT + \left(\frac{\partial S}{\partial P}\right)_T dp.$$

a. Show that

$$dS = \left(\frac{c_p}{T}\right)dT - \left(\frac{R}{M_a}\right)\frac{dp}{p}.$$

b. During adiabatic conditions ($dS = 0$) show that $d\Theta = 0$ and $\Theta = const\tan t \times \frac{T}{p^{\frac{(\gamma-1)}{\gamma}}}$.

2.5 In an urban area, the air temperature at 07:00 after a cloudless night in January is equal to $0°C$. The height of a temperature inversion was $h = 0$ and its depth $d = 100$ m. Calculate at which height from the inversion base (surface) it will be the same air temperature, if the value of the vertical temperature lapse rate is $-(\partial T/\partial z) = 5\,°C/km$ and the intensity of the inversion is equal to $0.1°C/10m$.

2.6 Calculate the concentration ($mole/cm^3$) and the mixing ratio (ppm) of water vapour at the Earth's surface for temperature equal to $T = 298K$ and relative humidity RH 50%, 60%, 70%, 80%, 90%, 95%, 99%. The water vapour pressure of pure water (saturation pressure) versus temperature is given by the expression

$$p^o_{H_2O}(T) = p_s \times \exp[13.3185\alpha - 1.976\alpha^2 - 0.6445\alpha^3 - 0.1299\alpha^4],$$

where:

$$p_s = 1013.25 mbar$$

and

$$\alpha = 1 - \left(\frac{373.15}{T}\right).$$

2.7 The prognosis of the daily atmospheric temperature structure under non-variable conditions is examined here. In this case it is necessary to examine the spatial temperature changes only in the vertical direction. It is assumed that the radiation absorption from the atmosphere is negligible and the dynamic temperature θ can be determined from the expression

$$\frac{\partial \theta}{\partial t} = \frac{\partial}{\partial z}\left(K\frac{\partial \theta}{\partial z}\right). \tag{2.49}$$

In the above expression it is assumed that K has a constant value. It is also assumed that at elevated heights the temperature profile is the same with an adiabatic rate equal to

$$\theta \to 0 \quad \text{when} \quad z \to \infty. \tag{2.50}$$

The temperature at the surface ($z = 0$) is dependent on the Sun's heat during the day and the cooling at night is due to radiation emission. Therefore $\theta(0,t)$ can be expressed as

$$\theta(0, t) = A \cos \omega t. \tag{2.51}$$

where A is the width of the daily variation and $\omega = 7.29 \times 10^{-5} \text{ s}^{-1}$.

A) Show that the solution which satisfies the equations $(1 - 3)$ is

$$\theta(z, t) = A e^{-\beta z} \cos(\omega t - \beta z),$$

where $\beta = \sqrt{\omega/2K}$. The above solution is called a steady-state solution and describes the temperature dynamics which correspond to the influence of equation 3.

B) Show that the height H which expresses the base or the top of a temperature inversion is given by the expression

$$\sin(\omega t - \beta H) - \cos(\omega t - \beta H) = \frac{g}{A\beta \widehat{c}_p} e^{\beta H}.$$

Is it possible to have more than one temperature inversion?
Show that $\frac{\partial \theta}{\partial z} = \frac{\partial T}{\partial z} + \Gamma$
where $\Gamma = \frac{g}{c_p}$.

2.8 It is observed in an area that the atmospheric temperature is decreased by 14 K between the Earth's surface and height of 2 km above it. What is the vertical lapse rate and how does it compare with the dry and wet lapse rate? Calculate the atmospheric stability conditions under the above conditions.

References

Ahrens, C. D. (1994). *Meteorology today – An introduction to weather, climate and the environment* (5th ed.). St. Paul: West Publishing Company.
European Union (2001). Air pollution research Report No 76, EU 2001.
Finlayson-Pitts, B. J., & Pitts, J. N. (1986). *Atmospheric chemistry: Fundamentals and experimental techniques*. New York: John Wiley & Sons.
Finlayson Pitts, B. J., & Pitts, J. N. (2000). *Chemistry of the upper and lower atmosphere*. San Diego: Academic Press.
Hanna, S. R., Briggs, G. A., & Hosker, R. P. (1981). *Handbook on atmospheric diffusion, technical information center*. Washington: U.S. Department of Energy.
Hemond, H. F., & Fechner-Levy, E. J. (2000). *Chemical fate and transport in the environment*. San Diego: Academic Press.
Housiadas, C., Drossinos, I., & Lazaridis, M. (2004). Effect of turbulent fluctuations on binary nucleation. *Journal of Aerosol Science, 4*(35), 545–559.

Lioy, P. J. (1990). Assessing total human exposure to contaminants. *Environmental Science and Technology, 24*, 938–945.

Seinfeld, J. H., & Pandis, S. N. (2006). *Atmospheric chemistry and physics* (2nd ed.). New York: John Wiley & Sons.

Stull, R. B. (1997). *An introduction to boundary layer meteorology* (pp. 199–202). London: Kluwer Academic Publishers.

Chapter 3
Atmospheric Circulation

Contents

3.1	Atmospheric Pressure and Pressure Gradient Systems	119
3.2	Atmospheric Pressure Changes	120
	3.2.1 Vertical Pressure Changes	120
	3.2.2 Non-canonical Pressure Changes	121
	3.2.3 Canonical Pressures Changes	121
3.3	Transfer of the Pressure to Its Mean Value at Sea Level	122
3.4	Isobaric Curves: Pressure Gradient Systems	122
3.5	Pressure Gradient Force	124
3.6	Movement of Air: Wind	126
	3.6.1 Forces Which Affect the Movement of Air	128
	3.6.2 Transport Equations of Air Masses in the Atmosphere	131
	3.6.3 Wind Categories	132
3.7	General Circulation in the Atmosphere	137
	3.7.1 Single and Three Cell Models	138
	3.7.2 Continuous Winds	140
	3.7.3 Periodic Winds	141
	3.7.4 Sea and Land Breeze	141
	3.7.5 Mountain and Valley Breezes	143
3.8	Vertical Structure of Pressure Gradient Systems	144
3.9	Equations of Atmospheric Circulation	145
	3.9.1 Equations of Circulation for a Compressible Fluid	145
References		149

Abstract Atmospheric circulation is studied in Chap. 3. General aspects of atmospheric pressure are examined in relation to pressure changes and isobaric curves. The transport equations of air masses in the atmosphere are derived in conjunction with the applied forces. These forces include the pressure gradient, the Coriolis force, and the centrifugal and friction forces. The geostrophic wind is also studied in conjunction with the wind at the surface level. The general circulation in the atmosphere is studied, with the help of a three-cell model, as well as the periodic and local wind systems. Finally the equations of circulation for compressible fluids are examined.

3.1 Atmospheric Pressure and Pressure Gradient Systems

Atmospheric pressure is the pressure which the atmospheric air exerts on the surface of objects due to its weight. Consequently every surface on the Earth's surface or inside the atmosphere receives the influence of the weight of the superincumbent air column. This act of a force per unit surface is called atmospheric pressure. Air pressure is the weight of air above a given level. The weight of the air column changes continuously due to complex atmospheric circulation and changes occurring within the air masses.

In Meteorology, atmospheric pressure is of significant importance since pressure changes are closely related to wind characteristics and weather conditions in the atmosphere. Atmospheric pressure is also the most important component for the illustration of weather conditions in weather maps.

Atmospheric pressure is displayed by the height of a mercury thermometer column within which it is balanced and is measured in millimeters (mmHg), inches (inHg), milibar (mb) or Hectopascal (Hpa). The relationships among the different units used for measurement of atmospheric pressure are:

$$1 \text{ atm} = 760.00 \text{ mmHg} = 29.92 \text{ in Hg} = 1013.25 \text{ mb} = 1013.25 \text{ Hpa},$$

$$1 \text{ mmHg} = 1.333 \text{ mb}, 1 \text{ in Hg} = 25.4 \text{ mmHg} = 33.86 \text{ mb},$$

$$1 \text{ mb} = 10^{-3} \text{bar} = 10^3 \text{dyn} \times \text{cm}^{-2} = 10^2 \text{Nt} \times \text{m}^{-2} = 0.75 \text{ mmHg},$$

$$1 \text{pa} = 1 \text{Nt m}^{-2}$$

The atmospheric pressure is measured with instruments which are called barometers.

3.2 Atmospheric Pressure Changes

Atmospheric pressure, similar to temperature and humidity of air, has several variations which can be resolved to vertical, non-canonical and canonical.

3.2.1 Vertical Pressure Changes

The atmospheric pressure at a specific geographical location is dependent on the weight of the air column above it. Therefore at higher elevations the weight of the air column above it decreases and the pressure also decreases. The pressure reduction inside the troposphere layer is close to 1 mmHg per 10–11 m.

In Chap. 1 the vertical pressure changes were described with the use of the hydrostatic equation. However, there exists also a semi-empirical approach which relates the difference of air pressure between two heights with the height difference Δh (m):

$$\Delta h = 18429 \times A \times (1 + aT) \times \log\left(\frac{H_o}{H}\right) \quad (3.1)$$

where, Ho and H are observed air pressures at the lower and upper levels of the atmosphere (mmHg), T is the mean temperature between the two levels (K), a is the factor of mercury expansion (m/K) and A is a factor which depends on the air humidity and geographical width and is close to unity.

3.2.2 Non-canonical Pressure Changes

The non-canonical pressure changes at a specific location are due to the passage of pressure systems above it. These changes are directly related to the weather conditions. The non-canonical changes can occur at a specific location for an extended period of time (up to 10 days). Significant changes may also occur also for time periods shorter than a day.

On a global level the values of atmospheric pressure at sea level range between 950 and 1,050 mbs and in exceptional cases between 925 and 1,070 mbs. The highest value of air pressure, equal to 1,084 mbs, occurred at the Agata of Siberia. The lowest value of air pressure occurred at the Mariana Islands of the Pacific Ocean and was equal to 876 mbs during the passage of a hurricane.

3.2.3 Canonical Pressures Changes

The canonical pressure changes are due to:

- The daily effect of heating and cooling of air which results from the Sun's radiation. These changes are known as semi-daily since they have a 12 h cycle.
- Atmospheric tides which result from the gravity attraction between the Sun and the moon.

The maximum values of pressure are observed between 09:00–10:00 and 21:00–22:00 local time, whereas smaller values (about 1–3 mbs lower than the maximum) between 03:00–04:00 and 15:00–16:00 local time. It is evident that the canonical changes have a stable periodicity and are much smaller than the non-canonical changes (see Fig. 3.1).

The change of the air pressure at a specific time period is called the pressure gradient inclination. In practice this is a 3-h time period before a meteorological observation. The estimation of the pressure gradient inclination helps in the determination of the pressure gradient system movement, since the air pressure decreases when a low pressure system is approaching and increases close to a high pressure system.

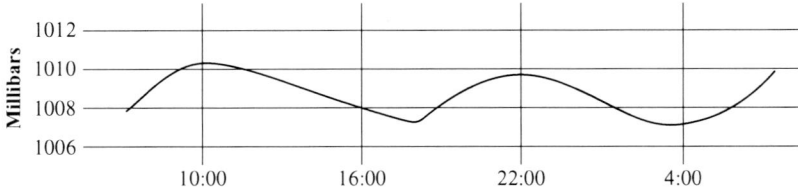

Fig. 3.1 Mean daily change of air pressure

3.3 Transfer of the Pressure to Its Mean Value at Sea Level

Measuring air pressure in reality is measuring the weight of a column of air. Consequently the pressure refers to the observation point. Since the pressure is decreased with height, the pressure at a higher elevation is lower than the pressure of zero elevation which is the mean sea level (MSL).

For the analysis of meteorological maps it is needed to compare pressure values from different meteorological stations at various elevations with a reference point. The common reference point is the mean sea level. Pressure values at each meteorological station are transferred to values which correspond to mean sea level. Figure 3.2 shows the transfer of pressure measurements at different locations to mean sea level values, taking into account that the atmospheric pressure is reduced at a rate close to 1 mb per meter.

3.4 Isobaric Curves: Pressure Gradient Systems

Every day and at short time intervals (every 3 h) at several places on Earth, measurements of atmospheric pressure are taken. Since the ground stations exist at various elevations, the measurements are translated into mean sea level values for comparison purposes. The observed pressure values are written on meteorological maps which are called surface maps. On these maps, lines are plotted which connect points with the same pressure. These lines are known as isobars. Every isobaric line has a specific pressure value and cannot intersect with another isobaric line since, in such a case, a point would have two different pressure values simultaneously. At surface maps, for reasons of resolution, the isobaric lines are plotted at intervals of 2 mbs or more usually at intervals of 4 mbs.

The isobaric lines divide areas of lower and higher pressure and have forms on surface maps which have high importance for meteorologists. Consequently some isobaric lines enclose areas with high pressures whereas others enclose areas with low pressures. These areas are called pressure systems and it is possible to determine five different shapes (Fig. 3.3):

- **Pressure Low or Cyclon (Low – L).** This shape is described with isobaric closed, usually cycles or elliptical curves, where the atmospheric pressure is

3.4 Isobaric Curves: Pressure Gradient Systems

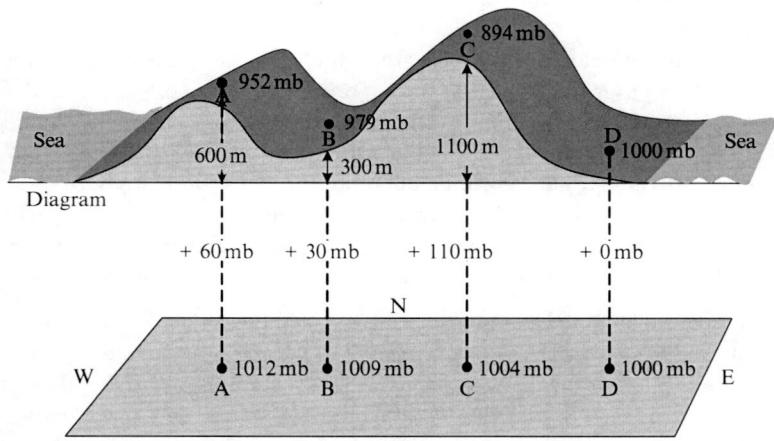

Fig. 3.2 Transfer of the atmospheric pressure at different locations (A, B, C, D) to the mean sea level pressure (Adapted from Ahrens 1994)

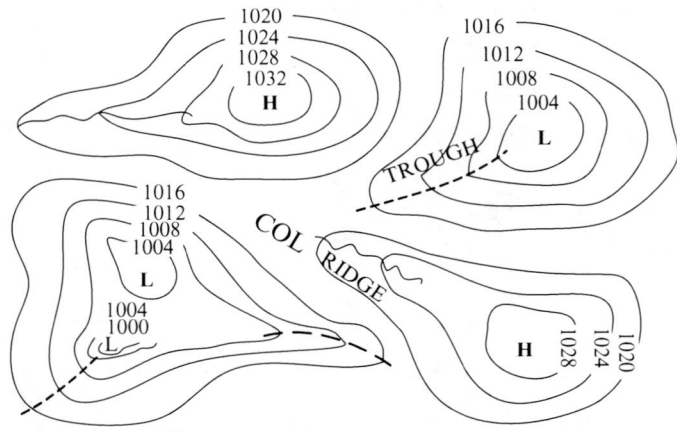

Fig. 3.3 Isobaric curves and pressure systems (*H* = high, high pressure system, *L* Low, Low pressure system, *RIDGE* Elongated high, *COL* Pressure saddle point, *TROUGH* Elongated low)

reduced from the circumference to its centre. On many occasions in the southeast direction of a cyclone system (at the northern hemisphere) the formation is observed of a secondary low, which tends to be isolated from the main centre of the low pressure system. The isobaric curves of this type tend to bend whether they have a centre or not. It is denoted by the letter L (Low) and coloured red.

- **Pressure high or Anticyclone (High – H).** This is described with isobaric closed, usually cycles or elliptical curves, where the atmospheric pressure

increases from the circumference to its centre. It is denoted the letter H (High) and coloured blue.
- **Pressure Saddle Point (COL).** The area which exists between two high (H) and two low (L) pressure systems that have a crosswise arrangement.
- **Elongated Low (TROUGH).** Elongation of a low pressure system with the minimum pressure along a line which is called a TROUGH LINE and depicts the positions of the maximum cyclonic curvature of the isobaric lines.
- **Elongated High (RIDGE).** Elongation of a high pressure system with the maximum pressure along a line which is called the RIDGE LINE and depicts the positions of the maximum anticyclonic curvature of the isobaric lines.

For the study of these systems inside the atmosphere, maps at different heights have been produced which are called maps of the upper atmosphere. The maps of the upper atmosphere are isobaric surfaces on which are written the actual heights at the different regions. For the production of these maps, data from radiosonde measurements are used. Furthermore, curves are produced which connect the regions which have the same height, distinguishing the upper from the lower heights. These curves are called contour lines and depict the variability of each isobaric level. The isobaric lines confine regions of higher heights, where there is depiction of high pressure systems. Isobaric lines which confine lower heights depict the low pressure systems. Putting the maps of the surface and the upper atmospheric levels vertically, one can depict the vertical structure of the pressure systems in the atmosphere (Fig. 3.4).

3.5 Pressure Gradient Force

The force which tends to move air molecules from high to low pressures in order to attain equilibrium is called pressure gradient force (F_B). The pressure gradient is

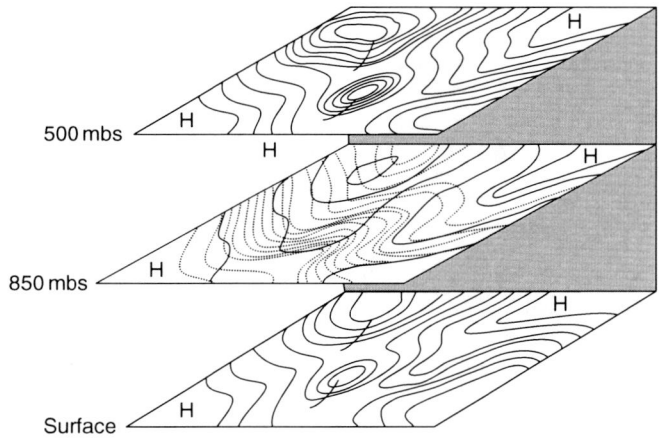

Fig. 3.4 Vertical structure of pressure systems

3.5 Pressure Gradient Force

dependent on the changes in atmospheric pressure and has a direction vertical to the isobaric lines. On a surface map the pressure gradient force is calculated as the ratio of the pressure difference between two successive isobaric lines to their distance. Therefore if the distance between two isobaric lines is longer, then the pressure gradient force is smaller. The pressure gradient force is given by the equation:

$$F_B = -\frac{\Delta P}{\Delta n} \qquad (3.2)$$

where Δn is the distance between two isobaric lines and Δp is the corresponding pressure difference. The sign $(-)$ is used to express the vector direction of the pressure gradient force which is from the high to low pressure.

Therefore when we calculate the pressure gradient force between two points AB and BΓ which have a distance between them of 60 and 90 km respectively and pressure difference 4 mb, we get that (Fig. 3.5):

$$AB : F_{B1} = -\frac{4\,\text{mb}}{60\,\text{km}} = -0.07\,\text{mb}\,\text{km}^{-1}$$

$$B\Gamma : F_{B2} = -\frac{4\,\text{mb}}{90\,\text{km}} = -0.04\,\text{mb}\,\text{km}^{-1} \qquad (3.3)$$

The minus signs $(-)$ denote the direction of the vectors and therefore $F_{B1} > F_{B2}$. The total pressure gradient force per unit mass is given by the expressions:

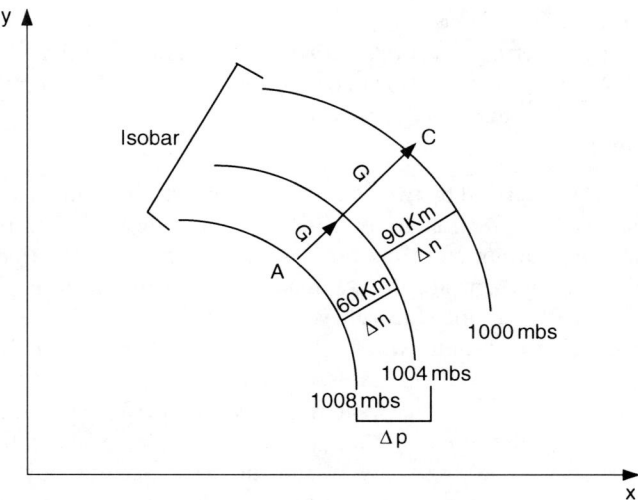

Fig. 3.5 Pressure change and determination of the pressure gradient force (the symbol G expresses the slope of the pressure gradient force with direction from the high to low pressure)

$$\frac{\vec{F}_B}{m} = -\frac{1}{\rho}\left[\frac{\partial p}{\partial x}\vec{i} + \frac{\partial p}{\partial y}\vec{j} + \frac{\partial p}{\partial z}\vec{k}\right] \tag{3.4}$$

where ρ is the density of the atmospheric air.

The value for the pressure gradient force F_B has a great importance in meteorology since it is closely related with the wind creation (direction and wind intensity).

3.6 Movement of Air: Wind

The changes of pressure and temperature result in two kinds of movement inside the atmosphere:

- The vertical movement (upwards and downwards air streams)
- The horizontal movement

The flux of air molecules gives the concept of wind. The flux is almost entirely horizontal since a very small percentage of the wind flux has a vertical direction. However, the vertical component of the wind transport is very important since it is related with the weather conditions above a specific area and the formation of cumulonimbus clouds (CuF) and storms.

The Earth's rotation, the inhomogeneity of the Earth's topography and the Sun's energy geographical distribution in the atmosphere and at the Earth's surface are the main factors which result in the atmospheric air being in constant movement. The most important forces which form and determine the wind dynamics are:

- The pressure gradient force
- The horizontal diverting force (Coriolis) which is due to the Earth's rotation
- The circular force (centrifugal) which appears when the wind is directed around a weather system of low or high pressure and
- The friction force

The wind is characterized by two components, its direction and its intensity. The wind direction is the direction from which the wind flows and is measured in degrees clockwise starting from the northern direction. The directions from the main points of the horizon are differentiated as northern, eastern, southern, or western and from the intermediate values (e.g. northwestern). The wind intensity is the measure of the horizontal wind velocity and is measured usually as m/s or knots (1 m/s = 3.6 km/h = 1.943 Knots = 2.237 m.p.h).

The calculation of the wind characteristics can be performed with the use of instruments and with a semi-empirical scale which is called a Beaufort scale. This scale was invented at the beginning of the nineteenth century by Admiral Sir Francis Beaufort and is based on the effects of the wind to the conditions over sea and at different objects on the Earth's surface. Table 3.1 shows the

3.6 Movement of Air: Wind

Table 3.1 Beaufort wind scale and its correspondence to other measurement scales of the wind velocity

Class beauf.	General description	Characterization	Wind velocity at height 6 m above the surface				Symbols
			m/s	km/h	mph	knots	
0	Calm	Smoke rises vertically	<0.6	<1	<1	<1	
1	Light air	Direction of wind shown by drifting smoke, but not by wind vanes	0.6–1.7	1–6	1–3	1–3	
2	Slight breeze	Wind felt on face; leaves rustle; wind vanes moved by wind; flags stir	1.8–3.3	7–12	4–7	4–6	
3	Gentle breeze	Leaves and small twigs move; wind will extend light flag	3.4–5.2	13–18	8–11	7–10	
4	Moderate breeze	Wind raises dust and loose paper; small branches move; flags flap	5.3–7.4	19–26	12–16	10–14	
5	Fresh breeze	Small trees with leaves begin to sway; flags ripple	7.5–9.8	27–35	17–22	15–19	
6	Strong breeze	Large tree branches in motion; whistling heard in power lines; umbrellas used with difficulty	9.9–12.4	36–44	23–27	19–24	
7	High wind	Whole trees in motion; inconvenience felt walking against wind; flags extend	12.5–15.2	45–55	28–34	24–30	
8	Gale	Wind breaks twigs off trees; walking is difficult	15.3–18.2	56–66	35–41	30–35	
9	Strong gale	Slight structural damage occurs (signs and antennas blown down)	18.3–21.5	67–77	42–48	36–42	
10	Whole gale	Trees uprooted; considerable damage occurs	21.6–25.4	78–90	49–56	42–49	
11	Storm	Winds produce widespread damage	25.5–29.0	91–104	57–67	49–56	
12	Hurricane	Winds produce extensive damage	>29.0	>104	>67	>56	

correspondence between the empirical classifications of Beaufort with the equivalent wind velocities.

It is very difficult to distinguish between the cause and the result in the process of wind formation in the atmosphere. In particular, there are correlations between the pressure, temperature and the wind inside the atmosphere. In reality the wind affects also its cause and there is a continuous movement in the atmosphere in order to reach equilibrium in the same manner as the ocean tends to keep a stable level.

The cause of the wind formation is the pressure difference between two different locations horizontally. The horizontal pressure differences at different locations are results of the unequal distribution of temperature in the atmosphere.

3.6.1 Forces Which Affect the Movement of Air

The force which can start the movement of the air molecules is the pressure gradient force which is a result of the pressure difference between two different locations. Furthermore, there are other forces which affect the air movement only if there is air movement inside the atmosphere.

3.6.1.1 Coriolis Force

The Coriolis force affects the movement of air molecules and is due to the Earth's rotation around its axis and affects every movement inside the atmosphere. The Coriolis force does not produce transport of air masses but changes the wind direction. Its name came from the French scientist G. Coriolis, who first studied this force in 1835.

If the Earth is assumed stationary, a missile ejected from the north pole having as target the point A at the Equator, will reach this point without a change of its trajectory (Fig. 3.6). In reality Earth is rotating clockwise with a stable angular velocity of rotation. Therefore the missile which will be ejected from the north pole with target the point A will reach in reality point B (at the west of point A) due to the Earth's rotation.

Therefore each movement inside the atmosphere affects this force. The Coriolis force pushes the bodies, and therefore also the air, which are moving inside the atmosphere clockwise at the northern hemisphere and counterclockwise at the south hemisphere, the deviation of the movement being counterclockwise.

The friction between the surface and the air decreases the effect of the Coriolis force at the lower layers. As discussed previously, the lower layer is called the boundary layer and extends close to 1,000 m above the Earth's surface where the effect of the topography is important. Above the boundary layer there is the geostrophic layer where the horizontal variations of the pressure and the Coriolis force affect the movement of the air masses.

3.6 Movement of Air: Wind

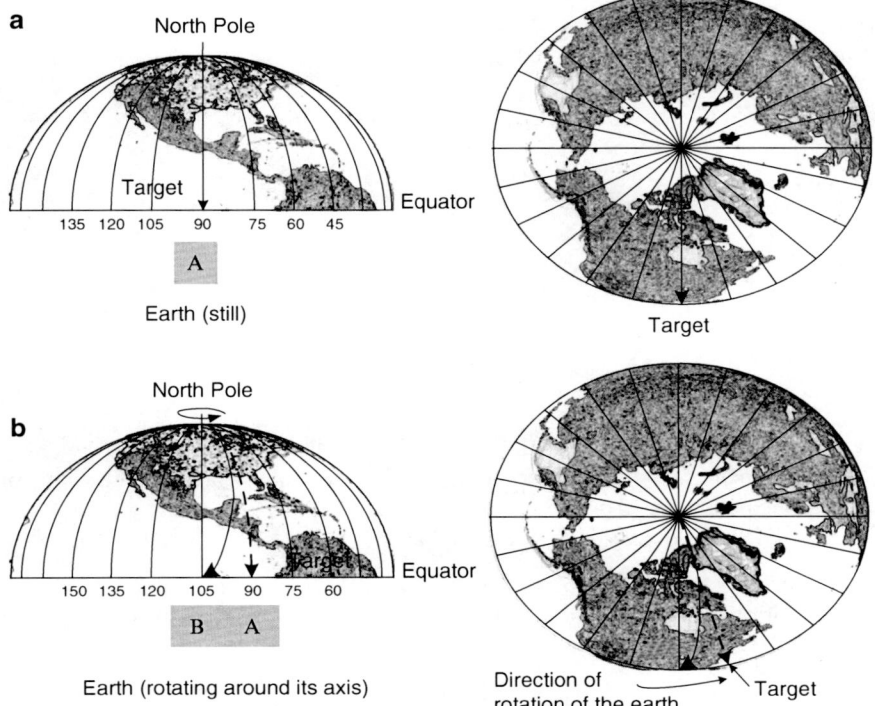

Fig. 3.6 The Coriolis force is an apparent force and is a result of the Earth's rotation

It can be calculated that the acceleration that an object encounters inside the atmosphere when it is moving with velocity u consists of two parts which are equal. The first part of the acceleration is equal to $-\vec{\Omega} \times (\vec{\Omega} \times \vec{r})$ and the second part equal to $-2(\vec{\Omega} \times \vec{V})$ where $\vec{\Omega}$ is the vector of the angular rotation of Earth and $\vec{r}$ is the radius from the Earth's centre to the position of the object.

The first term is the acceleration force in a direction vertical to the Earth's surface and is offset with the gravitational attraction. The second term $-2(\vec{\Omega} \times \vec{V})$ corresponds to the Coriolis force. The velocity vector must be different from zero in order for the Coriolis force to exist.

Expressing the Coriolis force from the equation $\vec{F}_c = -2(\vec{\Omega} \times \vec{V})$ results in the following components of the Cartesian system:

$$F_{cx} = -2\Omega(\omega \cos\phi - v \sin\phi) \tag{3.5}$$

$$F_{cy} = -2\Omega u \sin\varphi \tag{3.6}$$

$$F_{cz} = 2\Omega u \cos\varphi \tag{3.7}$$

At higher elevations it is possible to eliminate the vertical velocity component with respect to the u and v velocity components. In this case there are the following components of the Coriolis force per unit mass:

$$\frac{F_{cx}}{m} = 2\Omega v \sin \phi \Rightarrow \frac{F_{cx}}{m} = f_c v \quad F_{cx} = -2\Omega v \sin \phi \Rightarrow F_{cx}/m = f_c v \qquad (3.8)$$

and also:

$$\frac{F_{cy}}{m} = -f_c u \qquad (3.9)$$

where, f_c is called the Coriolis coefficient. The Coriolis coefficient is defined as follows:

$$f_c = 2\Omega \sin(\varphi) \qquad (3.10)$$

where $2\Omega = 1.458 \times 10^{-4}$ s^{-1} and φ is the Earth's latitude.

The parameter f_c is stable at the same latitude. At medium latitudes the value of the Coriolis parameter is on the order of $f_c = 10^{-4}$ s^{-1}.

From the above expressions it results that:

- The stronger the wind velocity, the more intense is the Coriolis force.
- The Coriolis force has a zero value at the equator and has its maximum value above the poles (Fig. 3.7).

As described, the Coriolis force is an apparent force which results from the Earth's rotation. Therefore there is no Coriolis force when there is no wind. The Coriolis force acts continuously vertically to the wind direction. In the northern hemisphere it impacts at the right of the wind direction, whereas at the south hemisphere it impacts at the left. This is the reason that the x constituent of the wind depends on the y constituent and vice versa.

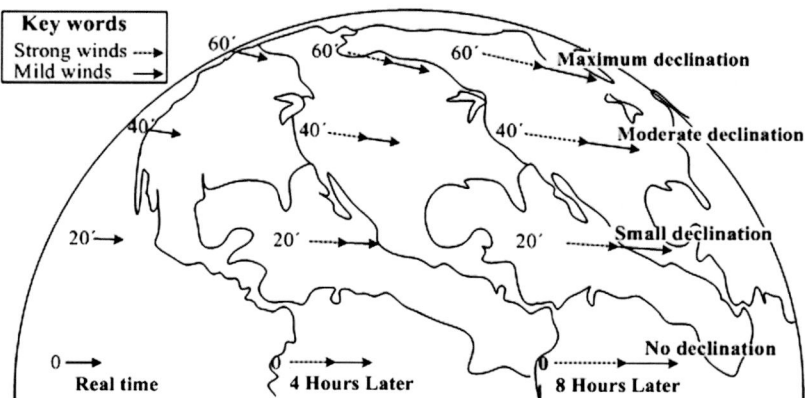

Fig. 3.7 The Coriolis force is increasing with increasing geographical width

3.6.1.2 Friction Force

The friction force F_τ is a result of surface friction, which is the force that opposes the wind movement above the surface. The friction force produces a slowdown to the air movement and therefore affects indirectly the Coriolis force which changes according to the wind velocity. Therefore a reduction of the wind velocity results in a reduction also of the Coriolis force.

The friction force depends on air velocity and surface roughness. The value of the force per unit mass is expressed as:

$$\frac{\vec{F_T}}{m} = -a\vec{V} \quad (3.11)$$

where α is a coefficient and its value depends on the surface roughness and V is the wind velocity. The friction force occurs inside the boundary layer which extends about 1,000 m from the Earth's surface and affects considerably the intensity and direction of the surface wind.

3.6.1.3 Centrifugal and Centripetal Forces

It is known from physics that the centrifugal and centripetal forces are equal and opposite and appear during curvilinear motion of bodies. The centrifugal force is regarded as a real force only for the observer who is connected with a moving reference system. Meteorology uses mainly the concept of centrifugal force (F_φ), which increases with the wind velocity as:

$$\frac{F_\Phi}{m} = \frac{V^2}{R} \quad (3.12)$$

where V is the wind velocity and R is the radius of the curvilinear movement. The centrifugal force has direction above the curvilinear track and tends to move away from the centre of the track.

3.6.2 Transport Equations of Air Masses in the Atmosphere

Starting from Newton's second law, and not taking account the vertical movement of air masses, the following equations are derived for the variability of the constituents of wind velocity:

$$\frac{du}{dt} = -\frac{1}{\rho}\frac{\partial p}{\partial x} + f_c v - F_{TX} \quad (3.13)$$

and

$$\frac{dv}{dt} = -\frac{1}{\rho}\frac{\partial p}{\partial y} - f_c u - F_{TY} \qquad (3.14)$$

The above equations are the well-known Navier–Stokes equations for the horizontal flux of a fluid.

3.6.3 Wind Categories

3.6.3.1 Geostrophic Wind

It is hypothesized that at the Earth's surface the isobaric lines are parallel lines and the high pressure is located in the south. It is assumed also that the friction force is negligible ($F_\tau = 0$) and therefore the forces which affect the horizontal movement are the pressure gradient (F_B) and the Coriolis (F_c) forces (Fig. 3.8).

At the beginning the molecules of atmospheric air under the influence of the pressure gradient force (F_B) are moving from a high pressure area to a low pressure area vertically to the isobaric charts. This happens in order to ensure an equilibrium of the pressure. As a movement of air begins, there starts also the influence of the Coriolis force which diverts the movement of air molecules clockwise in the northern hemisphere.

In reality the air is moving almost with a stable velocity parallel to the isobaric charts, having high pressure to the right during the movement at the northern hemisphere. On the contrary, in the southern hemisphere the high pressure occurs to the left in relation to the wind movement. Since the air flow occurs with stable velocity parallel to the isobaric charts, the Coriolis force (F_c) and the pressure gradient force (F_B) have opposite directions but the same value:

$$|F_c| = |F_B| \qquad (3.15)$$

The wind flow, which results from the equilibrium of the gradient pressure and the Coriolis force, is called Geostrophic wind (V_g). The intensity of the Geostrophic wind is analogous to the gradient pressure and the Coriolis forces. The Geostrophic wind is used for the approximate calculation of the real wind above areas where

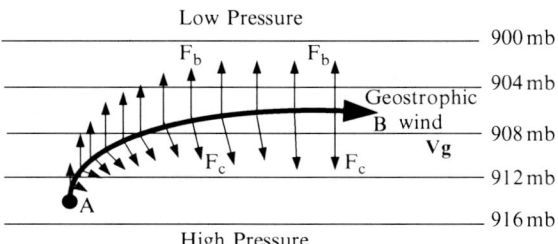

Fig. 3.8 A schematic representation of the geostrophic wind

3.6 Movement of Air: Wind

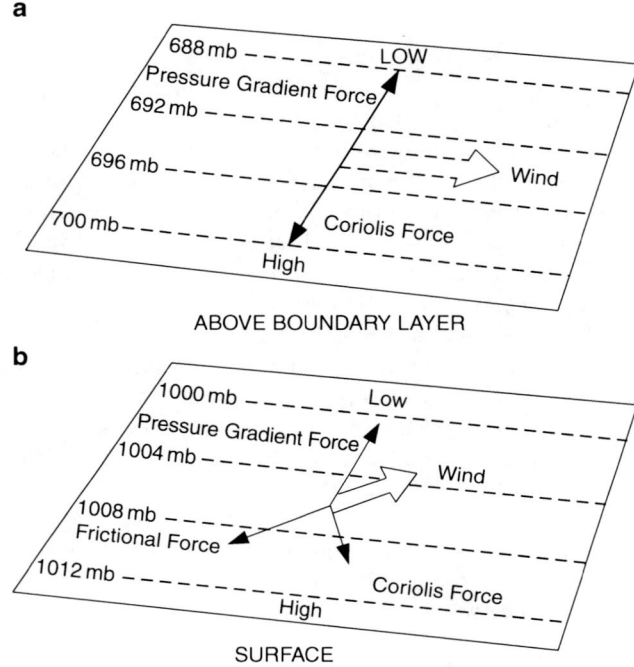

Fig. 3.9 Wind at the surface and at the upper atmospheric layers

there are no available meteorological data, such as above oceans and open seas. Furthermore, the Geostrophic wind is identical with the real wind above the boundary layer, where the wind flow between the constant pressure charts is almost straight and it is not dependent upon the friction force (Fig. 3.9).

From the balance of the pressure gradient force and the Coriolis force the following equations determine the velocity of the geostrophic wind at directions xx' and yy':

$$u = -\frac{1}{\rho f_c}\frac{\partial p}{\partial y} \tag{3.16}$$

and

$$v = -\frac{1}{\rho f_c}\frac{\partial p}{\partial x} \tag{3.17}$$

Therefore the velocity of the geostrophic wind will be equal to:

$$V = \sqrt{u^2 + v^2} = \frac{1}{\rho f_c}\sqrt{\left(\frac{\partial p}{\partial x}\right)^2 + \left(\frac{\partial p}{\partial y}\right)^2} \tag{3.18}$$

3.6.3.2 Pressure Gradient Wind

In the majority of cases, the constant pressure (isobaric) charts are not straight lines but curves. In many cases these are closed curves which pose centers of high or low pressures. In these cases, since the wind is directed around these centers, a centrifugal force (F_φ) develops with direction opposite to the centre of the pressure systems. Since the wind is moving with an almost constant velocity the pressure gradient force (F_B), the Coriolis force (F_c) and the centrifugal force (F_φ) must be in equilibrium, assuming that the friction force is very low ($F_\tau = 0$). The wind which results from these three forces is called the pressure gradient wind (V_α).

In the following paragraphs we will examine the pressure gradient force close to low pressure and high pressure systems. In the low pressure systems the isobaric charts are closed, circular or elliptical with the high pressure at the centre. For this reason the pressure gradient force (F_B) moves towards the centre, whereas the Coriolis (F_c) and the centrifugal forces have the same direction which is opposite to the direction of the pressure gradient force (Fig. 3.10a). Therefore, in the isobaric charts, the wind V_α moves along in parallel with the Coriolis force to move the air mass to the right in the northern hemisphere and to the left in the southern hemisphere. In a pressure low the wind direction is anticlockwise (known as cyclone circulation) at the northern hemisphere.

From the force balance we get:

$$\frac{1}{\rho}\frac{\partial p}{\partial R} = f_c V_\alpha + \frac{V_\alpha^2}{R} \tag{3.19}$$

The solution of the equation in relation to the wind velocity, taking into account only the solution with a physical meaning, gives us:

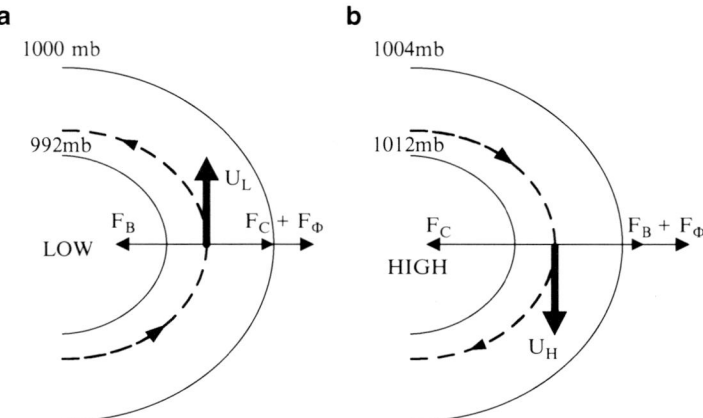

Fig. 3.10 Pressure wind, (a) at a pressure low and (b) at pressure high at the north hemisphere

$$V_a = -\frac{f_c R}{2} + \sqrt{\left(\frac{f_c R}{2}\right)^2 + \frac{R}{\rho}\frac{\vartheta p}{\vartheta R}} \qquad (3.20)$$

In high pressure systems the isobaric charts are also closed, circular or elliptical with the high pressure at the centre. Therefore the pressure gradient force (F_B) is directed opposite to the centre. The centrifugal force is in the same direction, whereas the Coriolis force goes in the opposite direction (Fig. 3.10b). The wind V_α is moving again in parallel with the Coriolis force to move the air mass to the right in the northern hemisphere and to the left in the southern hemisphere. At a pressure high, the wind direction is clockwise (known as anticyclone circulation) in the northern hemisphere. The equilibrium of forces results in:

$$\frac{1}{\rho}\frac{\vartheta p}{\vartheta R} + \frac{V_\alpha^2}{R} = f_c V_a \qquad (3.21)$$

The velocity of the pressure gradient wind is given from the expression:

$$V_a = \frac{f_c R}{2} - \sqrt{\left(\frac{f_c R}{2}\right)^2 - \frac{R}{\rho}\frac{\vartheta p}{\vartheta R}} \qquad (3.22)$$

under the condition that $\left(\frac{f_c R}{2}\right)^2 - \frac{R}{\rho}\frac{\vartheta p}{\vartheta R} \geq 0$

In the above equation the solution $V_a = \frac{f_c R}{2} + \sqrt{\left(\frac{f_c R}{2}\right)^2 - \frac{R}{\rho}\frac{\vartheta p}{\vartheta R}}$ results in an anticyclone circulation with unusually high velocities.

3.6.3.3 Friction Wind: Law of Buys-Ballot

Close to the Earth's surface and inside the boundary layer where the friction force (F_τ) is not negligible, the wind is not flowing parallel to the isobaric lines but is at an angle in relation to the isobaric lines (Fig. 3.11). Therefore due to the wind friction at the Earth's surface its intensity becomes slower and its direction changes. Therefore the wind friction velocity (V_T) above sea is close to two thirds of the geostrophic wind, whereas above land it is equal to one half of the geostrophic wind. Furthermore the wind direction is diverted with 10°–15° of the direction of the pressure gradient wind.

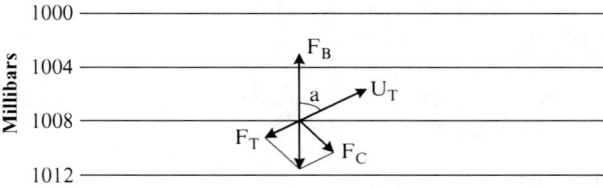

Fig. 3.11 Friction wind (V_T), inside the boundary layer, at horizontal and parallel isobaric lines

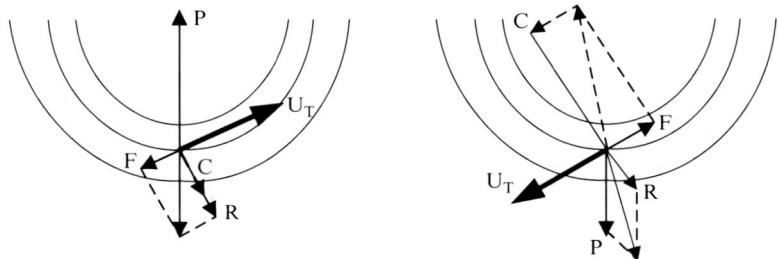

Fig. 3.12 Wind at the boundary layer at a pressure low and high systems at the north hemisphere

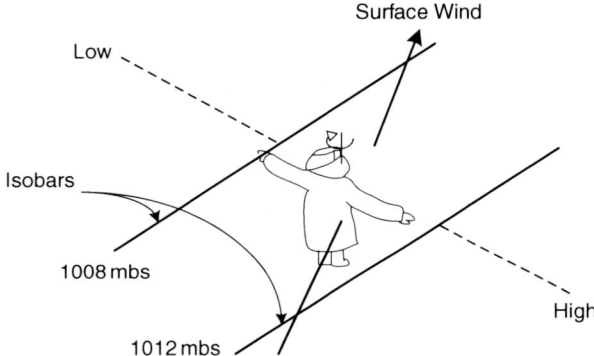

Fig. 3.13 Application of the law of Buys-Ballot for the northern hemisphere

With the help of the above facts it can be concluded that at a pressure low (LOW) the wind due to the friction force converges to the centre of the low pressure with an angle close to 20° above land (Fig. 3.12). At a high pressure area (HIGH) the wind diverges from the centre of the high pressure with almost the same angle.

The relation between the wind direction and the isobaric charts is derived from the law of the scientist Buys-Ballot. According to this law "if a person stands with his back to the wind at the northern hemisphere, then the low pressure is located to the left and the high pressure to the right" (Fig. 3.13). The contrary occurs in the southern hemisphere.

3.6.3.4 Wind at the Surface Layer

The wind at the surface layers of the atmosphere presents several abnormalities to its direction and intensity due to:

- Vertical lapse rate
- Heat distribution at land and sea

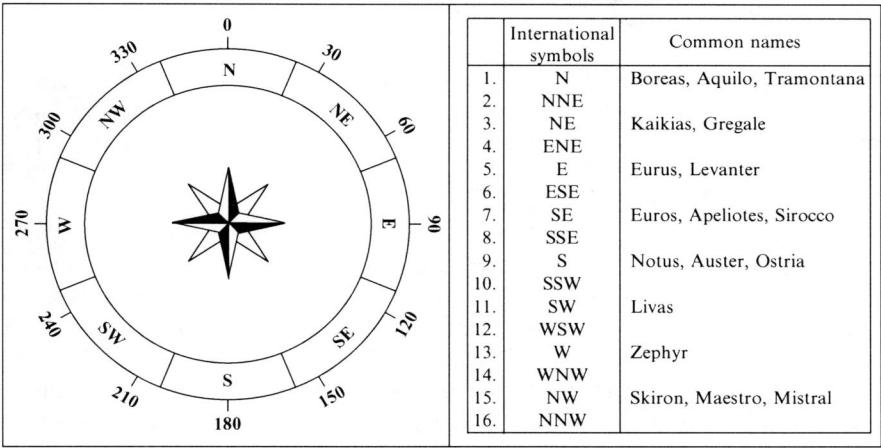

Fig. 3.14 Wind diagram for the determination of the wind direction

- Turbulent movements
- Friction with the Earth's surface

Therefore a wind which has on average the same intensity and direction for a few hours can change its direction and intensity for short time intervals. It is possible that a wind shear larger than 5 m/s from its average value may occur or even changes in the wind direction larger than 30°. The continuous change of the wind direction and especially of its intensity (for intensities larger than 10 m/s) results in wind shear which is observed when there are unstable conditions in the atmosphere and for clouds of vertical development (storms).

The wind variability is more intensive above land since the turbulent movement and the Earth's surface characteristics produce wind movements which may increase or decrease the wind intensity at specific time scales. With the increase of the wind velocity there is a consequent increase in the turbulent intensity which results in larger variability of the wind velocity and direction. With an increasing elevation the wind flow becomes more laminar and the wind becomes stable. For the determination of the wind direction it is customary to use a wind diagram (Fig. 3.14) in which are given also the common names of the winds.

3.7 General Circulation in the Atmosphere

Two simple models that describe the general circulation of the atmosphere are discussed in the following paragraphs.

3.7.1 Single and Three Cell Models

The fact that areas close to the Equator are heated more than the pole regions causes the air in these regions to become thinner due to heating and to ascend to the upper atmosphere. The upward movement of the warm air is continuous until its temperature becomes equal to the temperature of the surrounding atmosphere and then moves horizontally toward the poles.

The concentration of air masses above the poles results in their downward movement and the formation of high pressures at the pole surfaces. On the surface of the Equator there is low pressure due to upward movement and divergence of air to the poles. Therefore on the Earth's surface there is a formation of wind flow from the region of high pressure at the poles to the regions of low pressure at the Equator. The process of the movement of air masses from the Equator to the poles at the upper atmosphere and from the poles to the Equator at the Earth's surface denotes a simple form of the general circulation in the atmosphere.

The above model for a description of air circulation is called a single-cell model where it is supposed that the Earth's surface is covered by water and there is no difference between the temperature of land and sea. This cell is called the Handley cell, named for the English meteorologist who proposed this idea. In this theory the Sun is supposed to be vertical above the Equator in order not to be different in wind direction at different seasons. In order to study only force from the pressure differences, it is also assumed that the Earth does not rotate. The result from this hypothesis is shown in Fig. 3.15. The general circulation is shown as two separate cells, one per hemisphere.

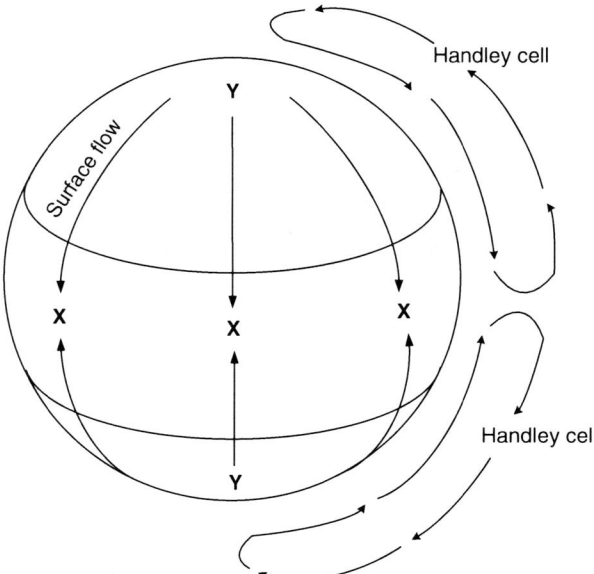

Fig. 3.15 The general circulation of air on the Earth which is covered totally by water and is not rotated as described in the single cell model by Handley

3.7 General Circulation in the Atmosphere

Such a simple circulation model does not exist in reality due to the Earth's rotation and the existence of the Coriolis force which moves the surface wind at the northern hemisphere to the right, with a result being formation of surface winds from the eastern directions at all latitudes. This results also in the formation of friction in relation to the Earth's rotation which however is not observed. It is also known that at medium latitudes the winds flow from the western directions. The above suggest that this simple single cell model cannot describe the real world due mainly to the Earth's rotation.

A more complex form of the general circulation of the atmosphere is the theory of a three-cell model. In this model, the Earth's rotation is incorporated as shown in Fig. 3.16. In these three cells per hemisphere there is re-distribution of the energy between the Equator and the poles.

According to this theory the Earth is divided into six zones at different latitudes, three in the northern and three in the southern hemisphere which are:

- 0°–30° Latitude, Hadley cell
- 30°–60° Latitude, Ferrel cell
- 60°–90° Latitude, Polar cell

The movement of air masses leads to the formation of zones of high pressure at latitudes 30° and 90° and, correspondingly, zones of low pressure at the Equator and at latitude 60°.

The air above the sea surface at the Equator is warm, the horizontal gradient of the temperature variation is small and the wind intensity is low. These regions are known as doldrums and are characterized as stable. Warm air masses ascend and

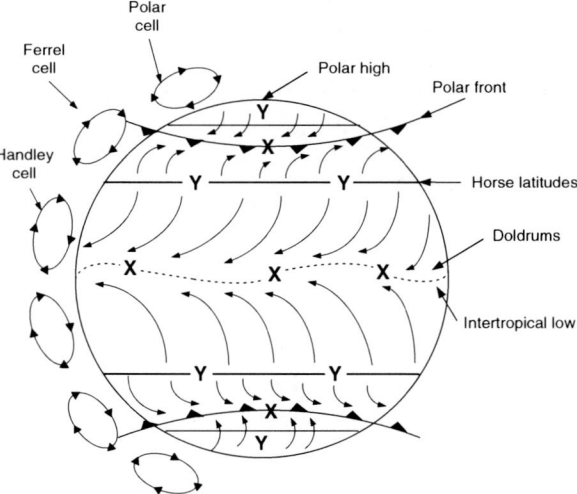

Fig. 3.16 Ideal temperature and surface pressure distribution above the rotational Earth with the three cell model

condense as extensive Cumulus clouds which further produce strong storms due to the large quantity of latent heat which is released. The heat which is included in these clouds transports them at higher elevations and produces a circulation in the Handley cell. The ascending air reaches the height of the tropopause which acts as an obstacle and diverts the air to the poles.

Some of the air at latitudes of 30° descends and is separated at the Earth's surface into two different currents. One is directed to the Equator and the other to the poles, whereas the major part of the wind at the upper atmosphere continues to flow to the poles. The wind flow from the Equator to the latitude of 30° at the upper atmosphere and the opposite current on the Earth's surface forms a closed circulation of wind, which is the first cell of the general circulation, known as a Handley cell.

The descent of humid air masses produces heating due to compression. These air masses produce clear skies and warm Earth's surface at the corresponding latitudes, and for this reason they produce mainly deserts. Above oceans the small pressure changes at the centre of the pressure high result in low wind speeds. These regions of the oceans are called horse latitudes. The remaining part of the descending air at latitude 30° moves to the poles, where at latitude 60° it meets the cold air which flows from the poles and forms a convergence zone which forces the warmer air to ascend, forming the polar fronts. Between latitudes 30° and 60° appears a second cell which is known as a Ferrel cell, whereas between 60° and 90° forms a third cell which is known as a polar cell.

The three-cell model is applicable under the condition that the Earth does not rotate and its surface is smooth and uniform. It is known that the Earth's rotation acts as an apparent force which affects the flow, diverting the wind to the right at the northern hemisphere and to the left at the southern hemisphere. Based on this principle and the fact that the air molecules move from the high to the low pressures, the continuous and periodic winds at the Earth's surface are formed.

Due to the uneven distribution of land and sea, the differences of the Earth's surface and many other thermal and dynamic parameters, the geographical distribution of the atmospheric pressure above the Earth's surface is quite complex. Studying the geographical distribution of atmospheric pressure at the height of mean surface level, we observe a distribution of pressure systems during winter and summer periods (Ahrens 1994).

3.7.2 Continuous Winds

Continuous winds are the winds which flow incessantly during the whole year and are due to the average conditions of atmospheric pressure at the Earth's surface, as it is formed from the general circulation in the atmosphere and the theory of the three-cell model. These include:

- Winds originating from the zone of high pressure at 30° to the Equator region under the influence of the Coriolis force. It gradually forms a system of

northeastern winds at the northern hemisphere and southeastern winds at the southern hemisphere which are called Trade winds.
- From the zone of 30° wind flow to the region of 60° under the influence of the Coriolis force. There is a formation of a system of western winds (westerlies) which are known also as anti-trade winds.
- At the polar zones, where there are high pressures, the wind flows from the region of 60° forming a system of eastern polar winds.

The trade winds and westerlies were known from the era of large explorations and therefore sailing boats made use of these winds.

3.7.3 Periodic Winds

The periodic winds flow due to changes of the zones of high and low pressure at specific time intervals which range from 1 year or lower time intervals. The major cause of the variation of the pressure zones is the Earth's surface topography and the distribution between land and sea, producing therefore an uneven heating of the different regions. Periodic winds include the yearly and daily periodic winds. Yearly winds include the monsoon and the annual (meltemi) winds.

The monsoon is a wind that blows in India 6 months during winter as northeastern cold and dry wind, and 6 months during summer as southern warm and humid air.

The annual winds prevail in the Aegean Sea region and are due to the combination of high pressure systems in southwest Europe with semi-permanent low pressures the southeastern Mediterranean. These winds blow during summer from mid-June to the end of September. They appear as northeastern at the north Aegean Sea, as northern at the central Aegean Sea and become northwestern in the area of Crete and the Dodecanese. Their intensity at midday hours exceeds many times 8 Beaufort, but decrease significantly during the night. Generally the annual winds are particularly beneficial for the island's climate during summer since they substantially lessen the prevailing high temperatures during summer.

Furthermore, the daily periodic winds include among others the sea and land breezes and the mountain and valley breezes. For the occurrence of these winds it is necessary for clear weather to prevail and the region not to be affected by a pressure system.

3.7.4 Sea and Land Breeze

The sea breeze is formed close to the seaside and is due to the temperature difference between the sea and the land surface. The mechanism of its formation

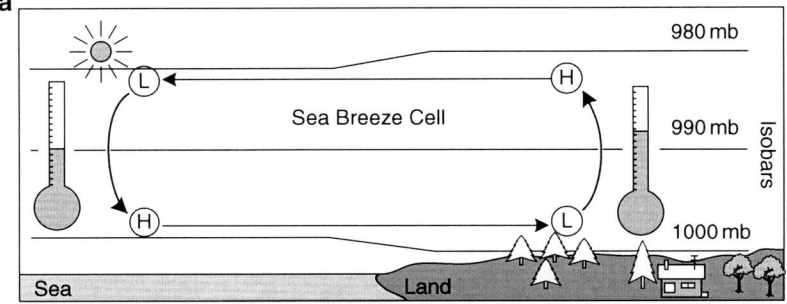

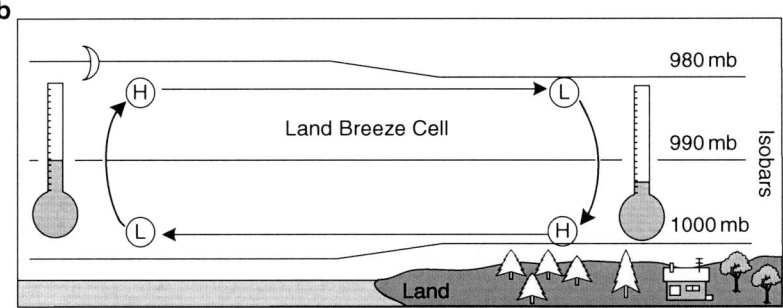

Fig. 3.17 Mechanism for the formation of (**a**) sea breeze. (**b**) land breeze

is a small cell of atmospheric circulation. Therefore during the morning hours the land surface warms more quickly than the sea surface which has larger heat capacity and is permeable at large depth from the Sun's rays. Therefore the air above the land receives heat and rises, expanding with the gradual formation of a system of low pressure. The colder and higher pressure air above the sea flows to the land to reach pressure equilibrium (Fig. 3.17). This wind flux from the sea to land is called sea breeze and its intensity increases gradually until the noon hours when it gets its maximum value. Finally the sea breeze gets weaker up to the sunset when it stops.

The direction of the sea breeze is vertical to the coastline and can develop up to distances of 15 km inland with intensity that can reach velocities up to 10 m/s. Its height can reach 500 m above the surface.

During the night there is cooling both for the land and the sea. However, the land is cooled quicker than the sea and therefore the pressure above the land during the night becomes higher than above the sea. The difference of the pressure produces a wind with a direction from the land to the sea. This wind is called land breeze and can occur during the whole night until the sunrise (Fig. 3.17).

The intensity and the height of the land breeze are lower than the intensity and the height of the sea breeze. Its height is lower than 100 m. This is due to the fact that the temperature differences between land and sea are greater during day than night.

3.7.5 Mountain and Valley Breezes

The breeze which appears on mountains and valleys arises due to the different air temperature above the surface in relation to the air temperature which exists at the same elevation of the free atmosphere. There is also the possibility of cold air transport from mountain top to lower elevations. Figure 3.18 shows a graphical representation of this phenomenon where the pressure difference between the plateau and the sea level results in wind formation through the slopes of the mountains to lower elevations. The descending wind is compressed during its descent since the atmospheric pressure increases close to the sea surface and therefore the air temperature increases.

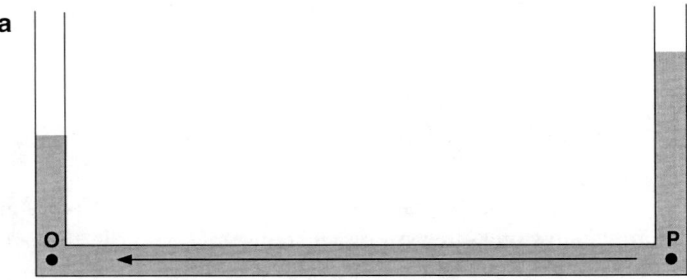

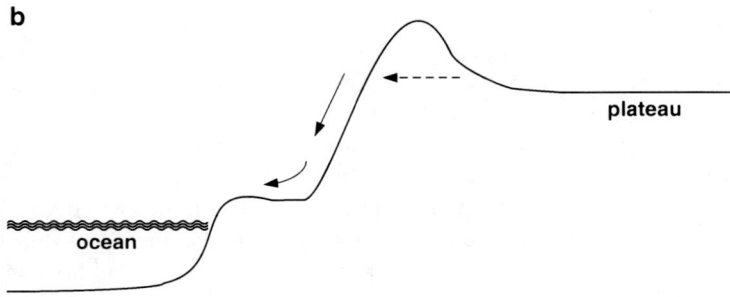

Fig. 3.18 Graphical representation of the pressure differences (**a**) Inside a tube the water height at the right side corresponds to higher pressure at point P in relation to the point O and therefore the water is moving accordingly to the arrow direction. (**b**) In relation to (**a**) the pressure gradient force is moving air from a cold *plateau*, where the pressure is high, to areas below where the pressure is low

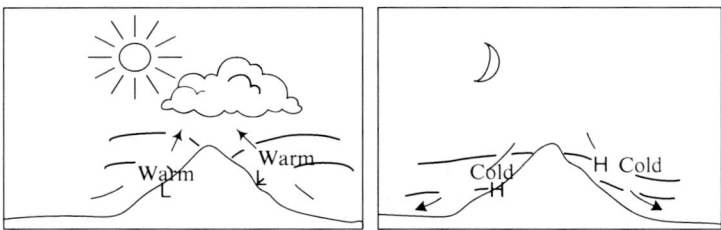

Fig. 3.19 Mechanism for the formation of the mountain and valley breeze

In the morning the air in valleys is heated more than the air at the mountain surface. Therefore the warmer air which is lighter starts to rise along the mountain slopes to the peaks (Fig. 3.20) where it is leveled and produces extended cloud layers along the mountain ridge. The valley breeze can start a little after sunrise and pauses completely after sunset in the same manner as the sea breeze.

In the case of mountain breeze the circulation mechanism is exactly the opposite from the valley breeze (Fig. 3.19). The night mountain breeze has lower intensity close to the surface except in specific areas due to topographic features.

3.8 Vertical Structure of Pressure Gradient Systems

It has been discussed previously that the wind inside the boundary layer moves anticlockwise around low pressure systems and converges towards the centre of the pressure low. Due to the convergence to the centre of the pressure low the wind is pressed to ascend inside the atmosphere until it reaches a height where it will be horizontal and will start to diverge. During this upward movement inside the pressure low, the ascending air mass is cooled due to expansion and reaches saturation, followed immediately by condensation processes which lead to the formation of clouds. These clouds furthermore can lead to rain. In the case that the upward movement of the air mass is abrupt, then it is possible that clouds of vertical development (storms) will form.

In the case of a pressure high the wind moves clockwise and diverges from the centre of the pressure high. Due to the divergence of the air at the lower layers of the pressure high, additional air is pressed to move vertically downwards which has originated from the convergence at a specific height in the upper atmosphere. The descending air becomes warmer due to its compression and its humidity percentage is reduced, forming clear weather conditions in the area close to a high pressure system.

Figure 3.20 shows the vertical structure of pressure systems inside the atmosphere. This structure resembles a closed system of wind circulation which shows symmetry between the systems at the surface and at the upper atmosphere.

3.9 Equations of Atmospheric Circulation

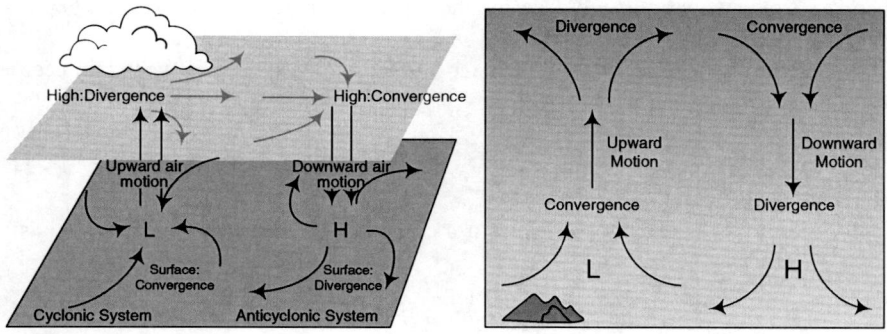

Fig. 3.20 Vertical form of the pressure gradient systems

The direction and intensity of the surface wind is dependent on the horizontal pressure distribution of the pressure systems as well as from the pressure movements in space. The upward movements of low pressure systems are related to the events of weather phenomena, whereas the downward movements in the high pressure systems favor good weather conditions.

3.9 Equations of Atmospheric Circulation

In Chap. 1 we established that the transport of air pollutants in the atmosphere occurs mainly in the boundary layer. In this section the variation of air at the lower atmospheric layers will be examined. At this elevation the flux is almost always turbulent and is needed to examine the basic principles of atmospheric turbulence. The study will be based on the equations which control the variations of air density, temperature and velocity.

3.9.1 Equations of Circulation for a Compressible Fluid

The equations of movement for a compressible fluid in a gravimetric field are expressed as:

$$\rho \left(\frac{\partial u_i}{\partial t} + u_j \frac{\partial u_i}{\partial x_j} \right) = \frac{\partial}{\partial x_k} \left[\mu \left(\frac{\partial u_i}{\partial x_k} + \frac{\partial u_k}{\partial x_i} \right) - \left(p + \frac{2}{3} \mu \frac{\partial u_j}{\partial x_j} \right) \delta_{ik} \right] - \rho g \, \delta_{3i} \quad (3.23)$$

where $u_i(x_1, x_2, x_3)$ is the air velocity at the direction i, μ is the air viscosity and δ_{ij} is the Kronecker delta which is expressed as $\delta_{ij} = 1$ for $i = j$ and $\delta_{ij} = 0$ for $i \neq j$.

The above equation is valid inside the boundary layer and at limited spatial and time areas. The Coriolis force is introduced into the above equation when the Earth's rotation has to be taken into account. The above equations are called

Navier–Stokes equations and describe the air transport in the environment expressing the momentum equilibrium.

For an incompressible fluid the Navier–Stokes equations can be written in vector form as:

$$\rho \frac{dv}{dt} = -\nabla p + \mu \nabla^2 v - \rho g \qquad (3.24)$$

Furthermore the energy equation for a compressible fluid can be expressed as:

$$\rho \left(\frac{\partial U}{\partial t} + u_j \frac{\partial U}{\partial x_j} \right) = k \frac{\partial^2 T}{\partial x_j \, \partial x_j} - p \frac{\partial u_j}{\partial x_j} + \Phi + Q \qquad (3.25)$$

where U is the internal energy per unit mass μάζας ($= \hat{c}_v T$ for an ideal gas), κ is the thermal diffusion, Φ is the produced heat per unit volume and time due to viscous dissipation and Q is the function which expresses the heat which is produced from each source inside the fluid.

The continuity equation for a compressible fluid can be written as:

$$\frac{\partial \rho}{\partial t} + \frac{\partial}{\partial x_i} (\rho \, u_i) = 0 \qquad (3.26)$$

Finally, for an ideal gas the pressure and the density are related through the equation of an ideal gas:

$$p = \frac{\rho R T}{M_a} \qquad (3.27)$$

Equations 3.23–3.26 are a system of six equations with six unknowns which are u_1, u_2, u_3, p, ρ and T. The equations can be solved if the appropriate initial and boundary conditions are chosen. Since there is strong interaction between the equations, the solution is usually calculated numerically. Under specific simplifications the above equation system can be solved analytically. In the following paragraphs the solution is presented for static atmospheric conditions and furthermore the solution is discussed for atmospheric motion conditions (Seinfeld and Pandis 2006).

For static atmosphere ($u_i = 0$) Eq. (3.23) can be written as:

$$\frac{\partial p_e}{\partial x_1} = \frac{\partial p_e}{\partial x_2} = 0 \quad \frac{\partial p_e}{\partial x_3} = -g \, \rho_e \qquad (3.28)$$

Equation 3.25 becomes also $\frac{\partial^2 T_e}{\partial x_3^2} = 0$, where it is supposed there no energy sources. The coefficient e denotes equilibrium values. From this expression after integration it results that $T_e = T_o \left(1 - \frac{x_3}{H}\right)$, where T_o is the temperature at the surface and H the height where the T_e is equal to zero.

From these equations it can concluded that the pressure, density and temperature in an atmosphere at equilibrium can be expressed as:

3.9 Equations of Atmospheric Circulation

$$\frac{\partial p_e}{\partial x_3} = -g\rho_e \quad (3.29)$$

$$p_e = \frac{\rho_e R T_e}{M_a} \quad (3.30)$$

$$T_e = T_o \left(1 - \frac{x_3}{H}\right) \quad (3.31)$$

With the integration of the above equations we get:

$$p_e = p_o \left(1 - \frac{x_3}{H}\right)^{gHM_a/RT_o} \quad (3.32)$$

$$\rho_e = \rho_o \left(1 - \frac{x_3}{H}\right)^{(gHM_a/RT_o)-1} \quad (3.33)$$

Therefore from the above results it can be concluded that when the atmosphere is static it can be written that:

$$\frac{p_e}{p_o} = \left(\frac{\rho_e}{\rho_o}\right)^n \quad \text{where} \quad \frac{n}{n-1} = \frac{gHM_a}{RT_o} \quad (3.34)$$

The lapse rate Λ can be defined as:

$$\Lambda = \frac{T_o - T_e}{x_3} \quad (3.35)$$

and as a result :

$$\Lambda = \frac{T_o}{H} = \frac{gHM_a}{R}\frac{n-1}{n} \quad (3.36)$$

When $n = \gamma = \frac{\hat{c}_p}{\hat{c}_v}$ the atmosphere is adiabatic and since for an ideal gas $\frac{R}{M_a} = \hat{c}_p - \hat{c}_v$ then:

$$\Lambda = \Gamma = \frac{g}{\hat{c}_p} \quad (3.37)$$

In the study of the atmosphere under motion, we examine the simple case that there is a very thin atmospheric layer close to Earth. With this assumption there are simplifications to the equations of continuity, motion and energy (Boussinesq simplifications). In this case the final form of the equation of motion can be written as:

$$\frac{\partial u_i}{\partial t} + u_j \frac{\partial u_i}{\partial x_j} = -\frac{1}{\rho_o}\frac{\partial \tilde{p}}{\partial x_i} + \frac{\mu}{\rho_o}\frac{\partial^2 u_i}{\partial x_j \partial x_j} + \frac{g\tilde{T}}{T_o}\delta_{i3} \quad (3.38)$$

where $\tilde{T}, \tilde{P}$ express the variations of the quantities due to the motion. After an analogous mathematical derivation the equation of energy is expressed as:

$$\rho_o \hat{c}_p \left(\frac{\partial \theta}{\partial t} + u_j \frac{\partial \theta}{\partial x_j} \right) = k \frac{\partial^2 \theta}{\partial x_j \partial x_j} + Q \tag{3.39}$$

where θ is the potential temperature and for the equation of continuity the result is $\frac{\partial u_i}{\partial x_i} = 0$.

Example. *Wind variation versus height in the atmosphere*

We examine the two-dimensional turbulent wind flux at the surface layer parallel to the surface for $x_3 = 0$, where x_1, x_2, x_3 correspond to the directions x,y,z respectively. It is assumed that $\bar{u}_1 = \bar{u}_1(x_3)$ and $\bar{u}_2 = 0$. In this example the quantity $\bar{u}_1(x_3)$ will be calculated for an adiabatic vertical temperature variation.

In this case $\theta = 0$ and it is enough to examine the x_1 part of the time dependent equation of motion which can be written as:

$$\frac{d}{dx_3} \overline{\rho u'_1 u'_3} = \mu \frac{d^2 \bar{u}_1}{dx_3^2} - \frac{\partial \bar{p}}{\partial x_1}$$

and furthermore can be expressed as:

$$\frac{\partial \bar{\tau}_{13}}{\partial x_3} = \frac{\partial \bar{p}}{\partial x_1}$$

where $\bar{\tau}_{13}$ is the total shear stress

$$\bar{\tau}_{13} = \tau_o + x_3 \frac{\partial \bar{p}}{\partial x_1}$$

After a mathematical derivation it can be concluded that

$$\frac{\bar{u}_1(x_3)}{u_*} = a \ln(x_3) + c$$

Measurements can give the values of the constant parameters and therefore:

$$\frac{\bar{u}_1(x_3)}{u_*} = \frac{1}{k} \ln \frac{u_* x_3}{v} + 5.5$$

where $u_* = \sqrt{\tau_o / \rho} \, \pi$ (friction velocity). Finally:

$$\frac{\bar{u}_1(x_3)}{u_*} = \frac{1}{k} \ln \frac{x_3}{z_o} \qquad x_3 \geqslant z_o$$

where z_o is the integration constant which is called friction length.

Problems

3.1 *Calculate the total derivative of the velocity u, when the wind flux is (a) stable, (b) unstable.*

3.2 *Calculate the Coriolis force at Equator and at the Earth's poles.*

3.3 If $u = 30$ m/s, $v = 10$ m/s and $\varphi = 45°$, calculate the value of the Coriolis force per unit mass.

3.4 If in the atmosphere there is a pressure gradient from east to west with a value of 1 kPa per 500 km, calculate the value of the Geostrophic wind. It is given by $\rho = 1$ kg/m^3 and $f_c = 10^{-4}$ s^{-1}.

3.5 If the value of the Geostrophic wind in an area of a pressure low is 10 m/s, calculate the value of the wind due to the pressure gradient at a distance of 500 km from the centre of the pressure low. It is given that $f_c = 10^{-4}$ s^{-1}.

3.6 In an area which is located east from the centre of a large typhoon at latitude $\varphi = 20°$ N, a pressure change of 50 mb per 125 km with wind velocity $\upsilon = 70$ m s is observed. What is the distance between the typhoon centre and the observation location ? Suppose that the temperature and the pressure at the observation location are Tv = 280 K and pa = 930 mb respectively. It is given $R' = \frac{R}{MB_{αέρα}} = 2.8704 \frac{m^3 mb}{KgK}$ and $\Omega = 7.29 \ 10^{-5}$ rad/s.

3.7 We are given a cell with dimensions $\Delta x = 5$ km, $\Delta y = 4$ km, and $\Delta z = 0.1$ km and wind velocity values of u_1=+2, u_1=+2, u_2=+3, v_3=+1, v_4=−3, w_5=+0.03 w_6=+0.04 m/s at the west, east, north, south, upper and lower sides of the cell respectively. Calculate the value of the divergence of the local acceleration $((v\bullet)v)$.

3.8 (a) Assume $u_1 = 1$ m/s at the location $x_1 = 0$, $u_2 = 2$ m/s at the location $x_3 = 5,000$ m and $K_{m,xx} = 2,5 \ 10^3$ m^2/s. If the density and $K_{m,xx}$ have stable values, calculate the variation of the velocity u, due to the diffusion process at the centre of the cell after 1 h. (b) Assume $w_1 = 0,02$ m/s at the location $z_1 = 0$, $w_2 = 0,02$ m/s at the location $z_2 = 50$ m, $w_3 = 0,04$ m/s at the location $z_3 = 1,000$ m and $K_{m,zz} = 2,5 \ 10^3$ m^2/s. If the density and $K_{m,zz}$ are constant, calculate the variation of the velocity w due to the diffusion to the centre of the cell after 100 s.

References

Ahrens, C. D. (1994). *Meteorology today – An introduction to weather, climate and the environment* (5th ed.). New York: West Publishing Company.

Seinfeld, J. H., & Pandis, S. N. (2006). *Atmospheric chemistry and physics* (2nd ed.). New York: Wiley.

Chapter 4
Atmospheric Chemistry

Contents

4.1	Chemical Components in the Atmosphere	151
4.2	Chemistry of the Troposphere	152
	4.2.1 Sulphur Components	155
	4.2.2 Nitrogen Components	155
	4.2.3 Carbon Components	156
	4.2.4 Halogen Components	157
4.3	Particulate Matter	157
4.4	Photochemistry in the Free Troposphere	158
	4.4.1 Photochemical Cycle of Ozone and Nitrogen Oxides	158
	4.4.2 Chemistry of Carbon Dioxide	160
	4.4.3 Chemistry of Hydrocarbons	161
	4.4.4 Chemistry of Sulphur Compounds	161
4.5	Components of Aquatic Chemistry in the Atmosphere	162
4.6	Chemistry of the Stratosphere - Ozone	163
References		167

Abstract The atmosphere contains myriads of chemical components and reactive species at low concentrations (ppm and ppb levels). Chap. 4 examines the chemical components in the atmosphere in conjunction with these chemical reactions. The sulphur, nitrogen, carbon and halogen compounds are examined together with the particulate matter. The photochemical cycle of ozone and nitrogen oxides in the troposphere is examined together with the chemistry of carbon dioxide, hydrocarbons and sulphur compounds. Furthermore, we explore the chemistry of the stratosphere in relation to ozone depletion.

4.1 Chemical Components in the Atmosphere

The atmosphere contains myriads of chemical components and reactive species at low concentrations (ppm and ppb levels) as discussed in the Chap. 1. The troposphere behaves as a container of gaseous and aerosols. The transport of pollutants from the troposphere to the stratosphere is very slow (in a range of years), whereas the mixing of pollutants in the troposphere occurs during a few weeks. It has been observed that the atmosphere's composition is changing this century with an increase of the concentration of specific gaseous pollutants such as CO_2, CH_4, N_2O and halogen-containing compounds as well as aerosols.

Chemical reactions include for example the formation of ozone and nitrogen dioxide and involve a large number of nonlinear chemical reactions (e.g. Seinfeld and Pandis, 2006). Furthermore, the chemical reaction rates depend also on the background concentration of the various chemical species which are determined from the emission and meteorological characteristics. The modelling of gaseous chemical reactions in the atmosphere is a difficult task because of the complex chemical reactions and the stochastic mixing processes due to turbulence. Several simplifications are adopted in describing the gaseous phase chemical reactions in air quality models.

The meteorological conditions are one of the major factors which determine the spatial and temporal distribution of pollutants in the atmosphere. Other factors which affect the chemical reactions in the atmosphere are the Sun's radiation and the cloud cover.

Table 4.1 presents the structure of selected chemical compounds which are abundant in the atmosphere. The number of lines between the atoms denotes the atom bonds, whereas the single dot adjacent to the atoms indicates a free electron. Chemical compounds with free electrons are called free radicals and are very reactive in the atmosphere. Moreover chemical compounds with single bond (e.g., O_3) are also reactive since the single bond is easier to break than double or triple bonds (Jacobson 2002).

Some chemical compounds have also positive or negative charge. The charge distribution is developed during the molecule formation and the transfer of charge between the atoms. Molecules with positive and negative charge have zero net charge but the charge can increase the reactivity since the positive charged atom can attract a negative charge from another compound and vice versa. However, molecules with a net charge (e.g., SO_4^{2-}) are ions.

The residence time of the chemical components in the atmosphere and the spatial scale of transport present significant characteristics for their dynamics (see Fig. 4.1). For example gaseous components with long lifetime such as the CH_4 play a significant role in the greenhouse effect in the atmosphere.

4.2 Chemistry of the Troposphere

In the atmosphere is present every element of the periodic table. The chemical components in the atmosphere can be classified in a few main groups which play a significant role in atmospheric chemistry (Seinfeld and Pandis, 2006). The main groups are:

1. Compounds containing sulphur
2. Compounds containing nitrogen.
3. Compounds containing carbon.
4. Compounds containing halogen.

4.2 Chemistry of the Troposphere

Table 4.1 Chemical structure of selected chemical components in the atmosphere (Adapted from Jacobson (2002))

Compound name	Structure	Chemical formula with free electrons	Chemical formula without free electrons
Molecular oxygen	O=O	O_2 (g)	O_2 (g)
Molecular nitrogen	N≡N	N_2 (g)	N_2 (g)
Ozone	$-O-O^+=O$	O_3 (g)	O_3 (g)
Hydroxyl radical	Ȯ–H	ȮH	OH (g)
Water vapor	O(H)(H)	H_2O (g)	H_2O (g)
Nitric oxide	Ṅ=O	ṄO	NO (g)
Nitrogen dioxide	$-O-\dot{N}^+=O$	$\dot{N}O_2$	NO_2 (g)
Sulphur dioxide	O=S–O	SO_2 (g)	SO_2 (g)
Carbon monoxide	C≡O⁺	CO (g)	CO (g)
Carbon dioxide	O=C=O	CO_2 (g)	CO_2 (g)
Methane	H–C(H)(H)–H	CH_4 (g)	CH_4 (g)

(continued)

Table 4.1 (continued)

Compound name	Structure	Chemical formula with free electrons	Chemical formula without free electrons
Sulphate ion	O=S(−O⁻)(−O⁻)=O	SO_4^{2-}	SO_4^{2-}

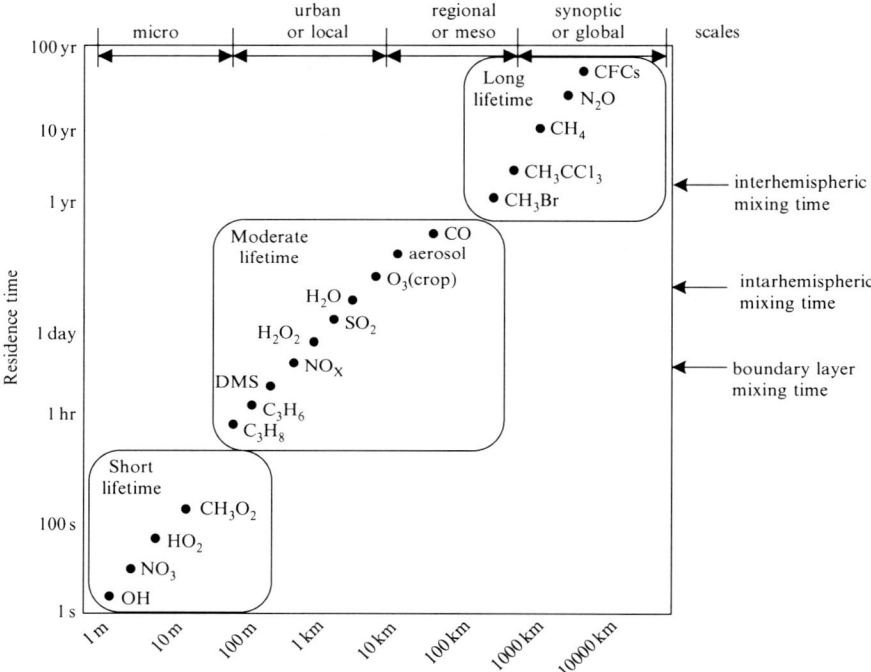

Fig. 4.1 Spatial scales and residence time for gaseous chemical species and aerosols in the atmosphere

The chemical compounds in the atmosphere after their emission participate in physico-chemical processes and are eventually removed to the Earth's surface. This cycle is called the *biogeochemical cycle* of the chemical compound. The *biogeochemical cycle* refers to the flux of a specific chemical compound between the Earth's surface (land, ocean and biosphere) and the atmosphere and its concentration in different compartments.

4.2 Chemistry of the Troposphere

In the following paragraphs a more detailed description of the four main categories of chemical components in the atmosphere is presented.

4.2.1 Sulphur Components

In the atmosphere the sulphur-containing compounds have concentration lower than 1 ppm and are originated from anthropogenic and natural emission sources. The sulphur dioxide (SO_2) concentration in areas far from urban settings is between 20 ppt and 1 ppb. In urban areas the concentration of sulphur components may reach levels of several hundred ppb.

A dominant sulphur compound that is emitted from the oceans is dimethyl sulfide (DMS) (CH_3SCH_3) which is produced from the biodegradation of marine phytoplankton. The concentration of DMS at the marine boundary layer ranges between 80 and 110 ppt and can reach the level of 1 ppb at eutrophying areas close to land. Other dominant sulphur compounds in the atmosphere include hydrogen sulfide (H_2S) and carbonyl sulfide (OCS). The OCS is produced from the oxidation of CS_2 which is emitted from the oceans and burning of the biomass. The OCS is removed by absorption into plants, by reactions with hydroxyl radicals and by deposition on the Earth's surface. The OCS is the most abundant form of sulphur in the atmosphere due to its low reactivity.

Table 4.2 presents measured mixing ratios of sulphur gasses and their atmospheric life times.

4.2.2 Nitrogen Components

The most abundant form of nitrogen is molecular nitrogen (N_2), which is chemically stable and does not participate in chemical reactions in the atmosphere.

Table 4.2 Average life time and measured mixing ratios of sulphur gasses in the atmosphere (Adapted from Lelieveld et al. 1997)

Species	Average lifetime	Mixing Ratio, ppt			
		Marine air	Clean continental	Polluted continental	Free troposphere
H_2S	2 days	0–110	15–340	0–800	1–13
OCS	7 years	530	510	520	510
CS_2	1 week	30–45	15–45	80–300	≤ 5
CH_3SCH_3	0.5 days	5–400	7–100	2–400	≤ 2
SO_2	2 days	10–200	70–200	100–10,000	30–260
SO_4^{2-}	5 days	5–300[a]	10–120	100–10,000	5–70

[a] non sea salt sulphate

Chemically reactive nitrogen compounds are nitrous oxide (N_2O), nitrogen monoxide (NO), nitrogen dioxide (NO_2),e nitric acid (HNO_3) and ammonia (NH_3).

Nitrous oxide is a colorless gas which is emitted mainly from natural sources and microbe activities in soil. It affects Earth's climate mainly due to its high thermal capacity and its large residence time in the atmosphere.

Nitrogen monoxide is emitted by combustion at high temperatures, whereas nitrogen dioxide is emitted during combustion at low quantities. NO_2 is mainly produced from the oxidation of NO in the atmosphere. The major part of NO_2 and NO is produced from bacteria inside soil during de-nitrification whereas nitric acid is produced from the oxidation of sulphur dioxide in the atmosphere and ammonia from natural sources.

The nitrogen oxides ($NO_x = NO + NO_2$) are very important molecules in atmospheric chemistry. High concentrations of NO_x in the atmosphere originate often from anthropogenic sources. Studies in the United States showed that 45% of NO_x anthropogenic emissions originate from transport, 30–35% from electricity production power stations and 20% from industry in general.

In addition, reactive nitrogen (NO_y) is called the sum of the nitrogen oxides ($NO_x = NO + NO_2$), as well as all products from the oxidation of NO_x such as HNO_3 (nitric acid), HNO_2 (nitrous acid), $NO_3^{\bullet}$ (nitrate radical), N_2O_5 (dinitrogen pentoxide), HNO_4 (peroxynitric acid), PAN (peroxyacetyl nitrate) (RC(O)OONO$_2$), alkyl nitrates ($RONO_2$), and finally peroxynitrates ($ROONO_2$).

4.2.3 Carbon Components

An atom of carbon has four free electrons and therefore can have free bonding with four other atoms. In case the carbon atom is connected only with other carbon and hydrogen atoms with simple bonding, molecules are form that are called alkanes with general formula C_nH_{2n+2} (e.g. methane (CH_4), propane (C_3H_8)). When the alkanes react and lose one hydrogen atom, then the resulting molecules are called alcylic radical (e.g. methyle ($CH_3\bullet$)).

If the alcylic radical is at a higher energetic level it can form a free radical which is denoted by a point. The free alcylic radical is denoted by R•. The alcylics are chemically reactive and play an important role in atmospheric chemistry.

Another category of chemical compounds which exist in the atmosphere are the alkenes, which have a double bond between two adjacent carbon atoms (e.g. ethylene $CH_2 = CH_2$). Molecules with two double bonds are named alkadienes, whereas with a triple bond are called alkynes (HC $\equiv$ CH).

Finally, an important category of hydrocarbons are the aromatic chemical components that are composed of six carbon atoms, which are connected among themselves with three simple and three double bonds and have a ring structure. The basic unit is the benzene:

benzene

Cyclic hydrocarbons which exist in the troposphere are the m-xylene, the toluene, the το o-xylene and the p-xylene:

m-xylene   toluene   o-xylene   p-xylene

Other basic forms of carbon compounds include the volatile organic compounds (VOC) and carbon monoxide (CO).

4.2.4 Halogen Components

The halogen components in the atmosphere form a quite wide category which includes, among others, the halogenated hydrocarbons and chlorofluorocarbons (CFCs) which have in their molecules atoms of carbon, chlorine and fluorine, as well as hydrochlorofluorocarbons (HCFCs). The anthropogenic emissions of the halogen compounds during 1990 were close to 2.5 Tg.

The attention paid to halogen compounds during recent years was mainly due to their role in ozone destruction in the stratosphere (especially trichlorofluorocarbon $CFCl_3$ (CFC-11) which is used as a blowing agent and CF_2Cl_2 (CFC-12) as a refrigerant). The chlorofluorocarbons are very stable chemical compounds in the troposphere with large residence time which are distributed homogeneously in the atmosphere. Their concentration decreases with height due to their photo-dissociation from the ultraviolet radiation with wave length 225 nm. Due to the important influence of CFCs on destruction of the stratospheric ozone layer, the Montreal protocol was signed in 1987 which resulted in considerable reduction of anthropogenic emissions of CFCs.

4.3 Particulate Matter

Particulate matter in the atmosphere is composed of stable suspensions of solid and liquid particles. Their diameter ranges between 2 nm up to 100 μm and is a complex mixture of many different chemical species originating from a variety of sources.

A detailed study of the particulate matter in the atmosphere and their dynamics is given in Chap. 5.

4.4 Photochemistry in the Free Troposphere

4.4.1 *Photochemical Cycle of Ozone and Nitrogen Oxides*

Ozone (O_3) belongs to the category of secondary pollutants in the atmosphere. It originates from photochemical reactions of primary gaseous pollutants, mainly from nitrogen oxides (NO_x), hydrocarbons and volatile organic compounds in the atmosphere (VOCs). These gaseous components have a great oxidation capacity and are studied globally for estimation of atmospheric pollution characteristics. Ozone has important health effects in the urban environment.

Under the influence of the Sun's radiation, photochemical pollutants such as nitrogen dioxide and ozone are produced. The ultraviolet radiation,[1] as a photochemical catalyst, contributes through breaking of the chemical bonds of NO_x and VOC to the production of atomic oxygen which furthermore produces ozone and reactive free radicals. This results in maximum ozone concentration in the atmosphere at late midday hours. The ozone concentration is variable in relation to the concentration of nitrogen oxides (NO_x) and hydrocarbons (HCs).

Ozone is produced in the atmosphere when molecular oxygen (O_2) reacts with atomic oxygen (O), whereas the main source of atomic oxygen in the troposphere is the photo-reaction of NO_2:

$$NO_2 + h\nu \ (\lambda < 420\,nm) \rightarrow NO + O \qquad (4.1)$$

photolysis of NO_2

$$O_2 + O + M \rightarrow O_3 + M \qquad (4.2)$$

production of O_3 where, M is a third chemical compound (usually N_2 or O_2), which absorbs the additional energy of vibration of the reaction and stabilizes the ozone molecule that is formed.

The elevated ozone concentrations in the atmosphere are removed with several mechanisms. These include (at secondary level) reactions with plant surfaces, the Earth's surface and different materials. The main removal mechanism consists of chemical reactions with NO_x, whereas NO is the main factor through the reaction:

$$NO + O_3 \rightarrow NO_2 + O_2 \qquad (4.3)$$

---

[1] The spectrum range between 10 and 380 nm.

4.4 Photochemistry in the Free Troposphere

The above three reactions give the photolytic cycle of the nitrogen oxides. At the same time, during the night there are reactions with NO_2 which destroy O_3 and produce HNO_3 through the production of N_2O_5. More specifically:

$$NO_2 + O_3 \rightarrow NO_3 + O_2 \tag{4.4}$$

$$NO_3 + NO_2 \rightarrow N_2O_5 \tag{4.5}$$

$$N_2O_5 + H_2O \rightarrow 2HNO_3 \tag{4.6}$$

The HNO_3 is removed in particulate matter in the form of acid rain. During daytime there is a production of HNO_3 with a rate of 0.5 ppb/h through the reaction:

$$NO_2 + HO \rightarrow HNO_3 \tag{4.7}$$

If there is enough ammonia in the atmosphere, NH_4NO_3 is formed. On the contrary nitric acid is produced in the gaseous phase. The oxygenated organic compounds under the influence of the Sun's radiation produce ozone. More specifically the concentrations of nitrogen oxide and hydrocarbons increase gradually early in the morning when the vehicular traffic increases in urban areas. After sunrise the small concentrations of nitrogen dioxide start to chemically react and produce ozone and oxygenated hydrocarbon compounds. In addition the influence of alkyl peroxy radical (RO_2) during the day starts to be important. The RO_2 starts to react with NO thus accelerating the production of NO_2 and decreasing the concentration of NO as depicted in the following reaction:

$$RO_2 + NO \rightarrow NO_2 + RO \tag{4.8}$$

The increased concentrations of nitrogen dioxide together with the more intense solar radiation accelerate the production of atomic oxygen (with the reaction of NO_2) and therefore the concentration increase of O_3 and other oxidation compounds. When the concentration of NO decreases considerably then the O_3 increases quickly. At the same time the alkyl peroxy radicals are oxidized to other organic compounds and therefore their concentration decreases as described in the Eq. (4.9) – (4.11):

$$NO_2 + h\nu \rightarrow NO + O \tag{4.9}$$

$$O + O_2 \rightarrow O_3 \tag{4.10}$$

$$RO_2 + O_2 + h\nu \rightarrow RO + O_3 \tag{4.11}$$

In summary, after the increase of the NO_x and HCs levels there is an increase of the O_3 levels as a result of the photochemical reactions which decrease the levels of the primary pollutants. These concentrations of NO_x are increased and remain

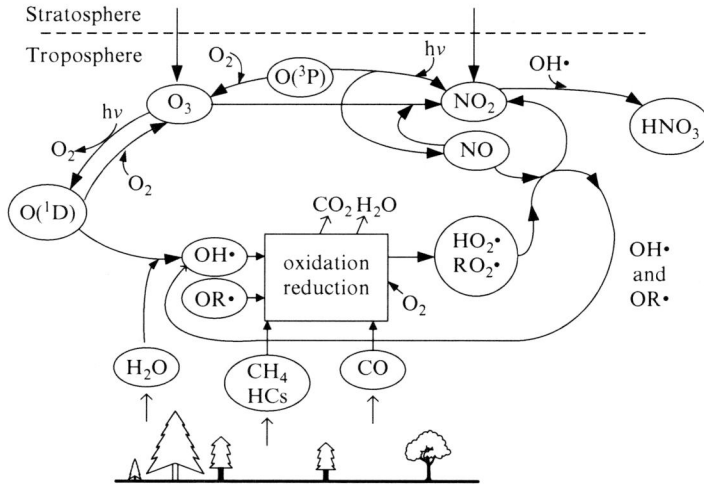

Fig. 4.2 Chemical reactions with the presence of solar radiation in the atmosphere

stable for a period of 1–4 h and furthermore there is a decrease as a result of their transport, their transformation to NO_2 and HNO_3 and their removal at the Earth surface. During the night the ozone production ceases.

In a polluted atmosphere the O_3 concentrations at a specific space and time are dependent on the light intensity, the NO_2/NO ratio, the concentration of reactive hydrocarbons (e.g. olefins) as well as from other compounds (e.g. aldehydes) and carbon monoxide react photochemically and produce RO_2. The increase of the ratio NO_2/NO, which is produced from atmospheric reactions including RO_2 radicals, result in important increases of the ozone levels. Fig. 4.2 shows the photochemical reactions in the troposphere of ozone and nitrogen oxides.

4.4.2 Chemistry of Carbon Dioxide

Carbon monoxide exists in the atmosphere and is emitted from combustion sources and is depleted with physical and chemical processes with a residence time in the atmosphere close to 36 days. Carbon monoxide is oxidised to carbon dioxide after a reaction with atoms of oxygen, ozone or nitrogen dioxide.

The atmospheric oxidation of carbon monoxide can be described with the following reactions:

$$(O_2)$$
$$CO + OH \rightarrow CO_2 + HO_2 \qquad (4.12)$$

$$HO_2 + NO \rightarrow NO_2 + OH \quad (4.13)$$

$$NO_2 + h\nu \rightarrow NO + O \quad (4.14)$$

$$O + O_2 + M \rightarrow O_3 + M \quad (4.15)$$

In total: $CO + 2 O_2 + h\nu \rightarrow CO_2 + O_3$

The above series of reactions produces ozone since the oxidation of NO to NO_2 is performed with the HO_2 radical and not with the ozone.

4.4.3 Chemistry of Hydrocarbons

There is a large number of hydrocarbons in the atmosphere especially in urban areas and as a result the chemical reactions of biogenic and anthropogenic hydrocarbons, as well as NO_x, dominate in comparison to the methane oxidation. There is a large number of chemical reactions for the hydrocarbons due to their large number of forms (e.g. alkanes, alkenes, aromatics).

Alkanes react mainly with hydroxyl radicals and are removed from the atmosphere. For example, methane reacts with hydroxyl radical through the reaction:

$$CH_4 + OH \rightarrow CH_3 + H_2O \quad (4.16)$$

Furthermore, the methylic radical which is formed reacts immediately with an oxygen molecule to form the methyl peroxy radical (CH_3O_2):

$$CH_3 + O_2 + M \rightarrow CH_3O_2 + M \quad (4.17)$$

In the troposphere, CH_3O_2 further reacts with NO, NO_2, HO_2 and RO_2. For the alkanes we can write the reactions with OH and NO_3, but the most important reaction is with OH ($> 90\%$):

$$RH + OH \rightarrow R + H_2O \quad (4.18)$$

$$RH + NO_3 \rightarrow R + HNO_3 \quad (4.19)$$

All hydrocarbons react with hydroxyl radicals. The higher reaction velocity is for alkenes and then come the aromatic and finally the alkanes.

4.4.4 Chemistry of Sulphur Compounds

Sulphur compounds in the atmosphere occur in different oxidation states, with those states being variable from -2 to $+4$. After the emission of sulphur components into the atmosphere, and especially the reduction of biogenics, there is a transformation of the oxidation state of sulphur to +4 (SO_2, CH_3SO_3H) and finally to the state +6 which is the oxidation state of sulphuric acid. The most important sources of sulphur in the atmosphere are from anthropogenic sources but a significant source is also the oceans.

Sulphur dioxide is the most important chemical form of sulphur in the atmosphere and is oxidized in both gaseous and particulate phases. In the gaseous phase there is a reaction of SO_2 with OH which can be described as:

$$OH + SO_2 + M \rightarrow HOSO_2 + M \quad (4.20)$$

$$HOSO_2 + O_2 \rightarrow HO_2 + SO_3 \quad (4.21)$$

The SO_3 which is formed reacts with water vapour in the gaseous and liquid phase after its absorption inside droplets and forms sulphuric acid as:

$$SO_3 + H_2O \rightarrow H_2SO_4 \quad (4.22)$$

The SO_2 concentration in the atmosphere shows also the pollution level and ranges at values close to 1 ppm in urban areas. At remote and background areas the concentrations are close to 0.2 ppb.

The chemical components with low oxidation state react with the radicals OH and NO_3. The H_2S reacts mainly with OH and the product of the reaction SH participates in a series of reactions which give finally SO_2:

$$H_2S + OH \rightarrow SH + H_2O \quad (4.23)$$

The main source of sulphur from the oceans originates from the oxidation of CH_3SCH_3 (DMS) with the radicals OH and NO_3. The residence time of DMS in the troposphere ranges from one to several days. The oxidation of DMS occurs through the NO_3 in the dark and cold regions and through OH at the lower latitudes.

4.5 Components of Aquatic Chemistry in the Atmosphere

Water is an inextricable component of life on our planet. Only 0.001% of the world's total water quantity exists in the atmosphere. Cloud formations pose the most obvious form of water processes in the atmosphere.

The equilibrium of a chemical component among the gaseous and liquid phases can be described with Henry's law. If there is equilibrium of a chemical component A in the gaseous (g) and liquid phase (aq) ($A(g) \leftrightarrow A(aq)$) it can be written that:

$$[A(aq)] = H_A \, p_A \quad (4.24)$$

where, p_A is the partial pressure of a component A in the gaseous phase (atm), [A(aq)] is the concentration of the component in the liquid phase (mol L^{-1}), and H_A is the Henry coefficient (mol L^{-1} atm^{-1}). Gasses with high solubility have high Henry coefficient.

The Henry law is applicable for light solutions. For thick solutions the solute concentration [A(aq)] in equilibrium at pressure p_A does not adhere to Henry's law. There is a large difference in the Henry values (mol L^{-1} atm^{-1}) for several gaseous species. At 298 K the following values are calculated for the Henry coefficients: $H_{O3} = 1.3 \times 10^{-3}, H_{SO2} = 1.23, H_{HCl} = 727, H_{CH3COOH} = 8.8 \times 10^3$ and $H_{NO3} = 2.1 \times 10^5$ mol L^{-1} atm^{-1}.

The temperature variation of an equilibrium coefficient such as Henry's constant can be described from the van't Hoff law:

$$\frac{d \ln H_A}{dT} = \frac{\Delta H_A}{RT^2} \qquad (4.25)$$

where, ΔH_A is the reaction enthalpy at constant temperature and pressure. The ΔH_A is a function of temperature and for small temperature changes is almost constant. The integration of the above equation results that:

$$H_A(T_2) = H_A(T_1) \times \exp\left[\frac{\Delta H_A}{R}\left(\frac{1}{T_1} - \frac{1}{T_2}\right)\right] \qquad (4.26)$$

Henry's constant increases when the temperature decreases which shows the high solubility of gas at lower temperatures. As an example the Henry constant for SO_2 increases from 1.24 to 3.28 mol L^{-1} atm^{-1} when the temperature decreases from 298 to 273K.

4.6 Chemistry of the Stratosphere - Ozone

Ozone is the most important gas present in the stratosphere and is responsible for the temperature increase in this layer as discussed in Chap. 1. If the ozone layer is transported to the Earth's surface under stable conditions of temperature and pressure, then the ozone column will have a width of 3 mm.

In recent years there has been a study made of the decrease of the ozone layer in the stratosphere at heights between 12 and 30 km. The ozone decrease is observed in September and October which is the beginning of spring in Antarctica. It is important to study the mechanisms responsible for the ozone destruction which is studied in this section.

It is known that the Sun's radiation which reaches the upper atmospheric layers includes also its ultraviolet spectrum which is composed of three regions in correspondence to the wave length of the UV radiation as follows (Fig. 4.3):

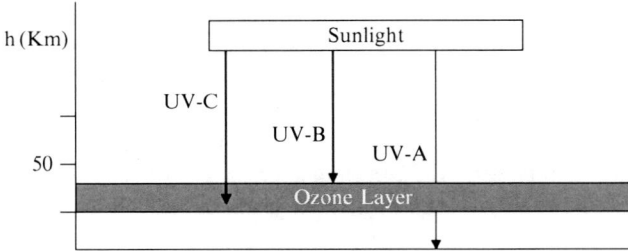

Fig. 4.3 Introduction of the Sun's ultraviolet radiation at the Earth's atmosphere and its absorption from the ozone layer at the stratosphere

- UV-A when $\lambda = 320\text{--}400$ nm,
- UV-B when $\lambda = 290\text{--}320$ nm,
- UV-C when $\lambda < 290$ nm.

The ozone is formed in the stratosphere at a height close to 30 km, where the UV radiation with wave length (λ) smaller than 242 nm (UV-C) breaks the molecular oxygen into two atoms of oxygen:

$$O_2 + UV - C \rightarrow 2O \tag{4.27}$$

Furthermore, the atomic oxygen reacts with molecular oxygen in the presence of a third molecule (N_2 or O_2) and there is the production of ozone:

$$O + O_2 + M \rightarrow O_3 + M \tag{4.28}$$

After the molecule of O_3 breaks with the absorption of UV-B and UV-C radiation as follows:

$$O_3 + h\nu \rightarrow O_2 + O \tag{4.29}$$

The O_3 can also react with atomic oxygen through the reaction:

$$O_3 + O \rightarrow O_2 + O_2 \tag{4.30}$$

The above photochemical theory was proposed by Chapman in 1930 but it was not able to explain the low ozone concentrations which have been observed lasting recent years. A study of the ozone destruction by other chemical compounds is proposed by other researchers. Chemical compounds which have the capacity to destroy ozone may occur at large concentrations or at low concentrations but cannot be destroyed through chemical reactions.

At the beginning of the 1970s, an important scientific finding from Crutzen (1970) and Johnston (1971) showed the effect of nitrogen oxides (NO_x) on the destruction of ozone without being consumed. Molina and Rowland (1974) showed

4.6 Chemistry of the Stratosphere - Ozone

the contribution of chlorofluorocarbons to the ozone destruction. Therefore an additional catalytic cycle has to be added to the Chapman mechanism which includes a free radical X (where X can be H, OH, NO, Cl or Br):

$$X + O_3 \rightarrow XO + O_2 \tag{4.31}$$

$$XO + O \rightarrow X + O_2 \tag{4.32}$$

$$\text{In total}: O_3 + O \rightarrow O_2 + O_2 \tag{4.33}$$

As can be seen in the above catalytic cycle the free radical is not consumed and the net result is the destruction of an ozone molecule and an oxygen atom and the production of two oxygen molecules. The importance of the catalytic cycle is dependent on the radical concentration and the velocity of the chemical reaction. In most cases the reaction (4-34) is very quick and therefore the cycle is controlled by the reaction (4-35).

Specifically chlorofluorocarbons (such as $CFCl_3$ and CF_2Cl_2) are destroyed in the stratosphere from ultraviolet radiation with the production of chlorine atoms:

$$CFCl_3 + h\nu \rightarrow CFCl_2 + Cl \tag{4.34}$$

$$CF_2Cl_2 + h\nu \rightarrow CF_2Cl + Cl \tag{4.35}$$

Furthermore the atomic Cl follows a catalytic cycle as described above.

Observations above Anarctica showed in 1985 a drastic reduction of the ozone layer, which is called the ozone hole. The destruction is higher than the expected destruction due to CFCs. It has been shown that the destruction is due to heterogeneous reactions which occur on the surface of aerosols. Above Antarctica there is a formation of clouds, which are called Polar Stratospheric Clouds (PSCs), at very low temperatures ($-80°C$). These polar clouds are composed of water ice crystals and dissolved nitric acid. The presence of these clouds contributes significantly to the ozone destruction.

First there is a photo dissociation of CFCs as described previously with the formation of a $ClONO_2$ molecule:

$$ClO + NO_2 \rightarrow ClONO_2 \tag{4.36}$$

With the presence of PSCs there are the following reactions:

$$ClONO_2 + HCl \rightarrow Cl_2 + HNO_{3(\text{solid phase})} \tag{4.37}$$

$$Cl_2 + h\nu \rightarrow 2\,Cl \tag{4.38}$$

$$Cl + O_3 \rightarrow ClO + O_2 \tag{4.39}$$

$$ClO + NO_2 \rightarrow ClONO_2 \tag{4.40}$$

In total: HCl + 2 O_3 + NO_2 → HNO_3 (solid phase) +2 O_2 + ClO

The above fast reactions occur on the surface of PSCs and allow the formation of reactive chemical species (Cl) which destroy ozone. The HNO_3 decreases the NO_2 concentration and is dissolved inside the aerosols in the stratospheric clouds. As a result the reactive forms of chlorine, such as ClO, remain free and are not allowed to form stable molecules such as $ClONO_2$.

The destruction of ozone in the stratosphere is an important consequent of the anthropogenic pollutant emission effects and an extensive survey is available in the scientific literature (e.g.. Finlayson-Pitts and Pitts 2000).

Problems

4.1 *The Chapman mechanism (1930) for the ozone removal in the stratosphere from the ultraviolet radiation ($\lambda < 242$ nm) can be described with the reactions:*

$$O_2 + h\nu \rightarrow O + O \tag{1}$$

$$O + O_2 + M \rightarrow O_3 + M \quad (very fast) \tag{2}$$

$$O_3 + h\nu \rightarrow O_2 + O \quad (very fast) \tag{3}$$

$$O_3 + O \rightarrow O_2 + O_2 \tag{4}$$

Show that for locally constant concentrations of oxygen atoms in the Chapman mechanism there is oxygen concentration equal to:
$[O] = \frac{J_{O_3}[O_3]}{K_2[O_3][M]}$, *where J_{O_3} is the photolysis rate of ozone and the production and removal rates of ozone are $3R_1$ and $2R_4$ respectively.*

4.2 *Calculate the analytical solution for the molecular ozone removal through the reaction:*

$$O_2 + h\nu \rightarrow O + O \quad (J)$$

If the oxygen is destroyed with the above reaction and is not further produced, what is the time for the concentration decrease to 10% from its initial value?

4.3 *The following reactions are given:*

$$A + h\nu \rightarrow B + C \quad (J_1)$$

$$B + h\nu \rightarrow C + C \quad (J_2)$$

where, J_1 and J_2 are the photolytic coefficients.

(a) *Write the reaction rate for the compounds A, B and C.*
(b) *Calculate the time dependent solution for each chemical compound with initial values $[A]_{t-h}$, $[B]_{t-h}$ and $[C]_{t-h}$. Suppose that J_1 and J_2 are stable versus time.*
(c) *When time reaches infinity ($h \rightarrow \infty$) how is the concentration of C ([C]) changing ? What is the limit $[C]_{h \rightarrow \infty}$, when $J_1 = 0$?*

4.4 *Calculate the mixing volume ratio of ozone, the partial ozone pressure and mass ratio of dry air when the arithmetic concentration of ozone is $N = 1.5 \times 10^{12}$ molecules/cm^3, the temperature $T = 285K$ and the partial pressure in the atmosphere $p_d = 980mb$.*

References

Crutzen, P. J. (1970). The influence of nitrogen oxides on atmospheric ozone content. *Q. J. R. Meteorol. Soc., 96*, 320–325.

B.J. Finlayson-Pitts, and J.N. Pitts *Chemistry of the upper and lower atmosphere*. Academic Press (2000)

Jacobson, M. Z. (2002). *Atmospheric pollution: History,science and regulation*. Cambridge: Cambridge University Press.

Johnston, H. S. (1971). Reduction of stratospheric ozone by nitrogen oxide catalysts from supersonic transport exhaust. *Science, 173*, 517–522.

Lelieveld, J., Roelofs, G. J., Ganzeveld, L., Feichter, J., & Rodhe, H. (1997). Terrestrial sources and distribution of atmospheric sulphur. *Phil. Trans. Roy. Soc. Lond. B., 352*, 149–158.

Molina, M. J., & Rowland, F. S. (1974). Stratospheric sink for chlorofluoromethanes: chlorine atom-catalyzed destruction of ozone. *Nature, 249*, 810–812.

Seinfeld, J. H., & Pandis, S. (2006). *Atmospheric chemistry and physics (2nd ed.)*. New York: John Wiley & Sons.

Chapter 5
Atmospheric Aerosols

Contents

5.1	Introduction	169
5.2	Size Distribution of Aerosols	171
5.3	Chemical Composition of Aerosols	180
5.4	Organic Aerosols	182
	5.4.1 Elemental Carbon- Primary Organic Carbon	184
	5.4.2 Secondary Organic Matter Formation (Secondary Organic Carbon)	185
5.5	Dynamics of Atmospheric Particulate Matter	186
	5.5.1 New Particle Formation	186
	5.5.2 Condensation and Evaporation	193
	5.5.3 Coagulation	196
5.6	Bioaerosols – Definition	197
References		198

Abstract Chapter 5 presents a general overview of the dynamics of atmospheric aerosols. The particle size is an essential parameter which determines the chemical composition, the optical properties, the deposition of particles and their inhalation in the human respiratory tract. Their size together with their chemical composition determine aerosol characteristics. Furthermore there is a focus on the organic aerosols in the atmosphere (elemental and organic carbon). The classical nucleation theory is presented in relation to new particle formation in the atmosphere together with condensation, evaporation and coagulation processes. Finally, general aspects of bioaerosols are studied.

5.1 Introduction

Aerosol is defined as a stable suspension of solid and liquid particles in the atmosphere. Atmospheric aerosols contain also the medium in which particles are suspended, which is the air. However, many times the terms aerosols and particles coincide in the literature. Aerosol particle sizes range from 0.001 μm (0.001 μm = 10^{-9} m = 1 nm = 10 Å) to 100 μm (10^{-4} m), hence, the particle sizes span over several orders of magnitude, ranging from almost macroscopic dimensions down to near molecular sizes. Aerosols are very important for public health and understanding of their dynamics is important for quantification of their effects on humans. Human exposure to aerosols occurs both in outdoor and indoor environments. The name is associated with Donnon (Whytlaw-Gray et al. 1923), although first publication of the term was due to Schmauss (1920) where it was used as an analogy to hydrosol, a stable liquid suspension of solid particles. Particles much greater than 100 μm do not stay airborne long enough to be measured and observed as aerosols. The lower limit is controlled by the size of a cluster of half a dozen or so molecules: this is the smallest entity of the condensed phase that can exist.

Particles in the atmosphere consist of a mixture of solid particles, liquid droplets and liquid components contained in the solid particles. Particles are variable in relation to their concentration, as well as to their physico-chemical and morphological characteristics. Particles can be products of combustion, suspension of soil materials, suspension of sea spray and can be also secondarily formed from chemical reactions in the atmosphere (Fig. 5.1). It can be concluded that airborne particulate matter is a complex mixture of many different chemical species originating from a variety of sources. Particulate matter can act as a transport medium for several chemical compounds, as well as, for biological materials which are absorbed or adsorbed on them. The role of particles in the atmosphere is very important both for effects on air pollution and climate. In the urban and semi-urban areas the air pollution due to ambient aerosols is significant with their concentration close to 10^7–10^8 particles/cm^3. The main characteristics of the airborne particles are their size, chemical composition and morphology.

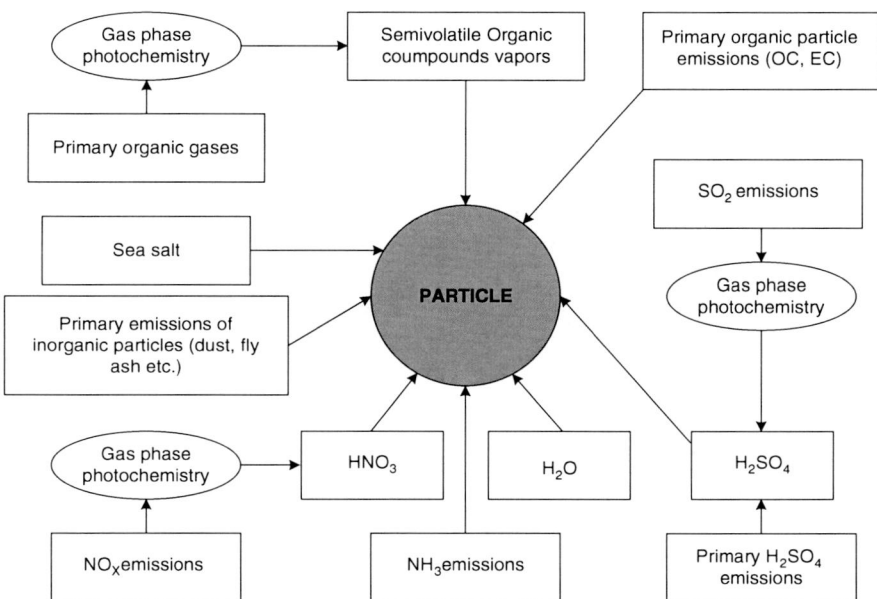

Fig. 5.1 Schematic representation of the chemical reactions and processes associated with the particulate matter (adapted from Meng and Seinfeld 1996)

Natural as well as anthropogenic aerosols have the potential to change the global radiation balance, an aspect highlighted in many studies of global climate change (IPCC 2007). Aerosols may influence the atmosphere in two important ways, direct and indirect effects. Direct effects refer to the scattering and absorption of radiation with a subsequent influence on the climatic system and planetary albedo. Indirect effects refer to the increase in available cloud condensation nuclei, due to an increase in anthropogenic aerosol concentration, and thus cloud albedo – the first indirect effect (or Twomey Effect). The microphysically-induced effect on the liquid water content, cloud height, and lifetime of clouds is the second indirect effect. The overall impact results in a cooling effect. The IPCC (2007) have estimated that the total direct aerosol radiative forcing combined across all aerosol types is -0.5 ± 0.4 W m^{-2} and that the indirect cloud albedo effect is -0.7 [-0.3 to -1.8] W m^{-2}. The role of anthropogenic aerosol is still a key factor of uncertainty in understanding and quantifying the present and future global climate change.

To quantify the impact of aerosols on climate and to assess, in turn, the feedback of climate change on aerosols requires a thorough understanding of the physico-chemical aerosol processes on a micro-scale and aerosol evolution in the context of regional and global scale circulation. This understanding can only be obtained by combining all available information from state-of-the-art experimental techniques and modelling tools. This is a challenging task, considering that tropospheric aerosol concentrations are nonlinearly dependent on the meteorological and chemical variables that govern their behaviour.

5.2 Size Distribution of Aerosols

Particle size is an essential parameter which determines the chemical composition, the optical properties, the deposition of particles and their inhalation in the human respiratory tract. Atmospheric particles can be found in a large size range, from a few tens of Angstroms (Å) to hundreds of micrometers (μm) (Hinds 1999). Figure 5.2 shows typical size distributions of particles in ambient air and Fig. 5.3 presents a typical number size distribution in an urban environment.

The categories of particles based on their size can be divided based on (1) their observed modal distribution, (2) the 50% cut-off diameter of the measurement instrument or (3) from dosimetric variables which are related to human exposure on atmospheric concentrations. In the last case the most common division is to $PM_{2.5}$ and PM_{10} particles where the index refers to the maximum aerodynamic diameter of the particles. The division is related with the possibility of $PM_{2.5}$ particles to enter the lower parts of the human respiratory tract. In the category (1) several sub-categories can be observed:

- *Nucleation mode*: Particles with diameter $<$ 10 nm, which are formed through nucleation processes. The lower limit of this category is not very well defined but it is larger than 3 nm.

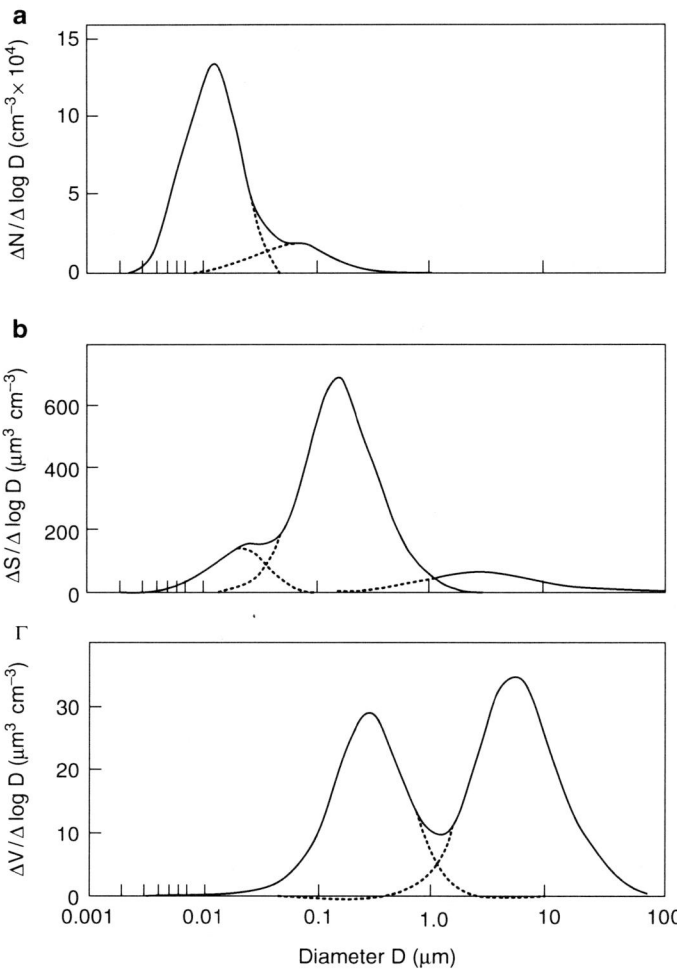

Fig. 5.2 Typical ambient aerosol distributions by (**a**) number (**b**) surface area (Γ) volume

- *Aitken mode*: Particles with diameter 10 nm < d < 100 nm. They originate from vapor nucleation or growth of pre-existing particles due to condensation.
- *Accumulation mode*: Particles with diameter 0.1 μm < d < 1 up to 3 μm. The upper limit coincides with a minimum of the total particle volume distribution. Particles in this mode are formed with the coagulation of smaller particles or the condensation of vapour constituents. The size of particles is not increased over this category with condensational growth. Furthermore, the removal mechanisms of particles in this category are very slow and as a result there is an accumulation of particles.
- *Ultrafine particles*: It contains particles in the Aitken and nucleation modes.

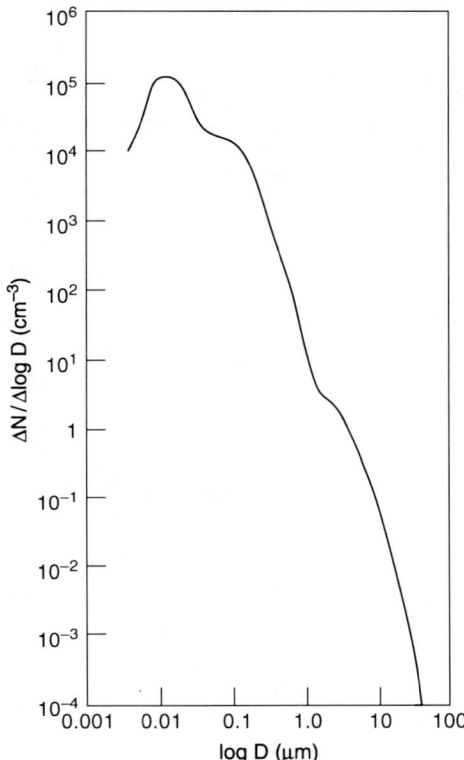

Fig. 5.3 Typical arithmetic distribution of particles in an urban environment (adapted from Hinds 1999)

- *Fine particles*: It includes the nucleation, Aitken and accumulation modes.
- *Coarse particles*: Particles with diameter between 3 and 10 μm.

Particles in the atmosphere have a distribution of sizes and lognormal distributions are used for their description (Hinds 1999). The aerosol size distribution is a crucial parameter that determines the dynamics of aerosols in the atmosphere, their transport, deposition and residence time. More specifically, Fig. 5.2 presents typical atmospheric aerosol distributions by number, surface area and volume.

The volume distribution has different features than the surface and number distributions. The volume distribution is usually bimodal with minimum ~1 μm (dividing limit between coarse and fine particles).

The arithmetic distribution has a maximum at the ultrafine mode, whereas the surface distribution has a maximum in the accumulation mode since there is a quite large number of particles in this region with considerable surface area. The volume distribution presents two logarithmic distributions, one at the accumulation mode and one at the coarse mode. In geographical regions with large production of new particles there have been observed three logarithmic distributions. On the

Table 5.1 Comparison between fine and coarse particles (adapted from Wilson and Suh 1997)

	Fine		Coarse
	Ultrafine	Accumulation mode	
Production process	Combustion High temperature processes Chemical reactions in the atmosphere		Brake of large particles
Formation	Nucleation Condensation Coagulation	Condensation Coagulation Chemical reactions of gasses inside/onto particles. Droplet evaporation at clouds/fogs. A solid core is existing inside the droplets	Mechanical breakup Spay evaporation Dust suspension Chemical reactions of gasses on/inside particles
Chemical composition	Sulphur	Sulphur, nitrogen, ammonium and hydrogen ions	Resuspended dust from roads or soil
	Elemental carbon	Elemental carbon	Suspended ash from incomplete burning of coal, oil and wood
	Metals	Large variation of organic chemical compounds Metals: compounds of Pb, Cd, V, Ni, Cu, Zn, Mn, Fe, etc.	Nitrates and chlorines
	Volatile organic compounds	water	Oxides of soil constituents (Si, Al, Ti, Fe) $CaCO_3$, NaCl, sea droplets, pollen, bioaerols, debris of tires, brakes
Solubility	More solubility than the accumulation mode	Very soluble and deliquescent	Insoluble and not hygroscopic
Sources	Combustion, secondary formation in the atmosphere from gaseous precursors,	Combustion of coal, oil, gasoline, diesel, wood, Secondary formation from gaseous precursors NO_x, SO_2 and organic compounds (e.g. terpenes) in the atmosphere.	Resuspension from industrial and soil dust, building construction, non controlled combustion of coal and oil, sea salt, biological sources.
	Processes at high temperature.	Processes at high temperature.	
Half life atmospheric time	Few minutes to hours	Days to weeks	Few minutes to hours
Removal processes	Growth to the accumulation mode, diffusion to rain droplets	Formation of cloud droplets, wet removal, dry deposition	Dry deposition, wet removal
Transport distance	< 1 up to several tens of km	Several 100 km up to few 1000 km	< 1 up to several tens of km

5.2 Size Distribution of Aerosols

contrary, at the urban environment a typical arithmetic distribution can be described from a single logarithmic distribution (Fig. 5.3).

The separation of fine and coarse particles is a determined factor since particles in these two regions are different in respect to their source, chemical composition, removal processes from the atmosphere, optical properties and affect on human health. There are therefore two different categories of particles which require individual study and need different legislation criteria for the protection of ecosystems and the population. In the region between 1 μm and 3 μm, where there is coexistence of the two categories, the distinction can be achieved due to particle origin.

The origin of fine particles is related mainly with anthropogenic activities (combustion sources, photochemical produced particles (urban haze)). The coarse particles are produced mainly from mechanical processes and can be distinguished as anthropogenic and natural aerosols (dust, sea spray). The coarse particles are characterized by quick removal due to gravitational settling and consequently relatively short residence time in the atmosphere. Since the removal mechanisms are more efficient for coarse and nuclei particles, the residence time of the aerosol accumulation mode in the atmosphere is larger than the other modes. Due to their relatively small residence time, the ultrafine and coarse particles are dominant close to their emission sources. On the contrary the accumulation mode particles are dominant far from the emission sources.

The fine particles are numerous compared to the coarse particles but their mass contribution is small compared to the total aerosol mass due to its small size and mass. On the contrary the coarse particles correspond to the major part of the aerosol surface and mass. The physical and chemical properties of aerosols are summarized in Table 5.1.

The sulphate, nitrate and ammonium particles have mainly anthropogenic origin and are observed in the fine mode. An exception is the NO_3^- ions, which adhere to NaCl particles and as the resulting particles are observed in the coarse mode. The resuspended dust and the sea salt aerosols are dominant in the coarse particle category. A small percentage of $SO_4^=$ particles are observed in larger sizes and has natural origin (oxidation of dimethyl sulfide (CH_3SCH_3, DMS) which is emitted from oceans). Trace species can be observed both in fine and coarse particles according to their sources.

A particle separation can also be performed due to its size as discussed previously. Airborne particulate matter (PM) is a complex mixture of different chemical components with its size ranging from a few nanometers to several 100 micrometers (Hinds 1999). It is apparent that particulate matter is not a single pollutant and its mass includes a mixture of many pollutants distributed differently at different sizes. The determination of the aerosol size distribution is one of the most important aspects involved both in measuring and modelling aerosol dynamics. Figure 5.4 shows size ranges of aerosols with their definitions.

In the atmosphere there is a continuous change of the size and the chemical composition of suspended particles. An example is given in Fig. 5.5, where particles are formed from combustion in a diesel engine and with increasing time

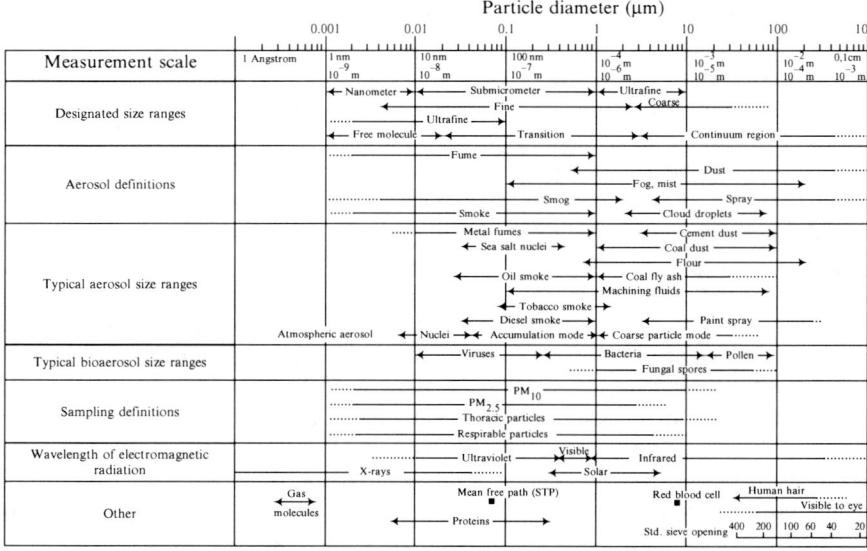

Fig. 5.4 Particle size range for aerosols (adapted from Hinds 1999)

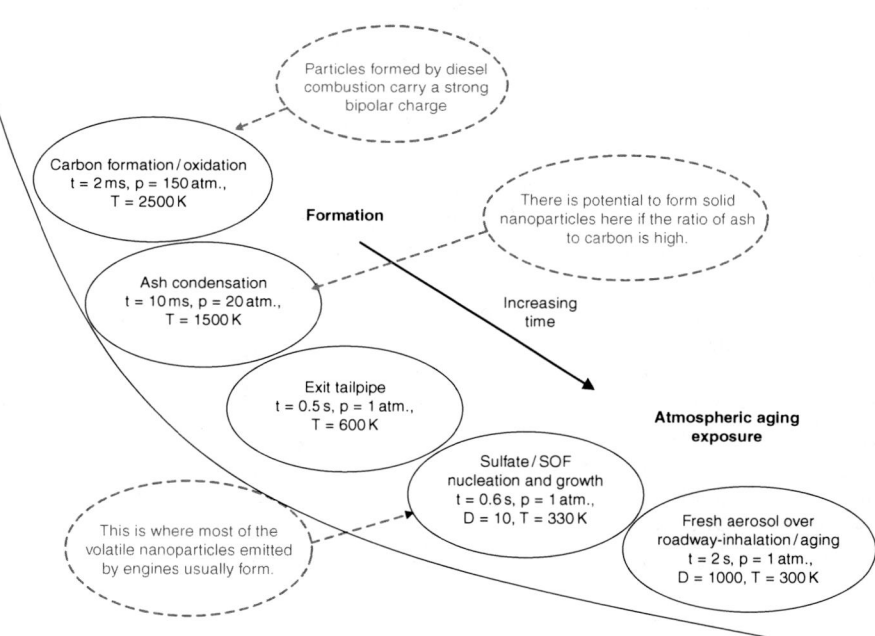

Fig. 5.5 Emission and secondary production of nanoparticles originating from a diesel engine (adapted from Ruzer and Harley 2005)

5.2 Size Distribution of Aerosols

they are diluted in the atmosphere, cooled with consequent alteration of their characteristics.

The majority of the aerosols in the atmosphere and the indoor environment are polydisperse. This means that the particles which comprise the aerosol have different sizes. Their size distribution is described through coefficients which for the logarithmic distribution are the Mass Median Diameter (MMD) or the Count Median Diameter (CMD), or the Mass Median Aerodynamic Diameter (MMςD) and the geometric standard deviation (σ_g).

For spherical particles the physical diameter can be used, whereas particles of other shapes can be characterized from the equivalent per volume diameter (d_e) which is the diameter of a spherical particle which has the same volume as the particles studied. The equivalent aerodynamic diameter is a very useful parameter which expresses the particle size with a homogeneous manner. This is the diameter of a sphere with density 1 g/cm^3 and the same settling velocity as the particle under study. The aerodynamic diameter is useful since it can be correlated with the residence time of particles in the atmosphere and their deposition in the human respiratory tract. It can be calculated in relation to the volume equivalent diameter with the following equation:

$$d_{ae} = d_e \sqrt{\frac{\rho}{\chi \rho_0} \times \frac{C(d_e)}{C(d_{ae})}} \qquad (5.1)$$

where ρ is the particle density, whereas $\rho_0 = 1$ g/m^3. The term χ is related with the shape of the particle under investigation and the values are varied between 1 (sphere) and 2. The terms $C(d_e)$ and $C(d_{ae})$ express a factor which is known as the Cunningham correction factor and expresses the deviation of the drag force from the Stokes law and is expressed in relation to the equivalent per volume diameter from the equation:

$$C(d_e) = 1 + (\lambda/d_e)\{2.514 + 0.8 \exp[-0.55(d_e/\lambda)]\} \qquad (5.2)$$

where $\lambda = 0.0683$ μm is the mean free path of air molecules at temperature 37 °C, relative humidity 100% and atmospheric pressure 76 cm Hg.

Particles with diameter smaller than 0.5 μm are better described with a thermodynamic equivalent diameter, which is the diameter of a spherical particle which has the same diffusion coefficient with the particle under study. The equivalent thermodynamic diameter can be calculated with the help of parameters which control the diffusion of particles with the equation:

$$d_{th} = \frac{kTC(d_e)}{3\pi\mu d_e} \qquad (5.3)$$

where T is the absolute temperature (310.15 K), k is the Boltzmann constant (0.014 × 10^{-23} J/grad) and μ is the dynamic viscosity of air (1.90 × 10^{-4} Poise).

Furthermore, the thermodynamic diameter is connected with the equivalent aerodynamic diameter of the particle with the expression:

$$d_{th} = d_{ae}\sqrt{\frac{\chi\rho_0}{\rho} \times \frac{C(d_{ae})}{C(d_{th})}} \qquad (5.4)$$

The expressions above for the particle diameter will be further used in Chapter 9 for the calculation of particle deposition in the human respiratory tract.

A logarithmic canonical distribution is used for the graphical distribution of particle mass for the determination of dose. However, the gravimetric instruments used for particle measurements use a number of stages for determination of the different size bins. The determination of parameters for the logarithmic distribution uses a specific methodology. First from the measurements a size distribution is obtained at specific size intervals (Fig. 5.6a) and the mass percentage at each size range is measured. Furthermore the cumulative distribution (Fig. 5.6b) is constructed, from which is obtained the mean mass diameter ($d_{50\%}$) and the diameter which corresponds to 84% (cumulative percentage) of mass.

The standard deviation is calculated from the expressions:

$$\sigma_g = \frac{d_{84\%}}{d_{50\%}} = \frac{d_{50\%}}{d_{16\%}} = \left(\frac{d_{84\%}}{d_{16\%}}\right)^{1/2} \qquad (5.5)$$

More specifically Fig. 5.6 shows the particle mass distribution versus the equivalent aerodynamic diameter with average mass diameter μ which is equal to 0.2 μm for fine particles and 3.2 μm for coarse particles and standard deviation σ_g, 2.2 and 2.1 respectively.

The solid line in Fig. 5.6b corresponds to coarse particles and the discontinuous line to fine particles.

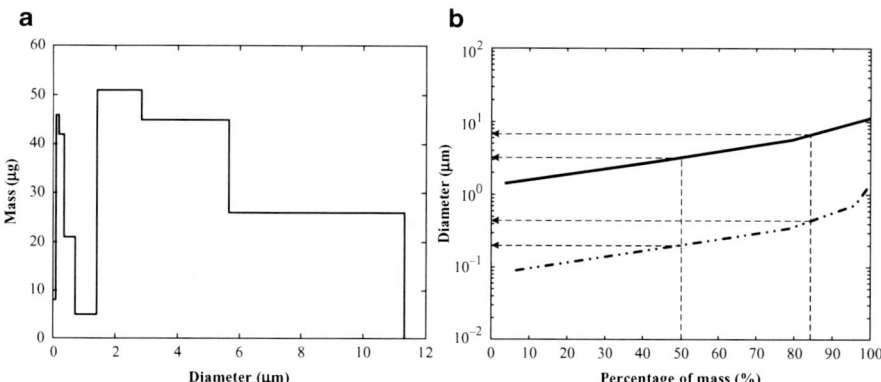

Fig. 5.6 (a) Particle mass distribution versus the equivalent aerodynamic diameter (b) Particle diameter versus the cumulative (adapted from Hinds 1999)

5.2 Size Distribution of Aerosols

The frequency function of the logarithmic canonical distribution can be expressed as:

$$f(d_{ae}) = \frac{1}{\sqrt{2\pi}\ln(\sigma_g)} \exp\left(-\frac{(\ln d_{ae} - \ln \mu)^2}{2\ln\sigma_g^2}\right) \quad (5.6)$$

and for a bimodal:

$$f(d_{ae}) = \frac{a}{\sqrt{2\pi}\ln(\sigma_{g,F})} \exp\left(-\frac{(\ln d_{ae} - \ln \mu_F)^2}{2\ln\sigma_{g,F}^2}\right) + \frac{1-a}{\sqrt{2\pi}\ln(\sigma_{g,C})}$$
$$\times \exp\left(-\frac{(\ln d_{ae} - \ln \mu_C)^2}{2\ln\sigma_{g,C}^2}\right) \quad (5.7)$$

where α is the fraction of the fine particles and μ is the average arithmetic diameter. The coefficients F and C refer to fine and coarse particles respectively.

The mean arithmetic diameter is given by:

$$\ln(\mu) = \frac{\sum \ln d_i}{N} \quad (5.8)$$

where N is the total number of particles and σ_g is the standard deviation which is expressed as

$$\ln \sigma_g = \left(\frac{\sum (\ln d_i - \ln \mu)}{N-1}\right)^{1/2} \quad (5.9)$$

There are different distributions which characterize specific properties of particles such as their number, surface, volume and mass. The number distribution describes the particle number at different sizes, whereas the mass distribution the particle mass at different particle sizes. The distributions are characterized from a mean diameter and geometric standard deviation.

The distributions are characterized from a mean diameter and geometric standard deviation. In logarithmic distributions there is the same geometric standard deviation. For characterized aerosols the mass concentration C_m is related with the arithmetic concentration C_N through the mean diameter, d_m, which is defined as:

$$d_m = \left(\frac{6}{\rho_p \pi N} m\right)^{1/3} = \left(\frac{\sum d_p^3}{N}\right)^{1/3} \quad (5.10)$$

Therefore it can be concluded that:

$$C_m = C_N \overline{m} = C_N \frac{\rho_p \pi}{6} d_m^3 \quad (5.11)$$

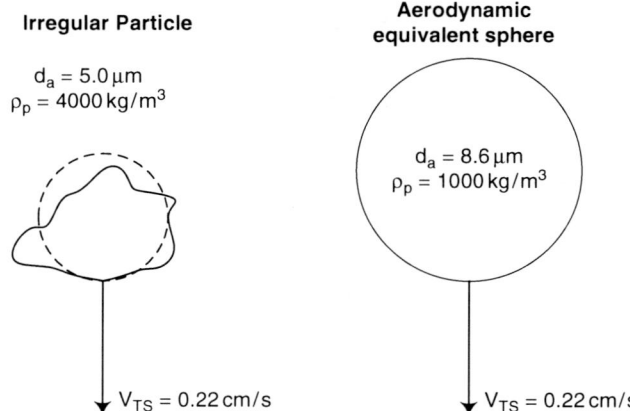

Fig. 5.7 An asymmetric particle and its aerodynamic equivalent sphere (adapted from Hinds 1999)

where $\overline{m}$ is the mass of a particle which has a mean diameter.

In most cases it is not necessary to know the real size, form or density of the particle but the knowledge of its equivalent aerodynamic diameter is adequate. Instruments for particle measurements in the atmosphere, such as impactors, are based on the measurement of the aerodynamic diameter. Figure 5.7 shows the definition of the equivalent aerodynamic diameter.

Furthermore, Fig. 5.8 shows the mass distribution of particles versus the aerodynamic diameter for an average mass diameter μ equal to 0.2 μm for fine particles and 3.12 μm for coarse particles and geometric standard deviation σ_g equal to 2.165 and 2.073 for fine and coarse particles respectively.

Figure 5.9 shows also the distribution of particles and the physico-chemical processes associated with the different sizes.

5.3 Chemical Composition of Aerosols

Atmospheric aerosols originate from either naturally occurring processes or from anthropogenic activity. Major natural aerosol sources include volcanic emissions, sea spray and mineral dust emissions, while anthropogenic sources include emissions from industry and combustion processes. Sources of PM can be either primary or secondary in nature. Primary sources are mainly of natural origin, whereas secondarily formed particles, in the atmosphere, are from both natural and anthropogenic origin as well and originate from chemical transformations of gaseous precursors such as sulfur dioxide, nitrogen oxides and VOCs. Recent research studies highlight also the importance of biogenic hydrocarbons (such as terpenes) in the formation of organic aerosols (Seinfeld and Pandis 2006). Sometime the distinction between natural and anthropogenic sources is not all clear. For example

5.3 Chemical Composition of Aerosols

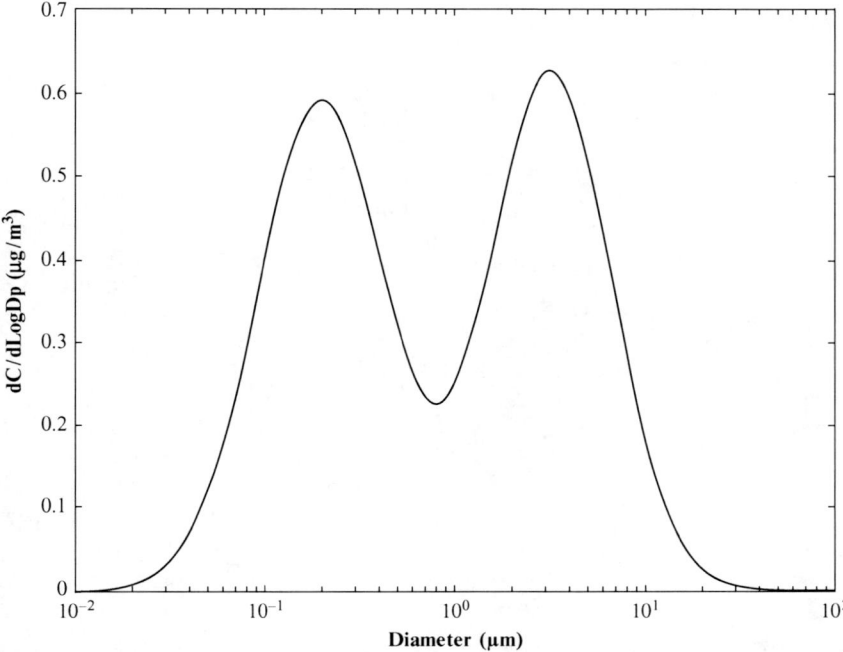

Fig. 5.8 Mass distribution of particles versus the aerodynamic diameter

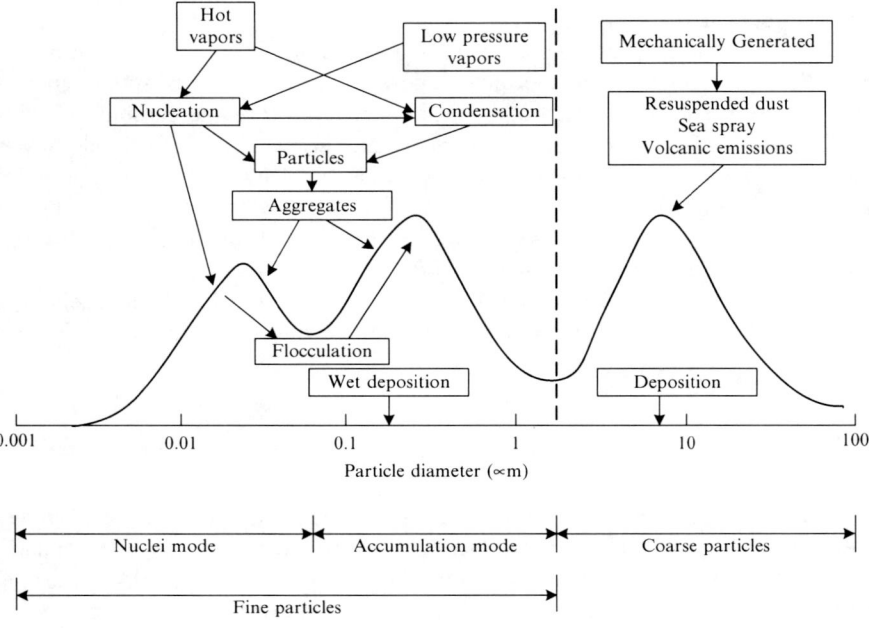

Fig. 5.9 Physico-chemical processes related to the aerosol particle size

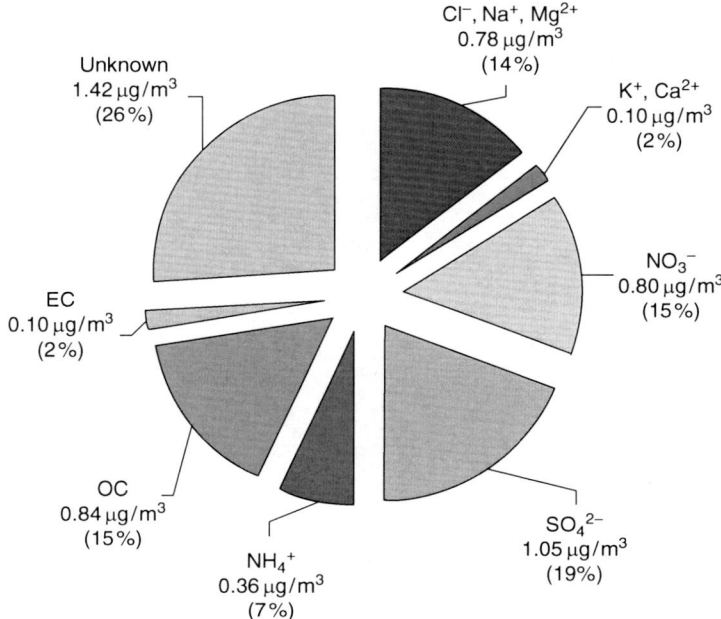

Fig. 5.10 Chemical mass closure of PM_{10} aerosols at the Birkenes station in Norway during 2004

smoke arising from natural wild fires is often categorized as anthropogenic in origin while mineral dust entrained into the atmosphere from agriculturally eroded regions has been considered as a natural source.

Crustal material, biogenic matter and sea-salt comprise the majority of natural aerosols. Anthropogenic aerosols are composed of primary emitted soot (elemental carbon) and secondary formed carbonaceous material (organic carbon) and inorganic matter (nitrates, sulfates, ammonium and water). Therefore modelling or measuring atmospheric aerosols involves many challenging tasks and is a fast evolving scientific area

Figure 5.10 shows as an example the chemical mass closure of PM_{10} aerosols at the Birkenes station in Norway during 2004.

In section 5.4 more details are given for organic aerosols which received an important scientific focus due to their health effects during recent years.

5.4 Organic Aerosols

The carbonaceous aerosols in the atmosphere can be identified as Primary Organic Aerosols (POA) and as Secondary Organic Aerosols (SOA) if they are primarily emitted or secondarily formed in the atmosphere respectively. The SOA forms as an oxidation product of gaseous organic precursors with low vapor pressure.

5.4 Organic Aerosols

There are several methodologies for the determination of elemental and organic carbon which result in an uncertainty in the distinction of these forms of carbonaceous aerosols. The determination of the organic aerosols based on a molecular level is also a challenging task since thousands of organic compounds are present in the particulate matter.

Soot is the only observed carbon particle form which includes all types of organic matter and is synonymous with combustion-generated primary carbonaceous aerosols. It consists of an array of organic compounds that are soluble in organic solvents and an insoluble part which is called Elementary Carbon (EC) or Black Carbon (BC) (Gelencsér, 2004; Cachier, 1998). The morphology of soot particles is in the form of aggregates which consist of several clusters. Combustion sources such as diesel engines emit soot particles at irregular agglomerate structures with an average diameter of primary spherules close to 22.6 nm ± 6.0 nm (Wentzel et al. 2003). Biomass combustion produces, during the smoldering phase, spherical particles which are compact and more stable than the irregular agglomerate structures, as shown in Fig. 5.11.

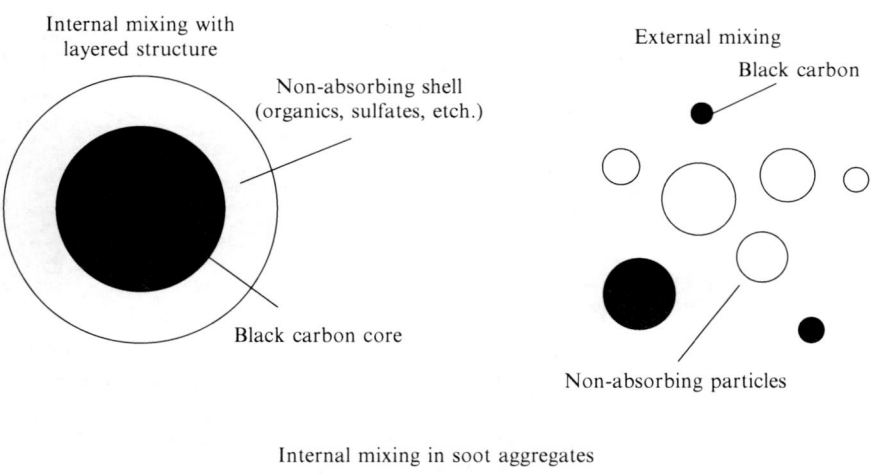

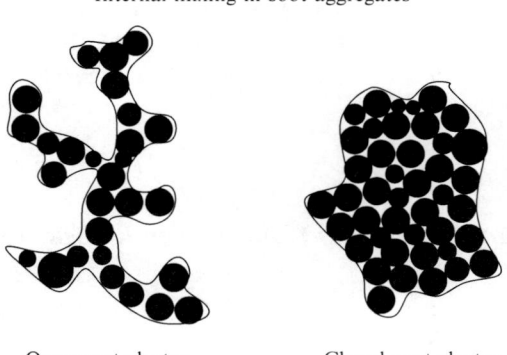

Fig. 5.11 Diagram on possible combinations between BC and non-absorbing materials in soot particles

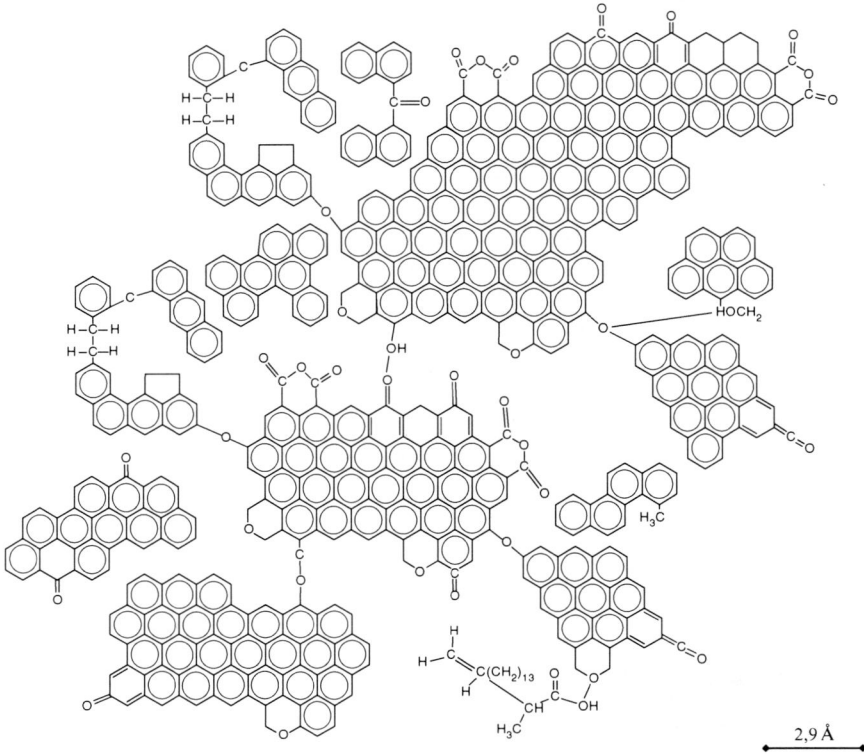

Fig. 5.12 Structure of hexane soot

The chemical composition of soot depends on its sources which can range from almost pure elemental carbon to a considerable contribution from organic matter. Measurements using analytical techniques proposed a structure for hexane soot (Akhter et al. 1985). It appears that combustion aerosols have a graphitic structure based on the C-O bonding with hydrocarbon segments at the particle surface. The transition between OC and BC fractions can be deduced from Fig. 5.12. There is a graphitic backbone and a gradual shift to OC. The main problem is to identify this shift from BC to OC.

5.4.1 Elemental Carbon- Primary Organic Carbon

Elemental carbon (EC) has a chemical structure similar to impure graphite and is emitted as primary particles mainly during combustion processes (wood-burning, diesel engines) (see Fig. 5.12). In Western Europe the contribution of diesel emissions to EC concentrations is estimated to be between 70 and 90%. Elemental carbon both absorbs and scatters light and contributes significantly to the total light extinction. Much higher concentrations of EC are found in urban areas compared to rural and remote locations. In rural and remote locations the EC concentration can vary between 0.2 and 2.0 μg m^{-3} and between 1.5 and 20 μg m^{-3} in urban areas, whereas the concentration of EC in remote ocean areas is in the range 5–20 ng m^{-3} (EPA, 2003). In polluted areas EC presents a bimodal size distribution with the first mode between 0.05 and 0.12 μm and the second mode between 0.5 and 1.0 μm. Average EC concentrations are around 1.3 and 3.8 μg m^{-3} for rural and urban sites respectively in the United States. The ratio of EC to total carbon is in the range 0.15–0.20 in rural areas and 0.2–0.6 in urban areas.

The organic carbon (OC) is a complex mixture of thousands of different organic compounds and a very small portion of it is molecular structure that has been characterized (around 10%). Combustion sources emit particles in the fine mode and the mass distribution has a mode at 0.1–0.2 μm. Gas chromatography combined with mass spectroscopy (GC/MS) as well as liquid chromatography combined with mass spectroscopy (LC/MS) has not, to date, provided a satisfactory solution for the molecular characterization of the major part of the ambient organic matter. Organic compounds that have been characterized include, amongst others n-alkanes, n-alkanoic acids and polycyclic aromatic compounds. Due to the difficulty in measuring organic compounds our current knowledge about organic matter is limited and incomplete. Primary emission sources for organic carbon includes combustion processes, geological (fossil fuels) and biogenic sources. Primary organic aerosol components include surface active organic matter on sea salt aerosol.

5.4.2 Secondary Organic Matter Formation (Secondary Organic Carbon)

An important part of secondary aerosol particles in the atmosphere is composed of secondary formed organic matter produced from oxidation of organic compounds. Partitioning of gas-particle organic compounds in the atmosphere is an important task for determining their association with the fine particulate matter. Understanding the mechanisms that control the conversion of organic matter from the vapor phase to particulate matter will provide valuable information for determining future control strategies to reduce the partitioning of organic matter in the particulate phase. However, there is a great complexity of the number of different chemical forms of organic matter and the absence of direct chemical analysis has resulted in the use of experimentally determined fractional aerosol yields, fractional aerosol coefficients and adsorption/absorption methodologies (Seinfeld and Pandis 2006) to describe the incorporation of organic matter in the aerosol phase.

5.5 Dynamics of Atmospheric Particulate Matter

5.5.1 New Particle Formation

Nucleation is one of the most fundamental processes that occur in the atmosphere and play an important role in processes such as condensation, new particle formation, cloud formation and crystallization. Nucleation can be defined as a phase transition (e.g. from the vapor to liquid phase). The phase change does not occur immediately but through the formation of small aggregations of molecules in the form of clusters. The conditions for the new particle formation directly from the vapor phase does not occur easily in the atmosphere due to low vapor pressure of gasses and the existence of suspended particles. This causes nucleation to occur on the surface of pre-existing particles or through the nucleation of vapor mixtures.

Nucleation which occurs without the presence of foreign (pre-existing) particles is called homogeneous nucleation, the process with foreign particles is called heterogeneous nucleation. When only one chemical compound participates in the nucleation process, it is called homomolecular nucleation, whereas when more than one chemical compound participates it is called heteromolecular nucleation.

In the atmosphere, homogeneous nucleation occurs mainly with the participation of two or more chemical compounds. A binary system which is important in the atmosphere is the system sulphuric acid//water. Homogeneous nucleation with the participation of only one component (e.g. water) is not feasible in the atmosphere. At relative humidity 200%, which is much higher than the conditions occurring in the atmosphere, and temperature of 20 °C the nucleation rate is 10–54 droplets cm^{-3} s^{-1}. Therefore, one would have to wait for 1,046 years under the above conditions for the formation of one water droplet.

Homogeneous nucleation occurs in a supersaturated vapour. The supersaturation of a chemical species A in air at temperature T is called the saturation ratio and is defined as:

$$S = \frac{p_A}{p_A^s(T)} \tag{5.12}$$

where p_A is the partial pressure of the gaseous species A and $p_A^s(T)$ is the saturation pressure of A which is in equilibrium with the liquid phase at temperature T.

The methodology of the classical nucleation theory is presented in the following paragraphs.

5.5.1.1 Nucleation Theory – Kinetic Method

The theory of nucleation is based on the solution of an equation system which describes the concentration variation of clusters with the addition or subtraction of molecules (Fig. 5.13). It is supposed that clusters impact with air molecules at a rate that is equilibrated thermally at time periods which are small compared with the

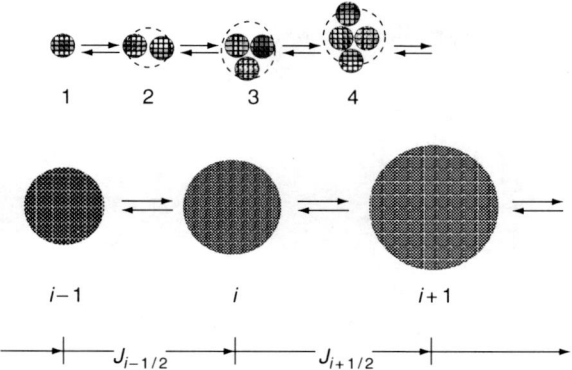

Fig. 5.13 Growth and evaporation of clusters (Adapted from Seinfeld and Pandis 2006)

time which is necessary for the addition or subtraction of a molecule. This denotes that the clusters have the same temperature as their environment.

The kinetic method of the nucleation theory assumes that clusters increase or decrease in size with the addition or removal of one molecule for non-associated vapors. Therefore if Ni(t) is the non-equilibrium number concentration of i-molecule clusters, the following equations describe the variation of the cluster concentration:

$$\frac{dN_i}{dt} = \beta_{i-1} N_{i-1}(t) - \gamma_i N_i(t) - \beta_i N_i(t) - \gamma_{i+1} N_{i+1}(t) \tag{5.13}$$

where β_i is the forward (condensation) flux per unit time and area at which an i-cluster gains a molecule, and α_i the backward (evaporation) flux at which the cluster loses a molecule.

The net rate (cm^{-3} s^{-1}) $J_{i+1/2}$ by which clusters change their size from i to $i+1$ is given by:

$$J_{i+1/2} = \beta_i N_i - \gamma_{i+1} N_{i+1} \tag{5.14}$$

It is assumed that the cluster concentration is in a steady state condition and therefore all the fluxes have to be equal to a stable flux J:

$$J_{i+1/2} = J, \text{ for all i} \tag{5.15}$$

A quantity f_i is defined from the equation:

$$\beta_i f_i = \gamma_{i+1} f_{i+1} \tag{5.16}$$

Dividing the Eq. (5.14) by (5.16) we get that:

$$\frac{J}{\beta_i f_i} = \frac{N_i}{f_i} - \frac{N_{i+1}}{f_{i+1}} \tag{5.17}$$

Putting $f_i = 0$ and obtaining the sum of the above equation from $i = 1$ to $i = i_{max}$ the following expression is derived:

$$J \sum_{i=1}^{i_{max}} \frac{1}{\beta_i f_i} = N_1 - \frac{N_{i_{max}}}{f_{i_{max}}} \tag{5.18}$$

The term f_i can be written as:

$$f_i = \frac{\beta_1}{\gamma_2} \frac{\beta_2}{\gamma_3} \cdots \frac{\beta_{i-1}}{\gamma_i} = \prod_{j=1}^{i-1} \frac{\beta_j}{\gamma_{j+1}} \tag{5.19}$$

The ratio f_i under nucleation conditions ($S > 1$) is higher than 1 since the rate of growth is higher than the evaporation rate. Furthermore $N_{imax} < N_1$ since under nucleation conditions the number of clusters is considerably lower than the number of the gas molecules. Therefore the second term in Eq. (5.18) can be omitted and has as a result:

$$J = N_1 \left(\sum_{i=1}^{i_{max}} \frac{1}{\beta_i f_i} \right)^{-1} \tag{5.20}$$

The above summation can be extended to infinity since the function f_i increases with the increase of i and the sum converges. Therefore the expression for the nucleation flux can be written as:

$$J = N_1 \left(\sum_{i=1}^{\infty} \frac{1}{\beta_i f_i} \right)^{-1} \tag{5.21}$$

The evaporation rate γ_i is difficult to be determined theoretically and therefore the kinetic problem for the evaluation of the nucleation flux becomes a problem of thermodynamics for evaluation of the equilibrium distribution of droplets. For this calculation it is necessary to examine the energy which is needed for the cluster formation.

The minimum energy which is necessary for the formation of a cluster with i molecules can be shown to be equal to (Debenedetti 1996):

$$W_{min} = \sigma A + \left(P - P' \right) V' + i \left(\mu'\left(T, P'\right) - \mu(T, P) \right) \tag{5.22}$$

where σ is the surface tension, A is the surface between the cluster and the gaseous phase, P is the partial air pressure and P' is the pressure inside the cluster, V' is the

5.5 Dynamics of Atmospheric Particulate Matter

cluster volume and μ, μ' are the chemical potentials for the liquid phase (cluster) and the gaseous phase respectively. For an incompressible cluster one can write that:

$$\mu'(T,P') - \mu'(T,P) = v'(P' - P) \tag{5.23}$$

where v' is the molecular volume in the liquid phase (it is assumed that the cluster occurs in the liquid phase). Therefore one can write that:

$$W_{min} = \sigma A + i\left(\mu'(T,P) - \mu(T,P)\right) = \sigma A + i\Delta\mu \tag{5.24}$$

where $\Delta\mu$, which is lower than 1, is the difference between the chemical potentials in the stable and metastable conditions in the liquid and gaseous phase. The A is analogous to $i^{2/3}$ (i the number of molecules in the cluster) and thus (Debenedetti 1996):

$$W_{min} = a\,i^{2/3} - bi = cr^2 - dr^3 \tag{5.25}$$

where a, b, c and d are positive numbers and r is the radius of the cluster. The minimum of the above expression results in:

$$i^* = \left(\frac{2a}{3b}\right)^3 \tag{5.26}$$

For spherical clusters the expression can be written as:

$$i^* = \frac{32\pi}{3}\left[\frac{(v')^{2/3}\sigma}{(-\Delta\mu)}\right]^3 \tag{5.27}$$

And the cluster radius which includes i^* molecules is given by:

$$r^* = \frac{2\sigma v'}{(-\Delta\mu)} \tag{5.28}$$

The minimum energy which is required for the formation of a cluster with i^* molecules can be expressed as:

$$W_{min} = \frac{4a^3}{27b^2} = \frac{16\pi}{3}\left[\frac{v'\sigma^{3/2}}{(-\Delta\mu)}\right]^2 \tag{5.29}$$

5.5.1.2 Calculation of the Nucleation Rate

It is supposed that there is an equilibrium distribution of clusters and therefore the cluster distribution can be expressed as:

$$N(n) \propto \exp\left[-\frac{W(n)}{kT}\right] \quad (5.30)$$

where W(n) is the reversible work which is required for the formation of a cluster with n moles. Since there is no necessary work for the formation of cluster with one molecule, the proportionality constant has to equilibrate the number density of the non connected molecules in the metastable phase. Therefore one can write that:

$$N(n) = N_{tot} \exp\left[-\frac{W(n)}{kT}\right] \quad (5.31)$$

where N_{tot} is the total density number in the metastable phase. Furthermore:

$$\begin{aligned} a_1 + a_1 &= a_2 \\ a_2 + a_1 &= a_3 \\ &\dots \\ a_{n-1} + a_1 &= a_n \end{aligned} \quad (5.32)$$

where α_n refers to a cluster which includes n moles. Therefore we can write:

$$\mu_n = n\mu_1 \quad (5.33)$$

where μ_n is the chemical potential of a droplet with n moles and μ_1 is the chemical potential per mole in the vapour phase. Therefore:

$$\mu_n = \lambda_n(T, v') + kT \ln \frac{N(n)}{\Sigma(N(n))} \quad (5.34)$$

where λ_n is the chemical potential of a cluster of size n when only this is present, v' is the molecular volume in the liquid phase and N(n) is the droplet concentration of n-moles. The solution of the Eq. (5.34) for the N(n) results (Debenedetti 1996):

$$N(n) = [\Sigma N(n)] \exp \frac{n\mu_1 - \lambda_n}{kT} \quad (5.35)$$

where the energy of cluster formation is written as:

$$W(n) = \lambda_n - n\mu_1 \quad (5.36)$$

5.5 Dynamics of Atmospheric Particulate Matter

During the condensation process the liquid embryo which is formed is assumed incompressible. Therefore:

$$\lambda_n = n\mu_1 + \sigma F(n) + n(\mu - \mu_1) = \sigma F(n) + n\mu \quad (5.37)$$

where, μ is the chemical potential per mole of liquid when it is in the gaseous phase. Therefore the chemical potential of a droplet which includes n moles is different than the chemical potential of n moles in the liquid phase.

The nucleation rate is given from the Eq. (5.21) (the f_i is denoted by F(n)) with the replacement of the summation with integration:

$$J = \beta N_{tot} \left[\int_{n \ll n*}^{n \gg n*} \exp\left(\frac{W(n)}{kT}\right) \frac{1}{F(n)} dn \right]^{-1} \quad (5.38)$$

where n^* is the critical number of molecules in the cluster.

Since the exponential is concentrated close to n^* the main contribution to the integral value comes from values close to n^*. Therefore the work which is needed for the cluster formation can be written as an expansion in relation with an unstable value at equilibrium and the F(n) is replaced with F(n^*). Therefore for an incompressible cluster it results that:

$$W(n) \approx W(n*) + \tfrac{1}{2}W''(n*)(n - n*)^2 = W(n*) + \tfrac{1}{2}W''(n*)(\delta n)^2 \quad (5.39)$$

where $W''(n^*)$ denotes the curvature of the function W(n) which is calculated at n^*. With a replacement of Eq. (5.39) to (5.38) we get that:

$$J = \beta N_{tot} F(n*) \exp\left(\frac{-W(n*)}{kT}\right) \left[\int_{-\infty}^{+\infty} \exp\left\{-\frac{1}{2kT}W''(n*)(\delta n)^2\right\} d(\delta n) \right] \quad (5.40)$$

where there is a replacement of the integration variables from n to δn and the exponential characteristics are used for the change of the integration limits from [1-n^*, Λ- n^* (Λ >> n^*)]. After the integration we get that:

$$J = \{\beta F(n*)\} \left(\sqrt{\frac{-W(n*)}{2\pi kT}}\right) \left[N_{tot} \exp\left(\frac{-W(n*)}{kT}\right)\right] = j(n*)ZN(n*) \quad (5.41)$$

The nucleation rate can be written as a product of three terms. The first term [j(n*)] is the product of β multiplied by the surface of the critical cluster and denotes the frequency of impingement of molecules on the critical cluster. The third term N(n*), is the equilibrium of the concentration of critical clusters. Therefore, the second term Z is a correction factor since the concentration of critical clusters is different than the equilibrium one. This term is referred to as a Zeldovich factor of non-equilibrium (Debenedetti 1996).

For incompressible clusters (e.g. droplets in a super-cooled vapour) it becomes (Debenedetti 1996):

$$J = 2\beta N_{tot} \left(\sqrt{\frac{\sigma(v')^2}{kT}}\right) \exp\left\{-\frac{16\pi}{3kT}\left[\frac{v'(\sigma)^{3/2}}{-\Delta\mu}\right]^2\right\} \quad (5.42)$$

If the dependence of the chemical potential in the liquid phase versus pressure is not taken into account, and assuming that the gaseous phase behaves as an ideal gas, we have the following expression for the homogeneous nucleation in the gaseous phase:

$$J = 2\beta N_{tot} \left(\sqrt{\frac{\sigma(v')^2}{kT}}\right) \exp\left\{-\frac{16\pi}{3kT}\left[\frac{\sigma(v')^{3/2}}{kT}\right]\left[\frac{1}{\ln S}\right]^2\right\}$$

$$= \frac{\rho_u^2}{\rho_l}\left(\sqrt{\frac{2\sigma}{\pi m}}\right) \exp\left\{-\frac{16\pi}{3kT}\left[\frac{v'(\sigma)^{3/2}}{-\Delta\mu}\right]^2\right\} \quad (5.43)$$

where: S is the supersaturation ($S = \frac{P}{P^e}$),

P^e is the saturation pressure of vapour at a specific temperature,
ρ_v ($= N_{tot}$) is the number density of vapour,
ρ_l ($= 1/u'$) is the number density for liquid,
m is the molecular mass,

and the rate of molecular impingement β from the vapour to liquid phase is given by the expression:

$$\beta = \frac{P}{\sqrt{2\pi mkT}} \quad (5.44)$$

5.5.1.3 Heterogeneous Nucleation

In the majority of cases, suspended particles exist in the atmosphere and on their surfaces occur phase transitions (nucleation) from the gaseous to the liquid phase. Suppose that a liquid embryo is formed onto a particle surface. The minimum work which is required for this process is given by the expression (Lazaridis et al. 1991):

$$W_{min} = \sigma_{gl}F_{gl} + (\sigma_{gs} - \sigma_{ls})F_{gs} + (P - P')V' + [\mu\prime(T, P\prime) - \mu(T, P)]n \quad (5.45)$$

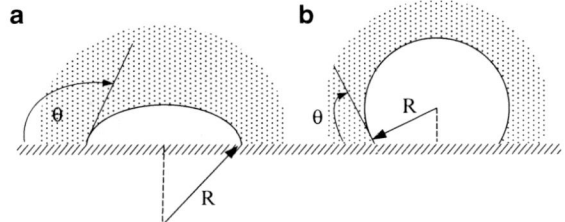

Fig. 5.14 Formation of a liquid droplet onto a solid surface under conditions of (**a**) small και (**b**) large wetting. The wetting refers to the contact solid-liquid

where σ refers to the interfacial tensions between the different phases, the indexes g, s and l refer to the applied phase (g refers to the gaseous, s to the solid and l to the liquid phase) and n the number of molecules in the cluster. At the triple interface between the liquid, solid and gas (Fig. 5.14) there is an equilibrium which satisfies the expression:

$$\sigma_{gl}\cos(\pi-\theta) + \sigma_{gs} = \sigma_{ls} \tag{5.46}$$

where, based on the geometry of Fig. 5.12, the following expression is satisfied:

$$\sigma_{gl}F_{gl} + (\sigma_{gs} - \sigma_{ls})F_{gs} = \pi R^2 \sigma[2(1+\cos\theta) + \sin^2\theta\cos\theta] \tag{5.47}$$

The minimum energy which is required for the formation of a critical nucleus is given by the expression:

$$W_{min} = \frac{16\pi\sigma^3}{3(P'-P)} \cdot \frac{(1+\cos\theta)^2(2-\cos\theta)}{4} \tag{5.48}$$

From the analysis which has been performed for the kinetics of the homogeneous nucleation, the expression for the heterogeneous nucleation is the following:

$$J = N_{tot}^{2/3} \alpha \sqrt{\frac{2\sigma}{\pi m B(2-\cos\theta)}} \exp\left[\frac{16\pi\sigma^3}{3kT(P^e-P)^2} \cdot \psi(\theta)\right] \tag{5.49}$$

where: α η is the available surface for the heterogeneous nucleation per unit volume in the liquid phase, $B \approx 1 - \frac{1}{3}(1 - \frac{P}{P^e})$ where P^e is the equilibrium vapour pressure and ψ(θ) the geometric factor $\cdot \frac{(1+\cos\theta)^2(2-\cos\theta)}{4}$ in the Eq. (5.49).

5.5.2 Condensation and Evaporation

The particles in the atmosphere can increase their size with water vapor condensation. The rate of size increase of the particles is dependent on the relative humidity, the particle size and the relative size of particles compared to the mean free path [1]. In the first phase of the particle size increase the particle diameter is smaller than the mean free path. At these conditions the rate of particle increase is dependent on the rate of the random molecular impacts between the particles and the water vapor molecules. The rate of particle diameter increase is given by the expression:

$$\frac{d(d_p)}{dt} = \frac{2 M a_c (p_\infty - p_d)}{\rho_p N_a \sqrt{2\pi mkT}} \text{ for } d_p < \lambda \tag{5.50}$$

where p_∞ is the partial water vapor pressure near the particle but at a distance from its surface and p_d is the partial water vapor at the particle surface, α_c is the condensation coefficient which denotes the percentage of molecules which are adhered to the particle surface after the impingement, M is the molecular liquid weight, ρ_p is the liquid density and N_α is the Avogadro number.

In the case that the particles are quite larger than the mean free path, the rate of particle increase is given by the diffusion rate of molecules to the particle surface. The rate by which the molecules impact with the particle can be calculated with the theory of coagulation but with a different coefficient since the molecules have much smaller diameter than the particle and the particle has much smaller diffusion coefficient than the molecules. The rate of particle increase can be written as:

$$\frac{d(d_p)}{dt} = \frac{4 M D_v}{R \rho_p d_p} \left(\frac{p_\infty}{T_\infty} - \frac{p_d}{T_d}\right) \phi \text{ for } d_p > \lambda \tag{5.51}$$

where R is the gas constant and φ is the correction coefficient of Fuchs. This equation is based on the molecular diffusion theory at the droplet surface. The diffusion equation, as well as the concept of difference, are not valid for a distance smaller than one mean free path from the droplet's surface. The correction factor of Fuchs is called also a Knudsen correction factor and is given as:

$$\varphi = \frac{2\lambda + d_p}{d_p + 5.33 \left(\lambda^2/d_p\right) + 3.42 \lambda} \tag{5.52}$$

In the case the growth of droplet size occurs slowly (when the relative humidity is between 100% and 105%), then the droplet temperature is the same as the air temperature and the pressure at the droplet surface can be calculated from the expression which gives the equilibrium pressure. In cases where there is quick increase of the droplet size then there is a temperature increase due to the latent heat release from the condensation process.

In the atmosphere there are mainly soluble particles and in the following paragraphs we will examine the effect of a soluble particle on the growth of a water droplet. A particle of sodium chloride will be examined as an example due to its presence in large quantities from ocean emissions. The introduction of salt inside water will lead to an increase of the water's boiling point with the decrease of the equilibrium pressure above the water surface. This results in an increase of the droplet size.

Salt dissolved inside water increases the boiling point of a solution and decreases the freezing point due to the decrease of water vapour pressure. The affinity of the dissolved salt with the water molecules enables the formation of stable droplets in saturated or under-saturated environments.

A droplet that contains dissolved constituents such as NaCl experiences two competitive phenomena which control the relationship between the Kelvin effect and the droplet size. First, the concentration of salt inside a droplet will increase with the decrease of its size. Secondly, a competitive phenomenon is the Kelvin effect, which brings an increase of the vapour pressure on the droplet surface with the decrease of its size. The relation between the Kelvin ratio and the droplet size which contains dissolved chemical components is written as:

$$K_R = \frac{p_d}{p_s} = \left(1 + \frac{6 i m M_w}{M_s \rho \pi d_p^3}\right) \exp\left(\frac{4 \gamma M_w}{\rho R T d_p}\right) \quad (5.53)$$

where m is the mass of the dissolved chemical constituent with mass M_s, M_w is the molecular weight of the solvent (usually water), ρ is the density of the solvent and I the number of ions of the molecule of salt which is dissolved. Figure 5.15 shows that the presence of dissolved salt changes dramatically the curve of pure water.

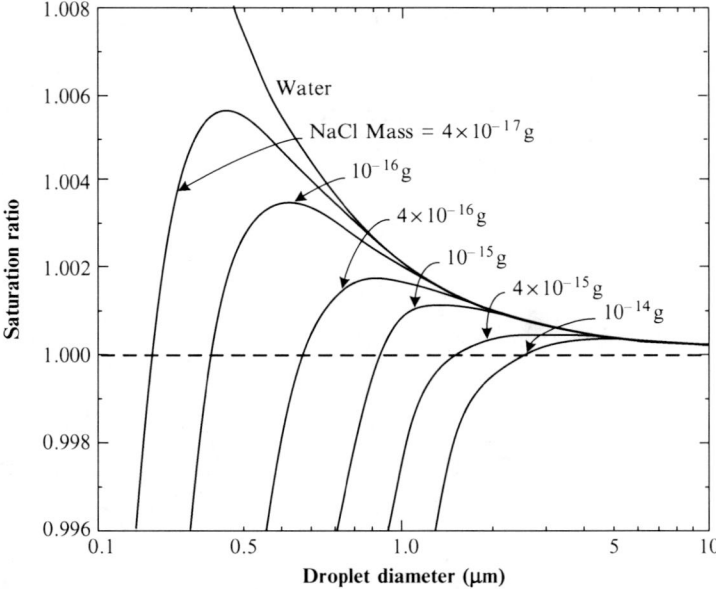

Fig. 5.15 Saturation ratio versus droplet size of pure water and of droplets which incorporate sodium chloride at temperature 293 K (20 °C). For each curve the region above it is the condensation region, whereas the region below it is the evaporation region

5.5.3 Coagulation

Coagulation is the process during which atmospheric particles impact together due to their relative movement and form larger particles. A result from coagulation is the decrease of the number of particles and increase of their size. When the relative movement of the particles is due to the Brownian motion, the process is called thermal coagulation. The relative particle motion can be due to external forces such as gravity, the aerodynamic phenomena or electrical forces. In this case the process is called kinematic coagulation.

We will study the simplest coagulation case which is the thermal coagulation for spherical particles with diameter larger than 0.1 μm. In the analysis which is performed it is supposed that the particles coagulate after each collision and the initial particle size change is small. At each collision there is a reduction of the particle number per unit volume.

The particle number ($N(t)$) at time t is given by the relation:

$$N(t) = \frac{N_o}{1 + N_o K_o t} \qquad (5.54)$$

where N_o is the initial particle number and K_o is the coagulation coefficient which is given by the relation $K_o = 4 \pi d_p D$ (m^3/s) (d_p is the particle diameter and D is the diffusion coefficient).

The coagulation rate, as shown in Eq. (5.54) is analogous to N^2, which denotes high rates for increased particle concentrations, whereas for lower concentrations the rate is smaller. The particle coagulation is restricted to particles larger than 0.1 μm since Eq. (5.54) does not describe in detail the concentration variation at distance of an average mean free path from the particle's surface. The deviation increases as the diameter decreases and is important for diameters smaller than 0.4 μm. A correction for the coagulation coefficient is proposed by Fuchs with the following relation:

$$K = K_o \beta \qquad (5.55)$$

At standard conditions values for the functions β, K_o and K are given in Table 5.2.

Table 5.2 Coagulation coefficients at canonical conditions (adapted from Hinds 1999)

Diameter (μm)	Correction coefficient β	K_o (m^3/s)	K (m^3/s)
0.004	0.037	168 10^{-16}	6.2 10^{-16}
0.01	0.14	68 10^{-16}	9.5 10^{-16}
0.04	0.58	19 10^{-16}	10.7 10^{-16}
0.1	0.82	8.7 10^{-16}	7.2 10^{-16}
0.4	0.95	4.2 10^{-16}	4.0 10^{-16}
1.0	0.97	3.4 10^{-16}	3.4 10^{-16}
4	0.99	3.1 10^{-16}	3.1 10^{-16}
10	0.99	3.0 10^{-16}	3.0 10^{-16}

5.6 Bioaerosols – Definition

Bioaerosols include all airborne particles of biological origin, i. e., bacteria, fungi, fungal spores, viruses and pollen and their fragments including various antigens. Particle sizes may range from aerodynamic diameters of ca. 0.5 to 100 µm (Nevalainen et al. 1991; Cox and Wathes 1995). Airborne micro-organisms become non-viable and fragmented over time due to desiccation. Indoor air contains a complex mixture of (i) bioaerosols such as fungi, bacteria and allergens, and (ii) non-biological particles (e.g., dust, tobacco smoke, cooking-generated particles, motor vehicle exhaust particles and particles from thermal power plants). Exposure to several of these biological entities as well as microbial fragments, such as cell wall fragments and flagella, as well as microbial metabolites e. s. endotoxin, mycotoxins and VOCs, may result in *adverse health effects*. In particular, increase in asthma attacks and bronchial hyper-reactivity has been correlated to increased bioaerosol levels.

More than 80 genera of fungi have been associated with symptoms of respiratory tract allergies. Cladosporium, Alternaria, Aspergillus and Fusarium are amongst the most common allergenic genera. Metabolites of fungi including toxins and volatile organic compounds are also believed to irritate the respiratory system. Furthermore, non-biological particles may serve as carriers of *fungal allergen molecules* into the lung independently of the whole fungal spore. In the case of non-viable combustion particles such as tobacco, smoke or cooking-generated particles, such an interaction could have serious implications. Allergic molecules could conceivably be carried into the lung at a greater depth than a fungal spore would be expected to penetrate. Bioaerosols are usually measured in standard Colony Forming Units per volume (CFU/m^3 counts).

Biological aerosols may also have considerable influence on the mass size distribution. Primary biological aerosol particles (PBAP) describe airborne solid particles (dead or alive) that are or were derived from living organisms, including microorganisms and fragments of all varieties of living things. PBAP includes viruses (0.005 µm $< r <$ 0.25 µm) (r, refers to particle radius), bacteria (r $\geq$ 0.2 µm), algae, spores of lichen mosses, ferns and fungi (r $\geq$ 0.5 µm), pollen (r $\geq$ 5 µm), plant debris such as leaf litter, part of insects, human and animal epithelial cells (usually $r > 1$ µm). Most biological aerosols arise from plants but secondary sources include industry (textile mills), agriculture (fertilizing) and sewage plants.

Bioaerosols may originate from almost all the natural and anthropogenic activity and every source has different emission bioaerosol characteristics. The bioaerosol concentration varies from relatively low levels ($\leq 10^2$ cfu/m^3) (e.g. at clean rooms and surgery rooms in hospitals) until very high (10^5–10^{10} cfu/m^3) during specific operations in industry and farming.

Recent findings by Winiwarter et al. (2009) suggest that the PBAB contribution to aerosol concentrations in Europe is between 2% and 3% with annual PBAB emissions of 233 Gg. This figure can be compared to global scale emissions for PBAPs in the range from 56 Tg annually to a value as high as 1,000 Tg.

Although indoor environments are considered to be protective, they can become contaminated with particles that present different and sometimes more serious risks than those related to outdoor exposures, when their concentrations exceed recommended maximum limits. These are TS > 1,000 CFUs/m^3 set by the National Institute of Occupational Safety and Health (NIOSH), 1,000 CFUs/m^3 (ACGC) with the culturable count for total bacteria not to exceed 500 CFUs/m^3 (Cox and Wathes 1995).

Problems

5.1 *Calculate the critical radius of a droplet which is formed from homogeneous nucleation when the saturation ratio is equal to (a) 4.0 and (b) 1.10. The particle density and temperature are equal in both cases ($\rho = 1.0$ g cm^{-3} and $T = 298.15$ K). Calculate the homogeneous nucleation rate in both cases. If the temperature decreases from $T = 298.15$ K to $T = 283.15$ K what is the new nucleation rate?*

5.2 *Calculate the variation of the number concentration and particle size which occurs for a time interval of 10 min for particles with diameter of 0.8 μm and with initial number concentration 10^8/cm^3. It is supposed that coagulation of particles occurs with a specific size under standard conditions of pressure and temperature.*

References

Akhter, M. S., Chughtai, A. R., & Smith, D. M. (1985). The structure of hexane soot I: Extraction studies. *Appl. Spectrosc.* 39,154–167.
Cachier, H. (1998) *Carbonaceous Combustion Aerosols.* In: Harrison, R. M., van Grieken, R. (eds.) Atmospheric particles. John Wiley, New York.
Cox, C. S., & Wathes, C. M. (1995). *Bioaerosols handbook.* NY: Lewis Publishers.
Debenedetti, P. G. (1996). *Metastable liquids: Concepts and principles.* NJ: Princeton University Press.
EPA, Air quality criteria for particular matter, (2003). EPA/600/p-99/002aD, U.S. Environmental Protection Agency.
Gelencer, A. (2004). *Carbonaceous Aerosol, volume Atmospheric and Oceangraphic Science Library Series, 30,* Springer ISBN 1–4020–2886–5.
Hinds, W. C. (1999). *Aerosol technology.* NY: Wiley-Inter science.
IPCC (2007). Climate Change 2007: *The Scientific Basis. Contribution of Working Group I to the Fourth Assessment Report of the Intergovernmental Panel on Climate Change.*
Lazaridis, M., Kulmala, M., & Laaksonen, A. (1991). Binary heterogeneous nucleation of a water-sulphuric acid system: the effect of hydrate interaction. *Journal of Aerosol Science, 22,* 823–830.
Meng, Z. Y., & Seinfeld, J. H. (1996). Time scales to achieve atmospheric gas-aerosol equilibrium for volatile species. *Atmospheric Environment, 30*(16), 2889–2900.
Nevalainen, A., Pastuszka, J., Liebhaber, F., & Willeke, K. (1991). Performance of bioaerosol samplers: collection characteristics and sampler design considerations. *Atmospheric Environment, 26A,* 531–540.

References

Schmauss, A. (1920). Die chemie des nebels der wolken und des regens. *Die Unschau, 24*, 61–63.

Seinfeld, J. H., & Pandis, S. N. (2006). *Atmospheric chemistry and physics* (2nd ed.). NY: John Wiley & Sons.

Wentzel, M., Gorzawski, H., Naumann, K.H., Saathoff, H., & Weinbruch, S. (2003). Transmission electron microscopial and aerosol dynamical characterization of soot aerosols. *Journal of Aerosol Science 34*, 1347–1370.

Whytlaw-Gray, R., Speakman, J. B., & Campbell, J. H. P. (1923). Smokes part I- a study of their behaviour and a method of determining the number of particles they contain. *Proceedings of the Royal Society A, 102*, 600–615.

Wilson, W. E., & Suh, H. H. (1997). Fine particles and coarse particles: concentration relationships relevant to epidemiologic studies. *Journal of the Air & Waste Management Association, 47*, 1238–1249.

Winiwarter, W., Bauer, H., Caseiro, A., & Puxbaum, H. (2009). Quantifying emissions of primary biological aerosol particle mass in Europe. *Atmospheric Environment, 43*, 1403–1409.

Chapter 6
Atmospheric Dispersion: Gaussian Models

Contents

6.1	Theories of Atmospheric Diffusion	201
6.2	Euler Description	202
6.3	Lagrange Description	203
6.4	Equations Describing the Concentration of Pollutants at Turbulent Conditions	204
	6.4.1 Diffusion Equation in Euler Description	204
	6.4.2 Diffusion Equation in Lagrange Description	205
	6.4.3 Solution of the Diffusion Equation for a Continuous Source with the Euler Methodology	206
6.5	Gaussian Model	207
	6.5.1 Limitations of the Gaussian Model	208
	6.5.2 Calculation of the σ_y and σ_z Coefficients. Stability Methodology	209
	6.5.3 Plume Rise	212
	6.5.4 Atmospheric Stability – Application to the Gaussian Models	216
6.6	Analytical Solutions of the Atmospheric Diffusion Equation	217
6.7	Two-Dimensional, Time Independent Line-Continuous Source with Changing Values of Velocity and Diffusion Coefficient	220
6.8	Characteristics of Plume Dispersion – Stability Conditions	223
6.9	Examples and Applications	226
6.10	Appendix 6.1 The Continuity Equation	230
References		232

Abstract A mathematical description of the concentration profile of different chemical compounds in the atmosphere is one of the applications of atmospheric models. The problem which one has to solve is to calculate the spatial and temporal concentration of air pollutants under specific emission conditions. In Chap. 6 the dispersion of air pollutants is examined with the analytical Gaussian approach. The Euler and Lagrange descriptions are used for solution of the continuity equation. Furthermore the limitations of the Gaussian approach are studied together with the calculation of Gaussian dispersion coefficients under different stability conditions. The plume rise from point emission sources is also studied together with the characteristics of plume dispersion.

6.1 Theories of Atmospheric Diffusion

The mathematical description of the concentration profile of different chemical compounds in the atmosphere is one of the applications of atmospheric models. The problem which someone has to solve is to calculate the spatial and temporal concentration of air pollutants under specific emission conditions. Of course in order to solve this problem it is necessary to have a detailed description of the pollutant emissions, the land topography, the reactivity of the pollutants and the transport conditions in the atmosphere. The accurate determination of the factors which control the concentration of pollutants in the ambient air such as the emission sources, the pollutant transport (meteorological conditions) and the surface topography define how realistic is the solution of the problem.

In the literature, several categories of models have been used such as deterministic, statistical and stochastic for the solution of problems for pollutant dispersion in the atmosphere. Numerical and analytical solutions have been applied for deterministic models. Numerical solutions have been used widely but require extensive input data and computational time. Stochastic Gaussian models have been used also for air pollution control.

There are two general approaches for the description of turbulent dispersion in the atmosphere, the Euler and Lagrange approaches. In the Euler approach the system is examined in fixed coordinates and is the most popular approach for the examination of mass and heat in the atmosphere. In the Lagrangian approach the concentration changes are described in relation to a non-fixed coordinate system (e.g. air). However, these two approaches can be related to each other as described in the current chapter.

6.2 Euler Description

The concentration of every chemical compound versus time has to satisfy the mass equilibrium inside the contained volume. This is the equation of continuity and its derivation is presented in Appendix 6.1 of this chapter. The change of mass inside the air volume due to convection is balanced with mass transfer which is produced from chemical reactions, molecular diffusion and emissions. The mathematical description of this procedure for each chemical species i is given by the following equation for the concentration c_i:

$$\frac{\partial c_i}{\partial t} + \frac{\partial}{\partial x_j}(u_j\, c_i) = D_i \frac{\partial^2 c_i}{\partial x_j \partial x_j} + R_i(c_1, ..., c_N, T) + S_i(\vec{x}, t) \qquad (6.1)$$

where u_j is the j constituent of the velocity of the medium (e.g. air), D_i is the coefficient of molecular diffusion of the chemical species i, R_i is the chemical reaction production rate of the chemical species i and S_i is the emission rate of the chemical species i at the location $\vec{x}$ at time t.

It has to be noted that the concentration c_i has to fulfill the continuity equation and the velocities u_j and the temperature T to fulfill the Navier–Stokes equations which are related through the functions u_j, c_i and T with the continuity equation and the ideal law of gasses. This matter has been studied in detail in Chap. 3.

It is supposed that the air pollutants do not change the meteorological conditions and the velocity can be expressed with two terms, the mean value ($\overline{u_j}$) and the stochastic part (u'_j) of the velocity, and therefore: $u_j = \overline{u_j} + u'_j$ (see Chap. 2). In the following subsections a solution of the continuity equation is constructed. By substituting the velocity in Eq. (6.1) we can derive:

$$\frac{\partial c_i}{\partial t} + \frac{\partial}{\partial x_j}\left((\overline{u_j} + u'_j)\, c_i\right) = D_i \frac{\partial^2 c_i}{\partial x_j \partial x_j} + R_i(c_1,\ldots,c_N,T) + S_i(\vec{x},t) \quad (6.2)$$

The concentration can also be expressed as two terms in the same way as the velocity: $c_i = \langle c_i \rangle + c'_i$. It is valid that $\langle c'_i \rangle = 0$ and therefore:

$$\frac{\partial \langle c_i \rangle}{\partial t} + \frac{\partial}{\partial x_j}(\overline{u_j}\langle c_i \rangle) + \frac{\partial}{\partial x_j}\langle u'_j c'_i \rangle = D_i \frac{\partial^2 \langle c_i \rangle}{\partial x_j \partial x_j} + R_i\left(\langle c_1 \rangle + c'_1,\ldots,\langle c_N \rangle + c'_N, T\right) + S_i(\vec{x},t)$$

(6.3)

Higher orders of Eq. (6.3) where higher terms such as $\langle u'_j u'_k c' \rangle$ are introduced can be made with multiplication of Eq. (6.3) with u'_j, subtracting the previous equation and then taking the mean value. In any case there will be more unknown parameters than the number of equations and this is referred to in the literature as the closure problem of the Euler description. Additional unknown terms appear when there are nonlinear chemical reactions.

6.3 Lagrange Description

In the Lagrange methodology the study is focused on the behavior of fluid particles. The movement of a particle is described through its trajectory. We introduce the probabilistic term $\psi(\vec{x},t)$, where $\psi(x_1,x_2,x_3,t)dx_1\,dx_2\,dx_3 = \psi(\vec{x},t)d\vec{x}$ is the probability of a particle to be located at time t in a volume which is located between x_1 until $x_1 + dx_1$, x_2 until $x_2 + dx_2$, and x_3 until $x_3 + dx_3$. From the definition of the function of the probability density it can be concluded that:

$$\int_{-\infty}^{\infty}\int_{-\infty}^{\infty}\int_{-\infty}^{\infty} \psi(x,t)\,dx = 1 \quad (6.4)$$

The probability density ($Q\left(\vec{x},t\mid \vec{x}',t'\right)$) can be expressed as a product of two terms, the probability density of a particle which is located at the point x', at time t' to be moved to the point x at time t, and the probability density of a particle which is located at the point x', at time t' to be integrated for all the initial points x'. As a result the probability is expressed as:

$$\psi(\vec{x},t) = \int_{-\infty}^{\infty} \int_{-\infty}^{\infty} \int_{-\infty}^{\infty} Q(\vec{x},t|\vec{x}',t') d\vec{x}' \tag{6.5}$$

Equation (6.5) for the probability density is expressed for one particle. For a number of particles m which are located initially in space the probability density is expressed with the summation of the probability density of all particles as:

$$\langle c(\vec{x},t) \rangle = \sum_{i=1}^{m} \psi_i(\vec{x},t) \tag{6.6}$$

Therefore expressing the $\psi(\vec{x},t)$ in relation to the initial particle concentration and the spatial and temporal source distribution $S(\vec{x},t)$ of particles we get that:

$$\langle c(\vec{x},t) \rangle = \int_{-\infty}^{\infty} \int_{-\infty}^{\infty} \int_{-\infty}^{\infty} Q(\vec{x},t|\vec{x}_o,t_o) \langle c(\vec{x}_o,t_o) \rangle d\vec{x}_o$$
$$+ \int_{-\infty}^{\infty} \int_{-\infty}^{\infty} \int_{-\infty}^{\infty} \int_{t_o}^{t} Q(\vec{x},t|\vec{x}',t') S(\vec{x}',t') dt' d\vec{x}' \tag{6.7}$$

The Eq. (6.7) is the basic equation of Lagrange which examines the mean particle concentration in a turbulent medium. The first term at the right side of Eq. (6.7) refers to the concentration at time t_o whereas the second term is the concentration resulting from emissions at time intervals between t_o and t. The problem arising from Eq. (6.7) is that it is difficult to calculate the probability density since there is no detailed knowledge of the turbulent properties except specific cases. In addition, the equation is valid in the case that the particles are not reacting chemically.

6.4 Equations Describing the Concentration of Pollutants at Turbulent Conditions

6.4.1 Diffusion Equation in Euler Description

As discussed previously the description of turbulent diffusion with the Euler method results in the closure problem due to the existence of new variables $\langle u'_j c'_i \rangle$, j = 1,2,3, as well as other variables which arise with the term $\langle R_i \rangle$ including nonlinear chemical reactions. It is necessary to introduce a relationship between terms $\langle u'_j c'_i \rangle$ with the average concentration $<c_i>$ and this is done with the method of mixing length which is described in Chap. 7.

6.4 Equations Describing the Concentration of Pollutants at Turbulent Conditions

In the mixing length methodology it is supposed that:

$$\langle u_j' c' \rangle = -K_{jk} \frac{\partial \langle c \rangle}{\partial x_k} \quad j = 1, 2, 3 \tag{6.8}$$

where the term K_{jk} is called eddy diffusivity and Eq. (6.8) is the mathematical expression of the K theory. Choosing the axis of the coordination system to coincide with the main axis of the diffusion strain K_{jk} has as a result only the terms K_{11}, K_{22}, K_{33} to be different to zero and therefore:

$$\langle u_j' c' \rangle = -K_{jj} \frac{\partial \langle c \rangle}{\partial x_j} \tag{6.9}$$

For the diffusion equation two main assumptions are made: (a) the molecular diffusion is negligible in relation to diffusion due to turbulence:

$$D_i \frac{\partial^2 \langle c \rangle}{\partial x_j \partial x_j} \ll \frac{\partial}{\partial x_j} \langle u_j' c' \rangle \tag{6.10}$$

and (b) that the atmosphere is incompressible:

$$\frac{\partial \overline{u_j}}{\partial x_j} = 0 \tag{6.11}$$

After the above assumptions the atmospheric diffusion equation can be written as:

$$\frac{\partial \langle c_i \rangle}{\partial t} + \overline{u_j} \frac{\partial \langle c_i \rangle}{\partial x_j} = \frac{\partial}{\partial x_j} \left(K_{jj} \frac{\partial \langle c_i \rangle}{\partial x_j} \right) + R_i(\langle c_1 \rangle, ..., \langle c_N \rangle, T) + S_i(\vec{x}, t) \tag{6.12}$$

It is possible to prove mathematically that Eq. (6.12) describes the atmospheric diffusion including chemical reactions, when the chemical reactions are slow (slower than the pollution transport) and the source distribution normal.

6.4.2 Diffusion Equation in Lagrange Description

Here we examine the solution of the diffusion equation in the Lagrange coordination system. The Eq. (6.3) is used without including the terms of molecular diffusion and chemical reactions. The solution of the equation is studied for one-dimensional flow. It is assumed that there is one source $S(x,t)$ and therefore the diffusion equation can be expressed as:

$$\frac{\partial c}{\partial t} + \frac{\partial}{\partial x}(u\,c) = S(x,t) \qquad (6.13)$$

The velocity u is a random quantity and the same applies for the value of the concentration c. Therefore we need to find an analytical solution for the concentration c for specific values of u. It is supposed that u is independent of x and is dependent only on t. It is supposed also that the probability density for u(t) is Gaussian and as a result can be expressed as:

$$p_u(u) = \frac{1}{(2\pi)^{1/2}\sigma_u} \exp\left(-\frac{(u-\bar{u})^2}{2\,\sigma_u^2}\right) \qquad (6.14)$$

with average velocity value $\bar{u}$ and dispersion σ_u. It is also assumed that u(t) is a stationary random process, where the correlation for the velocity is given by the expression:

$$\langle (u(t) - \bar{u})(u(\tau) - \bar{u}) \rangle = \sigma_u^2 \exp(-b\,|t-\tau|) \qquad (6.15)$$

The solution of Eq. (6.15) for variable source S(t) at x = 0 is given by:

$$c(x,t) = \int_0^t \delta(x - X(t,\tau))\,S(\tau)\,d\tau \quad \text{where} \quad X(t,\tau) = \int_\tau^t u(t')\,dt' \qquad (6.16)$$

where $\delta(x)$ is a delta function ($\delta(0) = 1$, $\delta(x) = 0$ when $x \neq 0$).

The above expression denotes that the maximum correlation between velocities at two different times occurs when the times are equal and is given by σ_u^2. When the distance is increased between the velocities u(t) and u(τ) then the correlation decreases exponentially with characteristic time 1/b. The concept of a stationary random process refers to the fact that the statistical properties of velocity u at two different times t and τ are dependent on the difference $t - \tau$ and not on the t and τ separately.

The solution of Eq. (6.16) after a series of mathematical calculations gives:

$$\langle c(x,t) \rangle = \frac{1}{(2\pi)^{1/2}\sigma_x(t)} \exp\left(-\frac{(x - \bar{u}t)^2}{2\,\sigma_x^2(t)}\right) \qquad (6.17)$$

In three dimensions the solution is expressed as:

$$\langle c(x,y,z,t) \rangle = \frac{1}{(2\pi)^{3/2}\sigma_x(t)\,\sigma_y(t)\,\sigma_z(t)} \exp\left(-\frac{(x-\bar{u}t)^2}{2\,\sigma_x^2(t)} - \frac{y^2}{2\,\sigma_y^2(t)} - \frac{z^2}{2\,\sigma_z^2(t)}\right)$$
$$(6.18)$$

6.4.3 Solution of the Diffusion Equation for a Continuous Source with the Euler Methodology

In this paragraph we will demonstrate a solution of the diffusion equation in case of a point source which emits pollutants with rate S at the beginning of the coordination system for an infinitive medium with velocity $\bar{u}$ in the direction x. From Eq. (6.12) and assuming K_{xx}, K_{yy} and K_{zz} as constant quantities we get that:

$$\frac{\partial \langle c \rangle}{\partial t} + \bar{u}\frac{\partial \langle c \rangle}{\partial x} = K_{xx}\frac{\partial^2 \langle c \rangle}{\partial x^2} + K_{yy}\frac{\partial^2 \langle c \rangle}{\partial y^2} + K_{zz}\frac{\partial^2 \langle c \rangle}{\partial z^2} \qquad (6.19)$$

under conditions $\langle c(x,y,z,0)\rangle = S\,\delta(x)\,\delta(y)\,\delta(z)$ and $\langle c(x,y,z,0)\rangle = 0$ when $x, y, z \to \pm\infty$.

The solution of the above equation is:

$$\langle c(x,y,z,t)\rangle = \frac{S}{8\,(\pi t)^{3/2}\,(K_{xx}\,K_{yy}\,K_{zz})^{1/2}} \exp\left(-\frac{(x-\bar{u}t)^2}{4K_{xx}t} - \frac{y^2}{4K_{yy}t} - \frac{z^2}{4K_{zz}t}\right)$$
$$(6.20)$$

and, examining Eqs. 6.20 and 6.18, it can be concluded that the solutions of Euler and Lagrange are identical in case that $\sigma_x^2 = 2K_{xx}\,t$, $\sigma_y^2 = 2K_{yy}\,t$ and $\sigma_z^2 = 2K_{zz}\,t$.

6.5 Gaussian Model

For a description of pollutant dispersion in the atmosphere we use in several cases models which assume that the pollutant concentration after their release from the source have a canonical distribution (Gaussian distribution) as examined in the previous paragraphs. These models are called Gaussian models.

The Gaussian models are extensively used since they describe realistically, based on comparison with field data, the pollutant dispersion at a local level for a stationary atmosphere. In a Cartesian orthogonal coordination system with origin at the bottom of the point source, with direction xx' the wind direction, yy' the direction vertical to the wind direction at the surface and zz' the vertical one, the concentration of pollutants at the position (x,y,z) can be described from the equation (Seinfeld and Pandis 2006):

$$c(x,y,z) = \frac{Q}{2\pi\sigma_y\sigma_z u} \exp\left[-\frac{y^2}{2\sigma_y^2}\right] \exp\left[-\frac{(H-z)^2}{2\sigma_z^2}\right] \qquad (6.21)$$

where, c(x,y,z) is the concentration of pollutant at point (x,y,z) expressed in σε μg/m^3, Q the emission rate expressed in μg/s, u the wind velocity (m/s), σ_y and σ_z the typical pollutant distribution deviations at axis yy' and zz' respectively and H the effective height of plume emission. The conditions which have to obtain in order to apply a Gaussian model to give realistic results are the following:

1. The pollutant emissions are continuous or at least the emissions occur for a time interval which is larger than the travel time of the pollutant from the source to the receptor point which the concentration has to be derived.
2. The pollutants are not reacting chemically in the atmosphere since the Gaussian model is not including chemical reactions.
3. At the wind direction the transport process is dominant to the turbulent dispersion.
4. The aerosol diameter to be smaller than 20 μm in order for their residence time in the atmosphere to be larger than the time intervals which are studied with the Gaussian models.
5. The atmosphere to be in a stationary condition in relation to the meteorological parameters for the time interval of transport from the pollution source to the receptors. This condition is satisfied in most cases. For example if there is transport of pollutants to distances smaller than 10 km and the wind velocity is 5 m/s, then the transport time is close to 35 minutes and for these time intervals the meteorological conditions are usually stable.

With the Gaussian model are calculated average pollutant concentrations at several points around the emission source. There is a difference in the pollutant concentration distribution between average and instant values.

The average concentration has a Gaussian distribution at the cross direction and to a lesser extent in the vertical direction. The question which arises is how large the time interval has to be at which the average concentration values have to be obtained in order for the average concentrations to have a canonical distribution. In the majority of cases average values of hourly concentrations are calculated. However, there are indications that the Gaussian distribution is observed also in cases where the average values are calculated for time intervals of ten's of minutes.

From the equation which describes the Gaussian plume it can be seen that the model cannot be applied in case the wind velocity tends to zero. In cases of very low wind velocity, it is set equal to 0.5 or 1 m/s based on the model being used.

6.5.1 Limitations of the Gaussian Model

6.5.1.1 Conditions of Intense Instability

When there are very unstable conditions in the atmosphere the air flux inside the boundary layer is turbulent and upward and downward air movements are present. This has as a result the pollutants emitted from a source to be transported quickly to the upper layers of the boundary layer or close to the surface. The upward movements have higher velocity than the downward transport and cover smaller area.

6.5 Gaussian Model

Therefore the pollutants have higher probability to be located in a downward air movement with a final result the main axis of the plume to move to the surface. In total the plume moves to the surface or to the base of the boundary layer. This fact is not considered in the Gaussian model and requires caution in application of the models in very unstable conditions.

6.5.1.2 Emissions Close to the Surface

When the emissions occur close to the surface, then difficulties arise in the application of Gaussian models due to the variation of wind velocity and the turbulence structure. More precisely the wind velocity close to the surface changes to logarithmic versus height and is difficult to use a characteristic velocity for the whole boundary layer. Furthermore the turbulent flux is not homogeneous in the vertical direction and deviates from the Gaussian distribution. The cross sectional distribution of emissions close to the surface continues to be canonical.

6.5.2 Calculation of the σ_y and σ_z Coefficients. Stability Methodology

The dispersion coefficients (σ) are determined from field measurements. The dispersion coefficients are dependent on the topography of the area of interest, the atmospheric stability and the distance and time from the start of the dispersion (Schnelle and Dey 1999).

The values in the scientific literature for the functions σ_y and σ_z versus distance from the source and stability conditions have been calculated during field experiments from the period between 1950 and 1960 (Hanna et al. 1981). The area of the experiment was homogeneous, the emissions were performed close to the surface and the measurements were done at distances smaller than 1 km from the source. These field experiments resulted in 1961 in the Pasquill curves (Fig. 6.1). The semi-continuous lines are due to the fact that the experiments were performed at a distance of 1 km away and the continuation of the lines is valid in ideal conditions. Probably the actual curves deviate from these ideal lines since the conditions in the atmosphere are almost never ideal.

Analytical expressions for the functions σ_y and σ_z (dispersion coefficients for the directions x and y) have been presented in the scientific literature. One of the first mathematical expressions was given during 1967 by Smith who showed hourly measurements at distances up to 10 km from a source with height of 108 meters. According to Smith's measurements the following expressions for σ_y and σ_z are given:

$$\sigma_y = a x^b \quad \text{and} \quad \sigma_z = c x^d \tag{6.22}$$

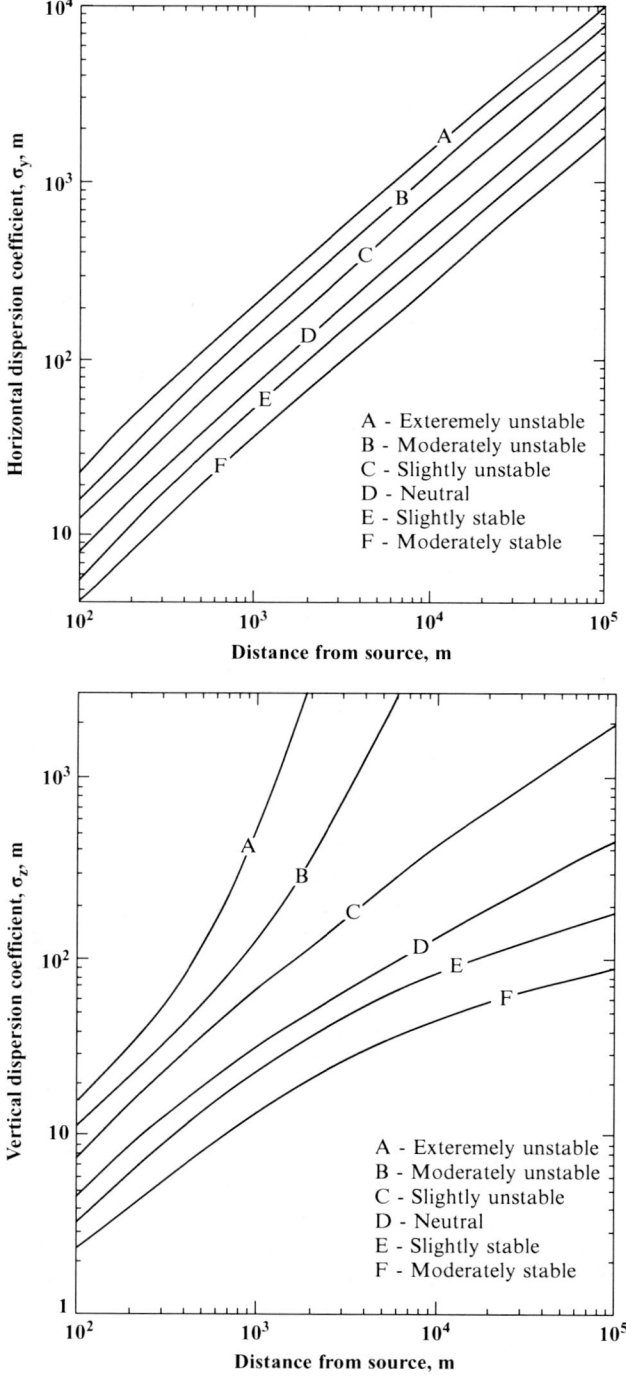

Fig. 6.1 Curves of the dispersion functions σ_y and σ_z based on the Pasquill formulation (adapted from Hanna et al. 1981)

6.5 Gaussian Model

More recent data have been presented by Briggs (Hanna et al. 1981) (Table 6.1). The values for the coefficients a, b, c, and d can be obtained from values given by the national laboratory at Brookhaven, USA (Table 6.2).

Pasquill during 1961 proposed a turbulent scheme which is known as the Pasquill-Gifford methodology. Calculations for the vertical and horizontal pollutant dispersion have been evaluated in relation to stability classes A up to F. Graphical representation of the Pasquill-Gifford coefficients are shown in Fig. 6.1, whereas mathematical functions are given in Table 6.1.

Lately the US Environmental Protection Agency (US EPA) has proposed an alternative methodology for the calculation of σ_y at an open field:

$$\sigma_y = \frac{1000 \, x \, \tan(TH)}{2.15} \tag{6.23}$$

where x is the distance along the wind direction (km) whereas σ_y is given in meters. The equations for the TH value from Eq. (6.23) are given in Table 6.3. For the function σ_z the expression $\sigma_z = a x^b$ is used where x is given in kilometers and σ_z in meters. Graphical representation of the Gaussian dispersion of pollutants from a chimney is shown in Fig. 6.2, where the horizontal and vertical dispersion of pollutants is shown.

Table 6.1 Proposed relationships by Briggs (Hanna et al., 1981) for the dispersion parameters $\sigma_y(x)$ and $\sigma_z(x)$ ($10^2 < x < 10^4$ m)

Stability conditions	$\sigma_y(x)$	$\sigma_z(x)$
Open field conditions		
A	$0.22 \times (1 + 0.0001 \, x)^{-1/2}$	$0.20 \times$
B	$0.16 \times (1 + 0.0001 \, x)^{-1/2}$	$0.12 \times$
C	$0.11 \times (1 + 0.0001 \, x)^{-1/2}$	$0.08 \times (1 + 0.0002 \, x)^{-1/2}$
D	$0.08 \times (1 + 0.0001 \, x)^{-1/2}$	$0.06 \times (1 + 0.0015 \, x)^{-1/2}$
E	$0.06 \times (1 + 0.0001 \, x)^{-1/2}$	$0.03 \times (1 + 0.0003 \, x)^{-1/2}$
F	$0.04 \times (1 + 0.0001 \, x)^{-1/2}$	$0.016 \times (1 + 0.0003 \, x)^{-1/2}$
Urban conditions		
A-B	$0.32 \times (1 + 0.0004 \, x)^{-1/2}$	$0.24 \times (1 + 0.001 \, x)^{1/2}$
C	$0.22 \times (1 + 0.0004 \, x)^{-1/2}$	$0.20 \times$
D	$0.16 \times (1 + 0.0004 \, x)^{-1/2}$	$0.14 \times (1 + 0.0003 \, x)^{-1/2}$
E-F	$0.11 \times (1 + 0.0004 \, x)^{-1/2}$	$0.08 \times (1 + 0.00015 \, x)^{-1/2}$

Table 6.2 Proposed parameters for the relationships $\sigma_y = a x^b$ and $\sigma_z = c x^d$

Stability conditions	Parameters			
	a	b	c	d
B_2	0.40	0.91	0.41	0.91
B_1	0.36	0.86	0.33	0.86
C	0.32	0.78	0.22	0.78
D	0.31	0.71	0.06	0.71

Table 6.3 Equations used for the calculation of the dispersion parameter σ_y for open field

Stability conditions	Equation for TH
A	TH = 24.1670-2.5334 ln(x)
B	TH = 18.3330-1.8096 ln(x)
C	TH = 12.5000-1.0857 ln(x)
D	TH = 8.3330-0.72382 ln(x)
E	TH = 6.2500-0.54287 ln(x)
F	TH = 4.1667-0.36191 ln(x)

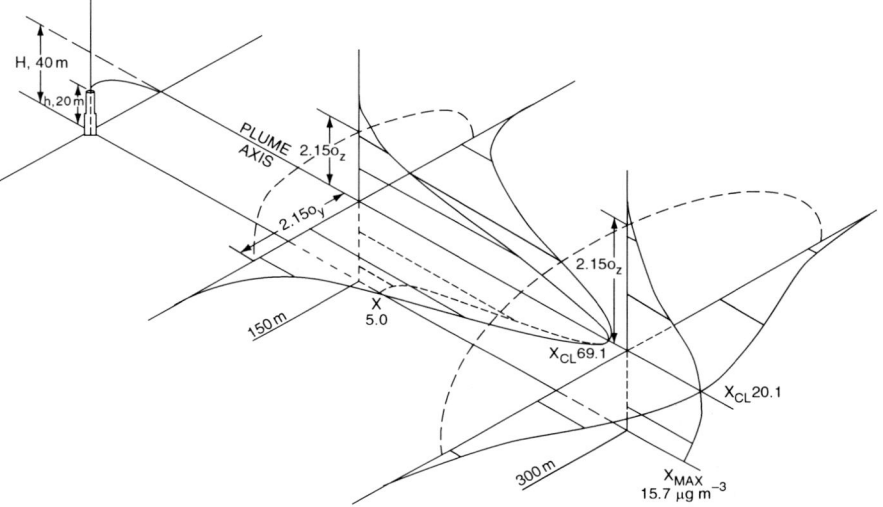

Fig. 6.2 Gaussian description of the emission of pollutants from a point source

6.5.3 Plume Rise

When the emissions from a stack have a higher temperature than the temperature of the environment, then the plume is rising due to thermal motion. In addition plume rise is also occurring if the emissions exit the stack with high velocity. Some plumes rise also due to different characteristics of the pollutants (density and composition). These gaseous species have positive upward transport (buoyant plume) if they are thinner than the air and a downward transport if they are heavier than the air

The rising of warm gaseous plumes can be described in the following stages (Fig. 6.3):

1. Thermal stage, which is characterized by:
   - Mixing due to initial turbulence,
   - Moderate ascending and planar shape,
   - Application of a linear thermal model and finally

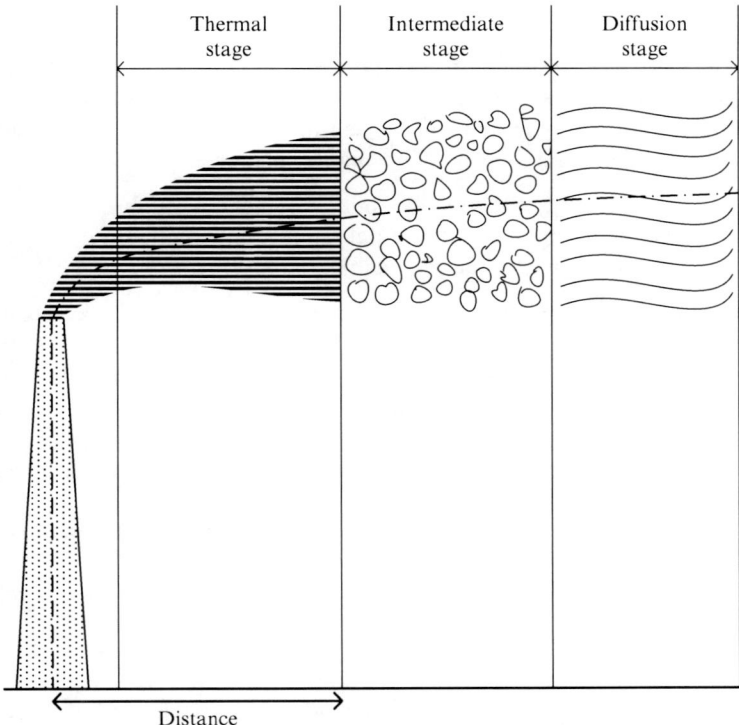

Fig. 6.3 Graphical description of plume rise. (Adapted from Schnelle and Dey 1999)

- Description of the phenomenon by the Briggs methodology which assumes that the plume reaches its maximum height at this stage.

2. Intermediate stage, which is characterized by:

   - Dominance of the atmospheric turbulence,
   - Breaking of the plume into small compartments and
   - A stepwise increase of the plume diameter.

3. Diffusion stage, which is characterized by:

   - Dominance of the atmospheric turbulence diffusion,
   - Plume formation– even larger diffusion and
   - Relatively slow development.

The objective to emit pollutants from an increased height has as a result to emit pollutants with temperatures higher than the ambient temperature and with high momentum. Therefore the effective stack height (h) is the sum of the normal stack height (h_s) and of the plume rise (Δh) and is expressed as:

$$h = h_s + \Delta h \tag{6.24}$$

The calculation of the plume rise is quite difficult task since it is dependent on many parameters such as meteorological conditions, the temperature difference between the plume and the ambient air and the velocity of the plume emission. Based on the initial conditions the plume can be characterized as:

Buoyant plume with initial rise $>>$ initial momentum,
Forced plume with initial rise $\cong$ initial momentum and finally,
Jet with initial rise $<<$ initial momentum.

The plume rise is expressed from the equation: $\Delta h = \frac{E x^b}{\bar{u}^a}$. As an example the coefficients for neutral and non-stable conditions are given as $\alpha = 1$, $b = 0$ and $E = 7.4 \, (F \, h^2)^{1/3}$ where F is the coefficient of flux buoyancy and is equal to $F = g \, d^2 \, V_s \, (T_s - T_a)/4 T_s \, (m^4 s^{-3})$ and g is the coefficient of the acceleration gravitational settling (m s^{-2}), d is the stack diameter (m), T_a is the air temperature at the height of the stack (T), T_s is the temperature of the emitted gas from the stack (T) and V_s is the exit velocity from the stack (m s^{-1}).

An extensively used expression for calculations of Δh is the Holland expression which for neutral conditions can be expressed as:

$$\Delta h = \frac{2 V_s r_s}{u} \left[1.5 + 2.68 \times 10^{-2} P \left(\frac{T_s - T_a}{T_s} \right) 2 r_s \right] \tag{6.25}$$

where r_s is the stack radius at the exit (m) and P is the atmospheric pressure (kP$_a$).

At atmospheric conditions, which are not neutral, the Δh is multiplied with a correction coefficient CF on which its value is dependent from the stability conditions and is expressed as:

$$\mathrm{CF} = (\mathrm{St}/10) + 0.70 \tag{6.26}$$

where the terms CF and St for different stability conditions of Pasquill is shown in Table 6.4.

A more detailed study on plume rise can be focused on the cause of the rise (momentum or thermal rise) as well as on the stability conditions of the atmosphere. In the following sections we will study the plume rise under variable conditions.

Table 6.4 Correction coefficients of the plume rise formula of Holland for different stability conditions in the atmosphere

Stability class	St	CF
A	5.0	1.20
B	4.0	1.10
C	3.5	1.05
D	3.0	1.00
E	2.0	0.90
F	1.0	0.80

6.5.3.1 Plume Rise due to Initial Momentum Under Neutral or Unstable Stability Conditions in the Atmosphere

This condition is dominant for values $V_S/U > 4$. The plume rise is calculated from the equation:

$$\Delta H_m = 3D_s V_S/U \qquad (6.27)$$

6.5.3.2 Thermal Rise Under Neutral or Unstable Stability Conditions in the Atmosphere

For the calculation of final plume rise it is necessary to know the pollutant flux due to the temperature difference:

$$F = g\ V_S D_S^2 (T_S - T)/4T_S \qquad (6.28)$$

where T_S is the temperature of the emitted gasses and T is the air temperature (K). The plume will obtain the final height at distance x_f from the stack where the turbulent dispersion prevails. For $F < 55$ it is $x_f = 0.049 \times F^{(5/8)}$ and $\Delta H_b = 21.425 \times F^{(3/4)}/U$, whereas for $F \geq 55$ it is $x_f = 0.119 \times F^{(2/5)}$ and $\Delta H_b = 38.71 \times F^{(3/5)}/U$. In the case that $\Delta H_m > \Delta H_b$, it is adopted the case (a).

6.5.3.3 Stable Conditions

Under stable conditions, the stability parameter s is calculated with the expression:

$$s = g/T\ (\delta\theta/\delta z) \qquad (6.29)$$

where $\delta\theta/\delta z = 0.02 K/m$ for a slightly stable atmosphere and $\delta\theta/\delta z = 0.035 K/m$ for stable atmosphere.

6.5.3.4 Plume Rise due to Momentum Under Stable Conditions

In the case the temperature of the emitted gasses is lower than the air temperature ($T_S < T$), the plume rise due to momentum is dominant and is calculated from the expression:

$$\Delta H_m = 1.5[(V_S^2 D_S^2 T)/(4\ T_S U)]^{(1/3)} s^{(-1/6)} \qquad (6.30)$$

In case $\Delta H_m > \Delta H_b$, the case (a) applies.

6.5.3.5 Thermal Rise Under Stable Atmospheric Conditions

In the case the temperature of the emitted gasses is higher than the air temperature ($T_S > T$), the thermal plume rise is dominant and is calculated from the expression:

$$\Delta H_b = 2.6[F/Us]^{(1/3)} \tag{6.31}$$

The distance from the plume x_f (km) is given by:

$$x_f = 0.002071 \; Us^{(-1/2)} \tag{6.32}$$

The value of ΔH_b for calm conditions is calculated from the expression:

$$\Delta H_b = 4F^{(1/4)} s^{(-3/8)} \tag{6.33}$$

6.5.4 Atmospheric Stability – Application to the Gaussian Models

The concept of potential temperature assists the examination of the atmospheric stability. The potential temperature θ is determined when adiabatically dried air is moved from its original condition to pressure of 1,000 mb. From Chap. 2 it is shown that:

$$\theta = T\left(\frac{P_o}{P}\right)^{\left(\frac{\gamma-1}{\gamma}\right)} \tag{6.34}$$

where $P_o = 1{,}000$ mb.

The relative importance of the pollutant transport and turbulent flux due to mechanical shear stress can be examined with the help of Richardson's number. A stability parameter s can be defined analogous to the rate by which the atmospheric stability decreases the turbulence:

$$s = \frac{g}{T}\left(\frac{\Delta\Theta}{\Delta z}\right) \tag{6.35}$$

The turbulence is formed from mechanical shear force with rate $\left(\frac{du}{dz}\right)^2$. The Richardson's number is the ratio of the stability s to the turbulence production and is a stability parameter of the atmosphere and also a parameter for the turbulence occurrence. The Eq. (6.36) defines the Richardson number (Schnelle and Dey 1999):

$$Ri = \frac{g\left(\frac{\Delta\Theta}{\Delta z}\right)}{T\left(\frac{du}{dz}\right)^2} \tag{6.36}$$

Table 6.5 The Richardson's number and stability conditions in the atmosphere

Stability conditions	Richardson number	Atmospheric conditions
Stable	Ri > 0.25	Absence of vertical mixing, low air velocity, decrease of mechanical turbulence, small plume dispersion
Stable	0 < Ri < 0.25	The mechanical turbulence is decreased and there is stratification of the atmosphere
Neutral	Ri = 0	Presence of only mechanical turbulence
Unstable	−0.03 < Ri < 0	Mechanical turbulence and transport
Unstable	Ri < 0.04	The transport is dominant, low wind velocity, important vertical transport, quick plume dispersion vertically and horizontally

The atmospheric stability can be expressed as stable, unstable or neutral with the help of the Richardson number. Negative Richardson numbers denote that the transport is dominant, the air velocity low and there is strong vertical transport in an unstable atmosphere. The plume which is released from a source diffuses vertically and horizontally. With an increase of mechanical turbulence, the Richardson number tends to zero and the plume dispersion decreases and tends to neutral stability where $(d\theta/dz) = 0$. Finally, when the Richardson number becomes positive, the vertical mixing ceases and the mechanical turbulence decreases. There is stratification in the atmosphere and there occur a low vertical dispersion of the plume. Table 6.5 denotes the above conditions.

6.6 Analytical Solutions of the Atmospheric Diffusion Equation

A study of the general form of analytical Gaussian solutions for pollutant dispersion in the atmosphere is presented in this section. The concentration is expressed in relation to the plume emission potential and is inversely proportional to the mean wind velocity. The Gaussian models have parameterized the pollutant dispersion through the coefficients σ and K which express the mixing process as measured by the turbulence intensity in the Lagrange and Euler coordination systems respectively. The dispersion model with an application of the diffusion coefficients K results from the continuity equation as discussed in the previous sections. Similar results can be obtained with an application of the statistical turbulent theory from which a model can result with the same characteristics as the model with the diffusion coefficients K (Schnelle and Dey 1999).

The Gaussian model will be applied here under different ideal assumptions for the mean pollutant concentration of pollutants which are emitted from a point source under Gaussian distribution. Analytical Gaussian solutions will be presented with the Lagrange methodology. The mean pollutant concentration is given by the expression:

$$\langle c(\vec{x},t)\rangle = \int_{-\infty}^{\infty}\int_{-\infty}^{\infty}\int_{-\infty}^{\infty} Q(\vec{x},t|\vec{x_o},t_o)\langle c(\vec{x_o},t_o)\rangle d\vec{x}_o$$
$$+ \int_{-\infty}^{\infty}\int_{-\infty}^{\infty}\int_{-\infty}^{\infty}\int_{t_o}^{t} Q(\vec{x},t|\vec{x}',t') S(\vec{x}',t')\, dt'\, d\vec{x}' \qquad (6.37)$$

The Q (transition probability density) expresses the probability that a hypothetical particle which is located at a point x', y', z' at time t' will be at point x, y, z at time t. Under conditions of stationary, homogeneous turbulence, Q has a Gaussian form. In the case of mean wind velocity at direction x with $\bar{v}=\bar{w}=0$ at an infinity medium, the expression for Q can be written as:

$$Q(x,y,z,t|x',y',z',t') = \frac{1}{(2\pi)^{3/2}\sigma_x\sigma_y\sigma_z}$$
$$\times \exp\left(-\frac{(x-x'-\bar{u}(t-t'))^2}{2\sigma_x^2} - \frac{(y-y')^2}{2\sigma_y^2} - \frac{(z-z')^2}{2\sigma_z^2}\right) \qquad (6.38)$$

Since the surface effect is of interest for the vertical dispersion of pollutants, the expression of Q can be divided into two terms in order to calculate the influence at the vertical direction z:

$$Q(x,y,z,t|x',y',z',t') = \frac{1}{2\pi\sigma_x\sigma_y}\exp\left(-\frac{(x-x'-\bar{u}(t-t'))^2}{2\sigma_x^2} - \frac{(y-y')^2}{2\sigma_y^2}\right) Q_z(z,t|z',t') \qquad (6.39)$$

The $Q_z(z,t|z',t')$ is calculated for different cases (a) total reflection, (b) total absorption, (c) partial absorption for two forms of the boundary layer (H): (a) $0 \leq z \leq \infty$, (b) $0 \leq z \leq H$ without diffusion above H. In the following cases mathematical expressions are given for $Q_z(z,t|z',t')$ from continuous point sources (Gaussian plume) for homogeneous turbulence.

<u>Total reflection at z = 0</u>

The presence of the surface at z = 0 can be accounted for with the addition of a hypothetical source at the point z = −z' to the real source which is located at the point z = z' for z ⩾ 0. It also supposed that $0 \leq z \leq \infty$. The final expression for $Q_z(z,t|z',t')$ can be written as:

$$Q_z(z,t|z',t') = \frac{1}{(2\pi)^{1/2}\sigma_z}\left[\exp\left(-\frac{(z-z')^2}{2\sigma_z^2}\right) + \exp\left(-\frac{(z+z')^2}{2\sigma_z^2}\right)\right] \qquad (6.40)$$

Figure 6.4 shows the plume shape during the pollutant dispersion from a point source with the hypothesis of total reflection.

6.6 Analytical Solutions of the Atmospheric Diffusion Equation

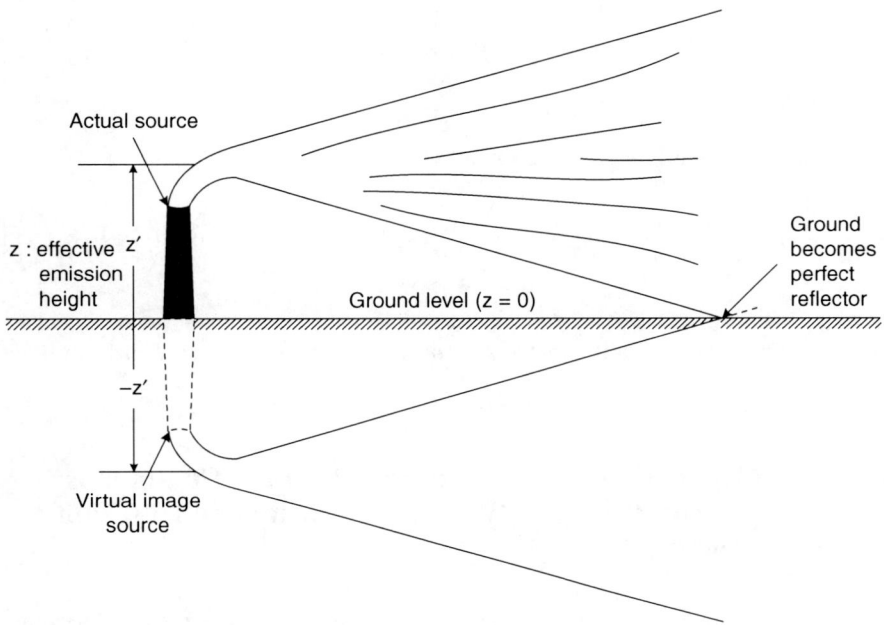

Fig. 6.4 Total reflection at $z = 0$

Total absorption at $z = 0$

In the case that the Earth's surface is a perfect absorber, the pollutant concentration at $z = 0$ is zero. The $Q_z(z,t|z',t')$ has a similar form as in the case of total reflection with the difference that subtraction of the hypothetical source at $z = -z'$ has to be performed. The result (for $0 \leqslant z \leqslant \infty$) is:

$$Q_z(z,t|z',t') = \frac{1}{(2\pi)^{1/2}\sigma_z}\left[\exp\left(-\frac{(z-z')^2}{2\sigma_z^2}\right) - \exp\left(-\frac{(z+z')^2}{2\sigma_z^2}\right)\right] \quad (6.41)$$

The mean pollutant concentration from a continuous source (total reflection) with intensity q at height h above the Earth's surface is given by the expression:

$$\langle c(x,y,z)\rangle = \lim_{t\to\infty}\int_0^t \frac{q}{(2\pi)^{3/2}\sigma_x\sigma_y\sigma_z}\exp\left(-\frac{(x-\bar{u}t')^2}{2\sigma_x^2} - \frac{y^2}{2\sigma_y^2}\right)$$
$$\left[\exp\left(-\frac{(z-h)^2}{2\sigma_z^2}\right) + \exp\left(-\frac{(z+h)^2}{2\sigma_z^2}\right)\right]dt' \quad (6.42)$$

The expression is calculated at the limit $\sigma_x \to 0$ (slender plume) with the following result:

$$\langle c(x,y,z)\rangle = \frac{q}{2\pi\bar{u}\sigma_y\sigma_z}\exp\left(-\frac{y^2}{2\sigma_y^2}\right)\times\left[\exp\left(-\frac{(z-h)^2}{2\sigma_z^2}\right)+\exp\left(-\frac{(z+h)^2}{2\sigma_z^2}\right)\right]$$
(6.43)

For a surface with total absorption at z = 0 we get:

$$\langle c(x,y,z)\rangle = \frac{q}{2\pi\bar{u}\sigma_y\sigma_z}\exp\left(-\frac{y^2}{2\sigma_y^2}\right)\times\left[\exp\left(-\frac{(z-h)^2}{2\sigma_z^2}\right)-\exp\left(-\frac{(z+h)^2}{2\sigma_z^2}\right)\right]$$
(6.44)

Table 6.6 shows several expressions for the pollutant concentration from point sources.

6.7 Two-Dimensional, Time Independent Line-Continuous Source with Changing Values of Velocity and Diffusion Coefficient

The diffusion from a continuous line source which emits vertically to the wind direction can be described by the following equation (Hanna et al. 1981):

$$u\frac{\partial C}{\partial x} = \frac{\partial(K_z\,\partial C/\partial z)}{\partial z}$$
(6.45)

with the following boundary conditions:

$$C \to 0 \quad \text{when} \quad x = z \to \infty$$
(6.46)

$$C \to \infty \quad \text{when} \quad x = z \to 0$$
(6.47)

$$K_z\frac{\partial C}{\partial z} \to 0 \quad \text{when} \quad z \to 0 \quad \text{and} \quad x > 0$$
(6.48)

$$\int_0^\infty u\,C\,dz = Q_l \quad \text{when} \quad x > 0.$$
(6.49)

where Q_l is the emission rate of the line source. The third boundary condition confirms that there is no flux of gaseous pollutant at the lowest region (e.g. surface).

The solutions from the above equation system refer only to sources which are located at the surface. In the literature the following expressions are given:

$$K_z = K_1\left(\frac{z}{z_1}\right)^n$$
(6.50)

$$u = u_1\left(\frac{z}{z_1}\right)^m$$
(6.51)

6.7 Two-Dimensional, Time Independent Line-Continuous Source

Table 6.6 Expressions for pollutant dispersion from point Gaussian sources (adapted from Seinfeld and Pandis 2006)

Mean concentration	Hypothesis
Gaussian puff formula $\langle c(x,y,z;t) \rangle = \dfrac{q}{(2\pi)^{3/2}\sigma_x\sigma_y\sigma_z} \exp\left(-\dfrac{(x-x'-\bar{u}(t-t'))^2}{2\sigma_x^2} - \dfrac{(y-y')^2}{2\sigma_y^2}\right)$ $\times \left[\exp\left(-\dfrac{(z-z')^2}{2\sigma_z^2}\right) + \exp\left(-\dfrac{(z+z')^2}{2\sigma_z^2}\right)\right]$	Total reflection at $z = 0$ $\bar{u} = (\bar{u}, 0, 0)$ $S = q\,\delta(x-x')\,\delta(y-y')\,\delta(z-z')\,\delta(t-t')$ $0 \leqslant z \leqslant \infty$
Gaussian puff formula $\langle c(x,y,z;t) \rangle = \dfrac{q}{(2\pi)^{3/2}\sigma_x\sigma_y\sigma_z} \exp\left(-\dfrac{(x-x'-\bar{u}(t-t'))^2}{2\sigma_x^2} - \dfrac{(y-y')^2}{2\sigma_y^2}\right)$ $\times \left[\exp\left(-\dfrac{(z-z')^2}{2\sigma_z^2}\right) - \exp\left(-\dfrac{(z+z')^2}{2\sigma_z^2}\right)\right]$	Total absorption at $z = 0$ $\bar{u} = (\bar{u}, 0, 0)$ $S = q\,\delta(x-x')\,\delta(y-y')\,\delta(z-z')\,\delta(t-t')$ $0 \leqslant z \leqslant \infty$
Gaussian puff formula $\langle c(x,y,z;t) \rangle = \dfrac{q}{2\pi H \sqrt{\overline{K}_{xx}\overline{K}_{yy}}}$ $\times \exp\left(-\dfrac{(x-x'-\bar{u}(t-t'))^2}{4\overline{K}_{xx}} - \dfrac{(y-y')^2}{4\overline{K}_{yy}}\right)$ $\times \left\{\underbrace{\dfrac{1}{2}}_{n=1} + \sum_{n=1}^{\infty} \cos\lambda_n z \cos\lambda_n z' \exp\left[-\lambda_n^2 \overline{K}_{zz}\right]\right\}$ $\lambda_n = \dfrac{n\pi}{H},\ \overline{K}_{xx} = \tfrac{1}{2}\sigma_x^2,\ \overline{K}_{yy} = \tfrac{1}{2}\sigma_y^2,\ \overline{K}_{zz} = \tfrac{1}{2}\sigma_z^2$	Total reflection at $z = 0$ $\bar{u} = (\bar{u}, 0, 0)$ $S = q\,\delta(x-x')\,\delta(y-y')\,\delta(z-z')\,\delta(t-t')$ $0 \leqslant z \leqslant H$
Gaussian puff formula	Total absorption at $z = 0$ $\bar{u} = (\bar{u}, 0, 0)$

(continued)

Table 6.6 (continued)

Mean concentration	Hypothesis
$\langle c(x,y,z,t)\rangle = \dfrac{q}{2\pi H \sqrt{\overline{K}_{xx}\overline{K}_{yy}}}$ $\times \exp\left(-\dfrac{(x-x'-\bar{u}(t-t'))^2}{4\overline{K}_{xx}} - \dfrac{(y-y')^2}{4\overline{K}_{yy}}\right)$ $\times \sum_{n=1}^{\infty} \dfrac{(\lambda_n^2+\beta^2)\cos[\lambda_n(H-z')]\cos[\lambda_n(H-z)]}{H(\lambda_n^2+\beta^2)+\beta}$ $\times \exp(-\lambda_n^2 \overline{K}_{zz})$ $\lambda_n \tan \lambda_n H = \beta \kappa_{ZI} \beta = v_d / K_{zz}$	$S = q\,\delta(x-x')\,\delta(y-y')\,\delta(z-z')\,\delta(t-t')$ $0 \leq z \leq H$
Gaussian plume formula $\langle c(x,y,z)\rangle = \dfrac{q}{2\pi \bar{u}\sigma_y \sigma_z} \exp\left(-\dfrac{y^2}{2\sigma_y^2}\right) \times \left[\exp\left(-\dfrac{(z-h)^2}{2\sigma_z^2}\right) + \exp\left(-\dfrac{(z+h)^2}{2\sigma_z^2}\right)\right]$	Total reflection at $z=0$ $\bar{u} = (\bar{u}, 0, 0)$ $S = q\,\delta(x)\,\delta(y)\,\delta(z-h)$ Slender hypothesis $0 \leq z \leq \infty$
Gaussian plume formula $\langle c(x,y,z)\rangle = \dfrac{q}{2\pi \bar{u}\sigma_y \sigma_z} \exp\left(-\dfrac{y^2}{2\sigma_y^2}\right) \times \left[\exp\left(-\dfrac{(z-h)^2}{2\sigma_z^2}\right) - \exp\left(-\dfrac{(z+h)^2}{2\sigma_z^2}\right)\right]$	Total absorption at $z=0$ $\bar{u} = (\bar{u}, 0, 0)$ $S = q\,\delta(x)\,\delta(y)\,\delta(z-h)$ Slender hypothesis $0 \leq z \leq \infty$
Gaussian plume formula $\langle c(x,y,z)\rangle = \dfrac{2q}{\sqrt{2\pi}\bar{u}\sigma_y H}\left\{\dfrac{1}{2}+\sum_{n=1}^{\infty}\cos\left(\dfrac{n\pi z}{H}\right)\cos\left(\dfrac{n\pi h}{H}\right) \times \exp\left[-\left(\dfrac{n\pi}{H}\right)^2 \dfrac{\sigma_z^2}{2}\right]\right\}\exp\left(-\dfrac{y^2}{2\sigma_y^2}\right)$	Total reflection at $z=0$ $\bar{u} = (\bar{u}, 0, 0)$ $S = q\,\delta(x)\,\delta(y)\,\delta(z-h)$ $0 \leq z \leq h$

Table 6.7 Gamma function[11]

M	n	S	$\Gamma(s)$
0.9	0.1	0.679	1.33
0.8	0.2	0.692	1.31
0.7	0.3	0.708	1.29
0.6	0.4	0.727	1.26
0.5	0.5	0.750	1.23
0.4	0.6	0.778	1.19
0.3	0.7	0.813	1.15
0.2	0.8	0.857	1.11
0.1	0.9	0.917	1.06
0	0	0.5	1.77

For the first nine lines it is supposed that m = 1-n whereas for the last line it is supposed that K and u are constant

Therefore the general solution can be expressed as:

$$C(x,z) = \frac{Q_l z_1^m (m-n+2)}{2 u_1 \Gamma(s)} \left[\frac{u_1}{(m-n+2)^2 z_1^{m-n} K_1 x} \right]^s$$

$$\times \exp\left[\frac{u_1 z^{m-n+2}}{z_1^{m-n}(m-n+2)^2 K_1 x} \right] \quad (6.52)$$

where: $s = \frac{m+1}{m-n+2}$, and Γ is the gamma function (see Table 6.7)

In the case that there are constant values for u and K ($n = m = 0$), the solution is Gaussian:

$$C = \frac{Q_l}{1,23 u_1} \left(\frac{u_1}{4 K_1 x} \right)^{1/2} \exp\left(-\frac{u_1 z^2}{4 K_1 x} \right) \quad (6.53)$$

6.8 Characteristics of Plume Dispersion – Stability Conditions

It is important to study the pollutant dispersion under different meteorological conditions with the use of Gaussian models. Table 6.8 presents the plume structure under different stability conditions. The coefficient p denotes a dimensionless number and the variability of the air velocity versus height can be written as:

$$\frac{\bar{u}}{\bar{u}_m} = \left(\frac{z}{z_m} \right)^p \quad (6.54)$$

where $\bar{u}$ is the mean wind velocity at height z, $\bar{u}_m$ is the mean wind velocity at height z_m and the value of p is dependent on the stability conditions and the surface turbulence.

Table 6.8 Values of parameters which correspond to the profiles of temperature and wind velocity at the models of plume rise

Plume shape	h (m)	ΔT (°C/100 m)	p (dimensionless)	Atmospheric stability
Looping	5–100	−1.8	0.145	Unstable
Conning	5–100	−1.0	0.170	Neutral
Fanning	5–20	+1.4	0.44	Stable (with temperature inversion)
	200–400	−0.7	0.25	Slightly stable
Fumigation	5–125 (≥ 22.6 °C)	−0.7	0.25	Slightly stable
	125–200	+1.4	0.44	Stable (with temperature inversion)
	200–400	−1.0	0.170	Neutral
Lofting	5–200	+1.4	0.44	Stable (with temperature inversion)
	200–400	+0.7	0.25	Neutral

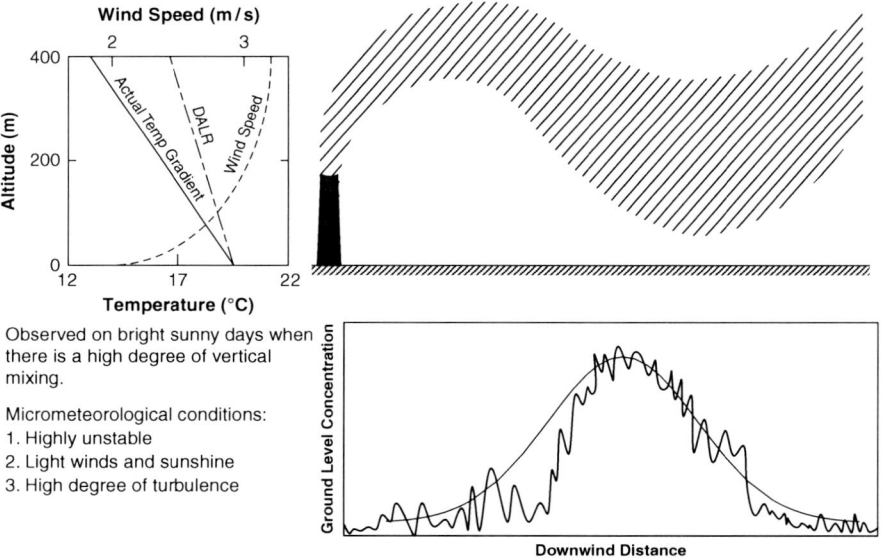

Fig. 6.5 Looping plume structure. (Adapted from Schnelle and Dey 1999)

As an example the characteristics of the plume dispersion will be studied for the looping plume structure (Fig. 6.5). The plume in this case is related to transport conditions and increased turbulence. This structure of the plume is observed on hot days with clear sky and low wind velocity. Sources with low stack height and low heat emissions can have a looping structure. The thermal instability of the atmosphere can transport the plume to the surface and have as a result increased surface concentrations. This can be changed quickly due to the presence of turbulence and

6.8 Characteristics of Plume Dispersion – Stability Conditions

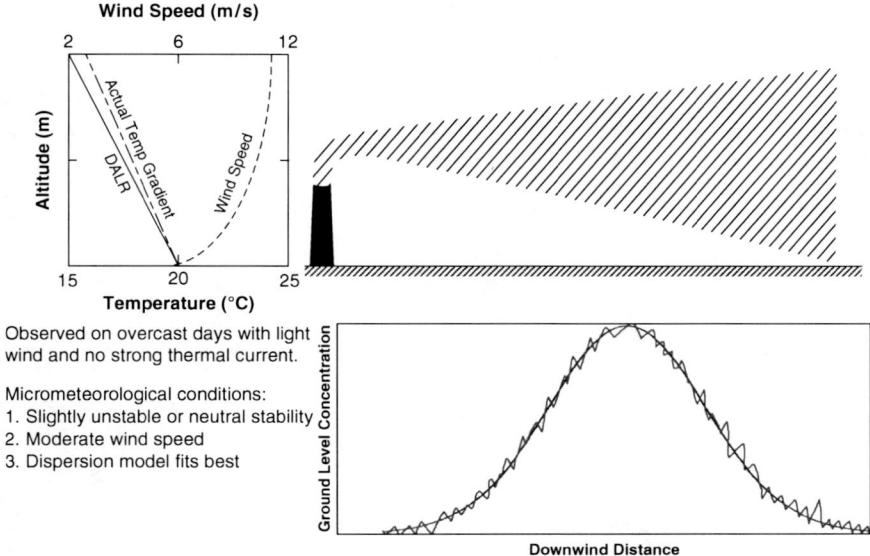

Fig. 6.6 Conical plume structure. (Adapted from Schnelle and Dey 1999)

can have as a result further plume transport upwards. The application of Gaussian models for plumes with looping structure does not give satisfactory results, as for example in the case of conical plume structure.

The conical plume structure is observed at neutral conditions and the Gaussian models are capable of describing the pollutant dispersion under these conditions (Fig. 6.6). Plumes that are characterized by vertical transport and low wind velocities present a high plume rise and therefore are lower compared to other conditions of surface concentration. Higher wind velocities result in increased pollutant dispersion. Neutral conditions are observed often in the atmosphere and in cases that stacks with low height exist, it may result in increased surface concentrations (Schnelle and Dey 1999).

Stable conditions are connected with fanning plume structure. This happens usually during inversion conditions in the morning when the vertical mixing is small. During inversion there is a cessation of the influence of vertical movements of unstable air masses from the surface. High mixing occurs under these conditions and the plume reaches the surface (fumigation). Under these conditions, high pollutant concentrations can exist at the surface. This structure is not observed for time intervals larger than half an hour. Therefore the fumigation is not important to the studies of the calculation of mean pollutant concentration from point sources for periods larger than 1 h. Furthermore, under conditions of stable plume inversion there is no penetration of the inversion and therefore there are higher concentrations at surface level. Finally, lofting conditions are observed usually late in the afternoon when an inversion starts to form and then moves aloft. The plume cannot penetrate the inversion downwards and so diffuses aloft.

6.9 Examples and Applications

Example 1. *Calculate the pollutant concentration resulting from a continuous surface source with a constant diffusion factor in one dimension and with air velocity equal to zero.*

The continuity equation at one dimension can be written as

$$\frac{\partial c}{\partial t} = K \frac{\partial^2 c}{\partial x^2}$$

The boundary conditions are: $c \to 0$ when $t \to \infty$ for all x. $c \to 0$ when $t \to 0$ for all x except the case $x = 0$ and we have also

$$\int_{-\infty}^{\infty} c \, dx = Q$$

where Q is the instantaneous source at units of mass per unit of surface with a solution:

$$\eta = \frac{x^2}{kt}$$

A hypothetical solution of the form: $u(x,t) = f(\eta)$ results in

$$\frac{\partial u}{\partial x} = \frac{df(\eta)}{d\eta} \frac{\partial \eta}{\partial x} = \frac{2x}{kt} \frac{df(\eta)}{d\eta}$$

$$\frac{\partial^2 u}{\partial x^2} = \frac{2}{kt} \frac{df(\eta)}{d\eta} + \left(\frac{2x}{kt}\right)^2 \frac{d^2 f(\eta)}{d\eta^2}$$

$$\frac{\partial u}{\partial t} = -\frac{x^2}{kt^2} \frac{df(\eta)}{d\eta}$$

With the replacement of the above expressions in the dispersion equation it results that:

$$4\eta \frac{d^2 f(\eta)}{d\eta^2} + (2 + \eta) \frac{df(\eta)}{d\eta} = 0$$

The above equation is an ordinary differential equation which can be easily solved (setting $f'(\eta) = \frac{df(\eta)}{d\eta}$) with the result:

$$\frac{f''(\eta)}{f'(\eta)} = -\frac{1}{2\eta} - \frac{1}{4} \Rightarrow \ln\left[\eta^{1/2} f'(\eta)\right] = -\frac{\eta}{4} + c \Rightarrow f'(\eta) = \frac{A}{\eta^{1/2}} \exp\left(-\frac{\eta}{4}\right)$$

$$\Rightarrow f(\eta) = A \int_{\eta_o}^{\eta} \mu^{-1/2} \exp\left(-\frac{\mu}{4}\right) d\mu$$

$$\zeta = \frac{\eta^{1/2}}{2} = \frac{x}{2(kt)^{1/2}} \Rightarrow d\zeta = \frac{1}{4}\eta^{-1/2} d\eta$$

$$u(x,t) = f(\eta) = g(\zeta) = B \int_{\zeta_o}^{\zeta} \exp(-v^2) dv$$

Replacement of value of η gives: $c = \frac{1}{(4kt)^{1/2}} \int_0^t \exp\left(-\frac{x^2}{4kt}\right) dx$

Application of the boundary conditions leads to: $c = \frac{Q}{(4\pi kt)^{1/2}} \exp\left(-\frac{x^2}{4kt}\right)$

The above expression is Gaussian with a standard deviation: $\sigma = (2kt)^{1/2}$
The expression between σ and k is used often for the determination of k from experimental values of σ. However, the assumption that k is constant is not valid since its value is variable versus time and space.

Example 2. Derive the plume rise height (Δh) from an electricity production station with the method of Holland. It is given that $r_s = 4$ m, $T_a = 293$ K, $T_s = 413$ K, $V_s = 15$ m/s, $P = 100$Kp$_a$, u $= 5$ m/s.

Using the equation for the plume rise height:

$$\Delta h = \frac{2 V_s r_s}{u}\left[1.5 + 2.68 \times 10^{-2} P \left(\frac{T_s - T_a}{T_s}\right) 2 r_s\right]$$

it is derived that $\Delta h = 185.5$ m

Example 3. There is a continuous linear source vertical to the wind direction at the surface level. It has length b S g Km^{-1} s^{-1}. Assume that the slender conditions are satisfied (see Table 6.4).

(a) If the origin of the coordination system is at the centre of the line, show that the mean pollutant concentration at the surface level downwind of the source is given by:

$$\langle c(x,y,0) \rangle = \frac{S}{(2\pi)^{1/2} \sigma_z \bar{u}}\left[erf\left(\frac{b/2 - y}{\sqrt{2}\sigma_y}\right) + erf\left(\frac{b/2 + y}{\sqrt{2}\sigma_y}\right)\right]$$

(b) For large distances from the source, show that the ground-level concentration at the plume axis ($y = 0$) is given by (Seinfeld and Pandis 2006)

$$\langle c(x,y,0) \rangle = \frac{S b}{\pi \sigma_y \sigma_z \bar{u}}$$

The linear source can be approximated as a sum of many point sources, each of them with emission Q dy'. Under this assumption the emissions from the linear source is homogeneous. Therefore:

$$TotalEmission = S = mass/(length) \times (time)$$

The dispersion along the wind direction (x) is very small compared to the transport and therefore the Gaussian solution for a point source can be written as:

$$\langle c(x,y,z) \rangle = \frac{q}{2\pi \bar{u} \sigma_y \sigma_z} \exp\left(-\frac{y^2}{2\sigma_y^2} - \frac{z^2}{2\sigma_z^2}\right)$$

$$\langle c(x,y,0) \rangle = \frac{S}{2\pi \bar{u} \sigma_y \sigma_z} \exp\left(-\frac{y^2}{2\sigma_y^2}\right)$$

Therefore for the entire linear source the solution is:

$$\langle c(x,y,0) \rangle = \frac{S}{\pi \bar{u} \sigma_y \sigma_z} \int_{-b/2}^{b/2} \exp\left(-\frac{(y-y')^2}{2\sigma_y^2}\right) dy'$$

For the above integration there is an exchange of variables as:
$\beta = \frac{(y-y')}{\sqrt{2}\sigma_y} \Rightarrow y' = y - \sqrt{2}\sigma_y \beta$ and also: $dy' = -\sqrt{2}\sigma_y d\beta$ *which results in:*

$$\langle c(x,y,0) \rangle = \frac{S}{\pi \sigma_y \sigma_z \bar{u}} \int_{\frac{y-b/2}{\sqrt{2}\sigma_y}}^{\frac{y+b/2}{\sqrt{2}\sigma_y}} \exp(-\beta^2) \left(\sqrt{2}\sigma_y\right) d\beta$$

The above integral can be expressed with the help of the error function (erf):
$erf(x) = \frac{2}{\sqrt{\pi}} \int_0^x e^{-\beta^2} d\beta = erf(-x)$ *and therefore:*

$$\langle c(x,y,0) \rangle = \frac{S}{\sqrt{2\pi} \sigma_z \bar{u}} \left[erf\left(\frac{b/2-y}{\sqrt{2}\sigma_y}\right) + erf\left(\frac{b/2+y}{\sqrt{2}\sigma_y}\right) \right]$$

β. *For* $y = 0$

$$\langle c(x,0,0) \rangle = \frac{S}{\sqrt{2\pi} \sigma_z \bar{u}} erf\left(\frac{b}{2\sqrt{2}\sigma_y}\right)$$

The error function can be expressed on a series and for small values of the terms the function can be approximated with the first term of the expansion:
$erf(x) = \frac{2}{\sqrt{\pi}} \sum_{n=0}^{\infty} \frac{(-1)^n x^{2n+1}}{n!(2n+1)} = \frac{2}{\sqrt{\pi}} \left[x - \frac{x^3}{3} + \ldots\right]$

And therefore for small x: $erf(x) = \frac{2}{\sqrt{\pi}} x$

$$\langle c(x \to \infty, 0, 0)\rangle = \frac{2S}{\sqrt{2\pi}\,\sigma_z \bar{u}} \left(\frac{2}{\sqrt{\pi}}\right) \left(\frac{b}{2\sqrt{2}\sigma_y}\right) = \frac{Sb}{\pi\,\sigma_y \sigma_z \bar{u}}$$

The above expression is equivalent with emissions from a point source with total emission Q b.

Problems

6.1 A cell has dimensions $\Delta x = 5$ km, $\Delta y = 4$ km and $\Delta z = 0.1$ km and the arithmetic velocities at west, east, south, north and the velocity at the lower z level are respectively:

$$u_1 = +3$$
$$u_2 = +4$$
$$v_3 = -3$$
$$v_4 = +2$$
$$w_5 = +2$$

If the atmosphere is incompressible, what is the value of the velocity w_6 at the top of the cell ?

6.2 Murphy and Nelson have shown that the Gaussian plume equation can be modified to include dry deposition velocity v_d by replacing the source strength q, usually taken as constant, with the depleted source strength $q(t)$ as a function of travel time t, where

$$q(t) = q_o \exp\left\{-\left(\frac{2}{\pi}\right)^{1/2} v_d \int_0^t \frac{1}{\sigma_z(t')} \exp\left(\frac{h^2}{2\sigma_z^2(t')}\right) dt'\right\}$$

Verify this result. Show that, if $\sigma_z = (2\,K_{zz}\,t)^{1/2}$ then (Seinfeld and Pandis 2006):

$$q(t) = q_o \exp\left\{\frac{2 v_d t^{1/2}}{(\pi K_{zz})^{1/2}}\right\}.$$

6.3 A power plant burns 10^4 kg/h of coal containing 2.5% sulphur. The effluent is released from a single stack of height 70 m. The plume rise is normally about 30 m, so that the effective height of emission is 100 m. The wind on the day of interest, which is a sunny summer day, is blowing at 4 m/s. There is no inversion layer. Use the Pasquill-Gifford dispersion parameters.

(a) Plot the ground-level SO_2 concentration at the plume centerline over distances from 100 m to 10 km. (Use log-log coordinates).

(b) Plot the ground-level SO_2 concentration versus crosswind distance at downwind distances of 200 m and 1 km.

(c) Plot the vertical centerline SO_2 concentration profile from ground level to 500 m at distances of 200 m, 1 km, and 5 km. (Seinfeld and Pandis 2006).

6.4 An instrument for SO_2 measurements has been located at distance 2,000 m from a point source downwind. The point source has height 100.7 m and the instrument is located on the top of a meteorological station (200 m). Wind measurements are performed at height 10 m. On a day with an overall cloud cover, the wind velocity has a value of 3.96 m/s and the temperature is equal to 18 °C. Calculate the SO_2 concentration which is measured by the instrument using the dispersion coefficients for an open field (Briggs). The characteristics of the point source are: $V_s = 12{,}0$ m/s, $d_s = 4{,}59$ m (stack diameter), $T_s = 478$ K and $Q = 232$ g/s SO_2.

6.5 At an electrical power station the stack has a height of 76.2 m. The total use of coal is 14,050 tons per year and the coal has on average 3.5% sulphur. Suppose that the total amount of SO_2 is emitted from the stack. Make a plot of the surface SO_2 concentration along the wind direction. The meteorological conditions and the characteristics of the stack are the following: $V_s = 14{,}7$ m/s, $d_s = 4.27$ m (stack diameter), $T_s = 416.15$ K, $h_s = 76.2$ m, $u = 5.0$ m/s, $T_a = 416.15$ K and stability (condition D).

6.6 A square source with surface 2.6×10^4 m^2 inside a landfill emits vinyl chloride. A measurement point is located 175° south and 250 m from the centre of the surface source. Winds with velocity of 3.08 m/s bring vinyl chloride concentrations to the measurement point A which is close to a populated area. During the measurements of a five day period, concentrations of vinyl chloride equal to 12, 5, 7, 12 and 9 ppb were measured. Calculate the emission rate of vinyl chloride (g/s).

6.10 Appendix 6.1 The Continuity Equation

In a volume cell the mass of air that enters inside the cell, minus the mass which flows out of the cell, equals the final mass remaining inside the cell minus the initial mass. The same relationship occurs for other atmospheric variables such as energy. Fig. 6.7 shows a volume cell with dimensions Δx, Δy and Δz (m). The concentration of air has boundary values N_1 and N_2 at the surface $\Delta y \times \Delta z$ (molecules m^{-3}) with corresponding values for the velocity u_1 and u_2 (m s^{-1}) respectively. The influx and outflux of air from the cell is $u_1 N_1$ and $u_2 N_2$ respectively (molecules m^{-3} s^{-1}).

The balance of the molecule concentration in the cell can be expressed as:

$$\Delta N \, \Delta x \, \Delta y \, \Delta z = u_1 N_1 \, \Delta y \, \Delta z \, \Delta t - u_2 N_2 \, \Delta y \, \Delta z \, \Delta t \qquad (6A.1)$$

Dividing both sides by Δt and with volume (Δx, Δy, Δz) it results that:

6.10 Appendix 6.1 The Continuity Equation

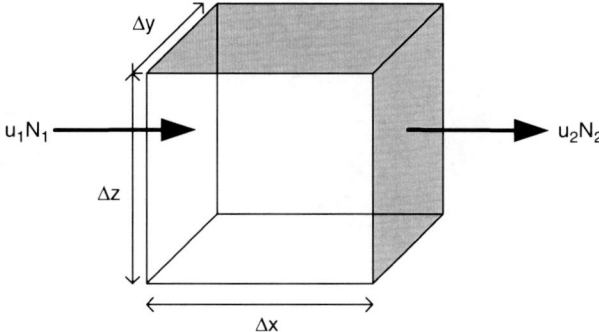

Fig. 6.7 An example of mass continuity. The number of molecules which enter minus the number which exits from the volume equals the number of molecules which remain inside the volume

$$\frac{\Delta N}{\Delta t} = -\left(\frac{u_2 N_2 - u_1 N_1}{\Delta x}\right) \tag{6A.2}$$

When $\Delta x \to 0$ and $\Delta t \to 0$ the above equation is expressed as:

$$\frac{\partial N}{\partial t} = -\frac{\partial (uN)}{\partial x} \tag{6A.3}$$

which is the continuity equation for a gas that is influenced by the velocity in one dimension. At three dimensions in a Cartesian coordination system the above equation can be written as:

$$\frac{\partial N}{\partial t} = -\frac{\partial (uN)}{\partial x} - \frac{\partial (vN)}{\partial y} - \frac{\partial (wN)}{\partial z} = -\nabla \bullet (vN) \tag{6A.4}$$

Furthermore:

$$\nabla \bullet (vN) = N(\nabla \bullet v) + v(\nabla \bullet N) \tag{6A.5}$$

And with replacement the Eq. (6A.5) to (6A.4) can be written:

$$\frac{\partial N}{\partial t} = -N(\nabla \bullet v) - v(\nabla \bullet N) \tag{6A.6}$$

and knowing:

$$v(\nabla \bullet N) = \frac{dN}{dt} - \frac{\partial N}{\partial t} \tag{6A.7}$$

it can be concluded that:

$$\frac{dN}{dt} = -N\left(\nabla \bullet v\right). \tag{6A.8}$$

A similar expression can be written for the density ρ_α of air:

$$\frac{d\rho_a}{dt} = -\rho_a\left(\nabla \bullet v\right) \tag{6A.9}$$

A more general form of the continuity equation, where there are emission sources and chemical reactions is given by the expression:

$$\frac{\partial N}{\partial t} = -\nabla \bullet (vN) + D\,\nabla^2 N + \sum_{n=1}^{N_{e,t}} R_n \tag{6A.10}$$

In the above equation the coefficient D is the coefficient of molecular diffusion of gas, which expresses the molecular kinetics due to their kinetic energy. The coefficient R_n expresses the variation of the gaseous concentration arising from chemical reactions. The term which is arising from molecular diffusion can be written as:

$$D\left(\nabla^2 N\right) = D\left(\nabla \bullet \nabla\right)N = D\left[\left(i\frac{\partial}{\partial x} + j\frac{\partial}{\partial y} + k\frac{\partial}{\partial z}\right) \bullet \left(i\frac{\partial}{\partial x} + j\frac{\partial}{\partial y} + k\frac{\partial}{\partial z}\right)\right]N$$
$$= D\left(\frac{\partial^2 N}{\partial x^2} + \frac{\partial^2 N}{\partial y^2} + \frac{\partial^2 N}{\partial z^2}\right)$$

$$\tag{6A.11}$$

References

Hanna, S. R., Metro Briggs, G. A., & Hosker, R. P. (1981). *Handbook on atmospheric diffusion*. Washington: Technical Information Center, U.S. Department of Energy.

Schnelle, K. B., & Dey, P. R. (1999). *Atmospheric dispersion modeling compliance guide*. New York: McGraw-Hill.

Seinfeld, J. H., & Pandis, S. N. (2006). *Atmospheric chemistry and physics* (2nd ed.). New York: John Wiley & Sons.

Chapter 7
Atmospheric Models: Emissions of Pollutants

Contents

7.1	Introduction	233
7.2	Dispersion Equations for Pollutant Transport at the Euler and Lagrange Coordinating Systems	235
	7.2.1 Model of a Single Volume in the Euler System	236
	7.2.2 Three Dimensional Models of Atmospheric Pollution	237
7.3	Statistical Evaluation of Atmospheric Models	238
7.4	Emissions of Atmospheric Pollutants	239
7.5	Emissions from the Biosphere	240
	7.5.1 Emissions of Volatile Organic Compounds from Vegetation	243
	7.5.2 Calculation of Biogenic Emissions	243
	7.5.3 Sea Salt Emissions	245
	7.5.4 Emissions of Air Pollutants from the Earth's Surface	246
	7.5.5 Emissions of Pollutants from Forest Fires	247
7.6	Examples and Applications	249
References		253

Abstract The atmosphere is a dynamic system where a large number of physical and chemical processes occur simultaneously. A comprehensive study of the atmosphere's dynamics and transport of pollutants is presented with the use of atmospheric models. In Chapter 7 the dispersion equations for pollutant transport in the Euler and Lagrange coordinating systems are examined. In particular we examine the model of a single volume. The emissions of atmospheric pollutants are also examined from a variety of sources including anthropogenic and biogenic sources. A specific methodology for the calculation of gaseous and particulate matter emissions is presented together with examples and applications.

7.1 Introduction

The atmosphere is a dynamic system where a large number of physical and chemical processes occur simultaneously. A comprehensive study of the atmosphere's dynamics and transport of pollutants can be carried out with the use of an

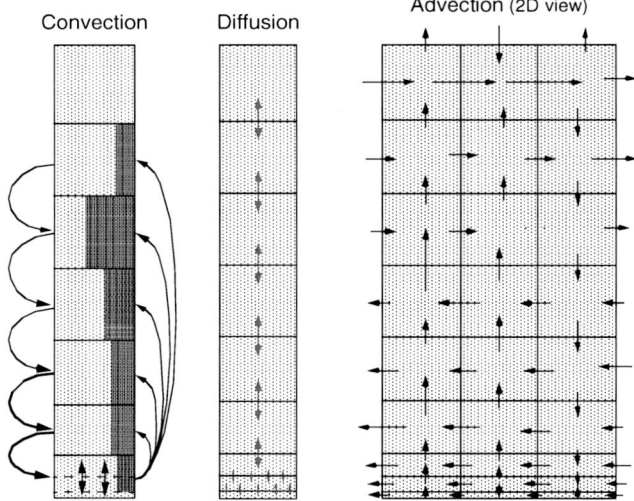

Fig. 7.1 Schematic description of the terms of convection and diffusion in one dimension model and advection in a two dimensional model

atmospheric model. Atmospheric models simulate the various processes occurring in the atmosphere and their interactions. The application of an atmospheric model involves also a comparison of the model's results with field and laboratory experiments. This assists in understanding of the chemical and physical processes occurring in the atmosphere as well as evaluation of the model's performance.

Laboratory measurements offer valuable information for specific processes and field measurements depict the atmosphere's characteristics and air composition at specific space and time intervals. An atmospheric model offers a complete picture of the spatial and temporal evolution of air pollutants inside the atmosphere at different heights. Schematic representation of the pollutant convection and diffusion in a one-dimensional cell model and advection in a two-dimensional cell model are presented in Fig. 7.1

An understanding of the atmosphere's dynamics can be achieved with a combination of measurements and an integrated modeling with the use of atmospheric diffusion models including chemical reactions. An atmospheric model may include also emission models and models for the physical and chemical processes of the gaseous and aerosol components. The model may also incorporate a number of computer modules which give the emissions of gaseous and aerosol components and the initial and boundary concentrations. Another set of models introduces meteorological data and the surface's topography. The main part of the model treats air transport and physical and chemical processes in the atmosphere. Finally, a part of the model analyzes the outputs of the model's simulations.

Fig. 7.2 depicts the main features of atmospheric models, originating from emissions to actual human exposure

7.2 Dispersion Equations for Pollutant Transport

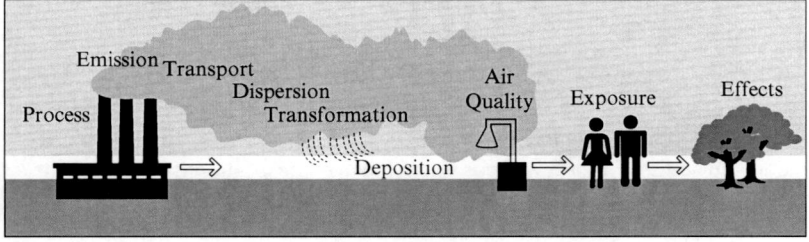

Fig. 7.2 Sequence of events in the atmosphere from pollutant emission – transport and transformation to human exposure and effects

Fig. 7.3 Cells of (**a**) zero, (**b**) one, (**c**) two and (**d**) three dimensions

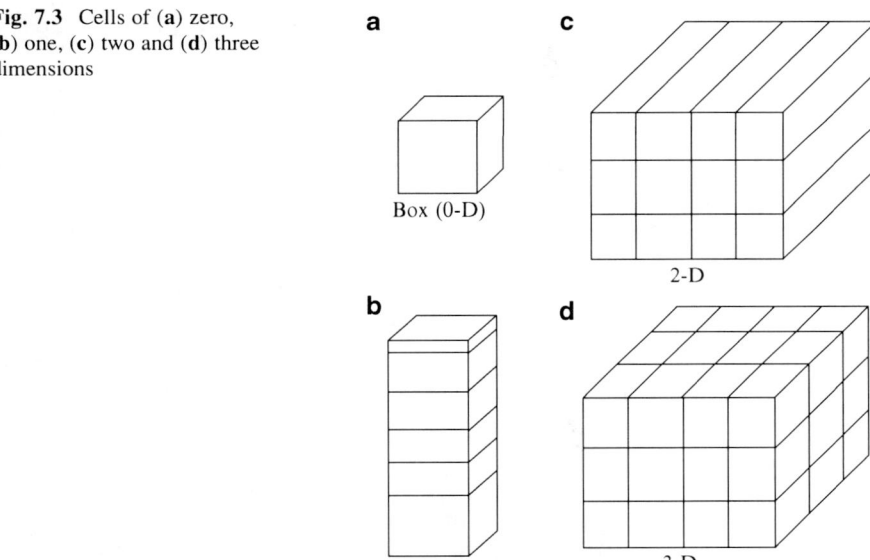

7.2 Dispersion Equations for Pollutant Transport at the Euler and Lagrange Coordinating Systems

As discussed previously the modelling of atmospheric processes can be performed generally in one of two coordinate systems, the Euler and Lagrange. Besides, the study area can be of zero dimensions (model of a single volume), one, two or three dimensions (Fig. 7.3).

The solution of the dispersion of pollutants with models of zero dimensions is treated in the current chapter. An interested reader can study the solutions in three-dimensional cell models in the literature (e.g. Seinfeld and Pandis 2006). In the following section a simple description of single-volume models is given with the Euler and Lagrange methods.

7.2.1 Model of a Single Volume in the Euler System

Figure 7.4 shows a model of a single atmospheric volume, where the height H(t) denotes the height of the boundary layer. In reality this height is variable during a day and is dependent directly on time.

Inside the volume is a mass conservation for the pollutants (c_i) which can be expressed for the volume H × Δx × Δy as:

$$\frac{d}{dt}(c_i \Delta x \Delta y H) = Q_i + R_i \Delta x \Delta y H - S_i + u H \Delta y \left(c_i^o - c_i\right) \quad (7.1)$$

where, Q_i is the emission rate of mass for the pollutant i (Kg h^{-1}), S_i is the rate of removal of the component i (Kg h^{-1}), R_i is the production rate (Kg m^{-3} h^{-1}) due to chemical reactions, c_i^o is the background concentration and u is the wind velocity which is supposed to have a constant direction. Division of the above expression with Δx Δy results in:

$$\frac{d}{dt}(c_i H) = q_i + R_i H - s_i + u \frac{H}{\Delta x}\left(c_i^o - c_i\right) \quad (7.2)$$

where, q_i and s_i denote the emission and deposition rates per unit surface (Kg m^{-2} h^{-1}). The deposition rate can be expressed in relation to the velocity of dry deposition of the component i, $v_{d,i}$, and can be expressed as:

$$s_i = v_{d,i} c_i \quad (7.3)$$

The result is an equation which describes the concentration variation in the volume cell for a stable height of the boundary layer and can be expressed as:

$$\frac{dc_i}{dt} = \frac{q_i}{H} + R_i - \frac{v_{d,i}}{H} c_i + \frac{u}{\Delta x}\left(c_i^o - c_i\right) \quad (7.4)$$

In the scientific literature the ratio of the cell length to the average wind velocity in the main direction is called residence time of the pollutant in the cell and is expressed as:

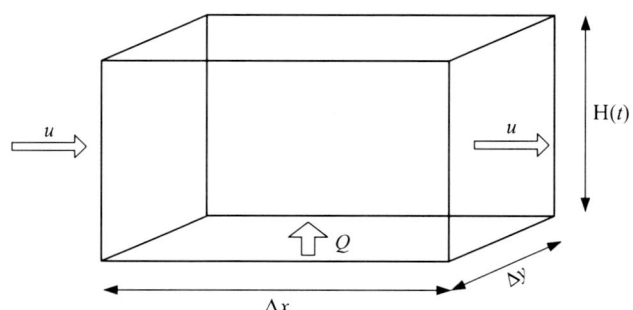

Fig. 7.4 Model of a single volume

7.2 Dispersion Equations for Pollutant Transport

$$\tau_r = \frac{\Delta x}{u} \quad (7.5)$$

The above problem is more interesting when the boundary layer height H(t) is changing versus time. It is supposed that the pollutant has a concentration c_i^a above the initial height of the boundary layer H. It is supposed that the time interval Δt the height of the boundary layer is increased from the value H to H + ΔH, whereas the pollutant concentration is changing from c_i to $c_i + \Delta c_i$. From the above conditions the mass conservation of the chemical component i results in

$$(c_i + \Delta c_i)(H + \Delta H) = c_i H + c_i^a \Delta H \quad (7.6)$$

And, omitting terms of second order, it results that

$$H \Delta c_i = (c_i^a - c_i) \Delta H \text{ for } \Delta t \to 0, \quad (7.7)$$

and also that $\frac{dc_i}{dt} = \frac{c_i^a - c_i}{H} \frac{dH}{dt}$

Finally Eq. (7.4) in the Euler coordination system can be expressed as:

$$\frac{dc_i}{dt} = \frac{q_i}{H(t)} + R_i - \frac{v_{d,i}}{H(t)} c_i + \frac{(c_i^o - c_i)}{\tau_r} \text{ for } \frac{dH}{dt} \leq 0 \quad (7.8)$$

$$\frac{dc_i}{dt} = \frac{q_i}{H(t)} + R_i - \frac{v_{d,i}}{H(t)} c_i + \frac{(c_i^o - c_i)}{\tau_r} + \frac{c_i^a - c_i}{H(t)} \frac{dH}{dt} \text{ for } \frac{dH}{dt} > 0 \quad (7.9)$$

An important hypothesis in the Lagrange coordination system for one cell is that there is no transport component in the horizontal direction. The cell in the Lagrange coordination system behaves as a point which follows the wind direction. This has as a result that the equations have a similar form as the Euler coordination system but without the term which describes the transport of pollutants. The result is that Eq. (7.8) and (7.9) in the Lagrange coordination system can be written as:

$$\frac{dc_i}{dt} = \frac{q_i}{H(t)} + R_i - \frac{v_{d,i}}{H(t)} c_i \text{ for } \frac{dH}{dt} \leq 0 \quad (7.10)$$

$$\frac{dc_i}{dt} = \frac{q_i}{H(t)} + R_i - \frac{v_{d,i}}{H(t)} c_i + \frac{c_i^a - c_i}{H(t)} \frac{dH}{dt} \text{ for } \frac{dH}{dt} > 0 \quad (7.11)$$

7.2.2 Three Dimensional Models of Atmospheric Pollution

The general formula for the dispersion equation of the concentration of a pollutant i (c_i) for an atmospheric model in three dimensions can be written as:

$$\frac{\partial c_i}{\partial t} + u_x \frac{\partial c_i}{\partial x} + u_y \frac{\partial c_i}{\partial y} + u_z \frac{\partial c_i}{\partial z} = \frac{\partial}{\partial x}\left(K_{xx}\frac{\partial c_i}{\partial x}\right) + \frac{\partial}{\partial y}\left(K_{yy}\frac{\partial c_i}{\partial y}\right) + \frac{\partial}{\partial z}\left(K_{zz}\frac{\partial c_i}{\partial z}\right)$$

$$+ R_i(c_1, c_2, \ldots, c_n) + E_i(x, y, z, t) - S_i(x, y, z, t) \tag{7.12}$$

where R_i, E_i and S denote the chemical reactions, emissions and removal respectively. The velocity is decomposed into three dimensions in the same way as the diffusion coefficients.

7.3 Statistical Evaluation of Atmospheric Models

The evaluation of atmospheric models is carried out by comparison of the model results with field data. The statistical evaluation of the model's results presents a powerful methodology for the evaluation of a model. The most efficient comparison can be performed with an analysis of field time series measurements with model results at different time intervals and locations.

For the statistical evaluation there is a need to have measured data from monitoring stations. Assume that there are n measurement stations each performing m measurements during the model evaluation period. The predicted and observed concentrations are defined as $PRED_{i,j}$ and $OBS_{i,j}$ respectively for the station i at time j. Useful evaluations of the model performance can be attained with a series of comparisons (Seinfeld and Pandis 2006):

a. The Mean Normalized bias (MNB):

$$MNB = \frac{1}{nm}\sum_{i=1}^{n}\sum_{j=1}^{m}\frac{PRED_{i,j} - OBS_{i,j}}{OBS_{i,j}} \tag{7.13}$$

b. The mean bias (MB):

$$MB = \frac{1}{nm}\sum_{i=1}^{n}\sum_{j=1}^{m} PRED_{i,j} - OBS_{i,j} \tag{7.14}$$

c. The mean absolute normalized gross error (MANGE):

$$MANGE = \frac{1}{nm}\sum_{i=1}^{n}\sum_{j=1}^{m}\frac{|PRED_{i,j} - OBS_{i,j}|}{OBS_{i,j}} \tag{7.15}$$

d. The mean error (ME):

$$MB = \frac{1}{nm}\sum_{i=1}^{n}\sum_{j=1}^{m}|PRED_{i,j} - OBS_{i,j}| \tag{7.16}$$

Besides the above comparisons for evaluation of an arithmetic model, one can also perform a diagnostic analysis based on a simple scenario such as zero emissions or zero removal in order to examine the sensitivity of the model and the study of possible arithmetic errors.

7.4 Emissions of Atmospheric Pollutants

The most important air pollutants are carbon monoxide and carbon dioxide (CO_2, CO), the Non-Methene Volatile Organic Compounds (NMVOCs), methane (CH_4), the nitrogen oxides and nitrous oxide (NO_x, N_2O), ammonia (NH_3), sulphur dioxide (SO_2), the particulate Matter (PM), the Heavy Metals (HM) and the Persistent Organic Pollutants (POPs). Emissions originate from anthropogenic and natural sources and have important consequences ranging from effects on ecosystems, urban air quality, human health and climate.

The construction of an emission inventory is an important tool in air quality management. It is a key element in air quality modelling which is used as a tool by policy makers for the development of air pollution abatement strategies. The quantification and assessment of pollutant emissions and their spatial and temporal variation is very important for the protection of the environment and the health of citizens. Together with air quality modelling, an emission inventory is used for assessment of the impact of specific human activities and of the main sources responsible for air quality deterioration in areas violating air quality standards and also for assessment of the results of specific mitigation strategies.

Data on land cover, population density, location and emissions of large point sources (LPS) are necessary to be used for the emission inventory from anthropogenic sources. For example, emissions from stationary sources including industrial combustion, mineral extraction and domestic heating has to be mapped to the "continuous urban fabric" and "discontinuous urban fabric" categories whereas agriculture emissions were apportioned to agricultural areas.

More specifically, for disintegration of emissions based on population data, equations of the following form can be used:

$$E_k(cell) = E_k(cell_{EMEP}) \frac{wf_{k,l}}{\sum_l (wf_{k,cell_{EMEP}})} \cdot \frac{pop(cell,l)}{pop(cell_{EMEP},l)} \qquad (7.17)$$

where $E_k(cell)$ is the emission (Mg/year) from sector k from the 100×100 m^2 grid cell and $E_k(cell_{EMEP})$ is the emission (Mg/year) from sector k from the EMEP grid cell (EMEP/CORINAIR 2002) for which emissions are spatially disaggregated, $wf_{k,l}$ is the weighting factor for emissions from sector k for the land cover class l which is deviated by the sum for every land cover class of the weighting factors for emissions from sector k in the specific $cell_{EMEP}$ ($wf_{k,cell_{EMEP}}$) (dimensionless), and pop(cell,l) is the population in the cell with land cover class l deviated with the total population living in cells with land cover class l in the whole $cell_{EMEP}$ ($pop(cell_{EMEP}, l)$) (in number of inhabitants). In the case where there are no weighting factors specific to each land cover class, the first fraction is omitted from the equation and the population of each cell is deviated with the total population of the $cell_{EMEP}$. The later approach was used for emissions from the use of solvents.

For emissions disintegrated based on land cover data we used equations of the form:

$$E_k(cell) = E_k(cell_{EMEP}) \frac{wf_{k,l}}{\sum_l (wf_{k,cell_{EMEP}}) \sum_l cell} \quad (7.18)$$

where $\sum_l cell$ is the count of cells with land cover class l in the $cell_{EMEP}$ for which emissions are spatially disaggregated. The derived emissions were then mapped to the grid of the area under study. Table 7.1 shows the main anthropogenic sources and their emissions. Furthermore Figs. 7.5–7.6 show the annual CO and $PM_{2.5}$ emissions during 1999 in Greece as an example application. Higher emissions are located close to main urban and industrial areas.

7.5 Emissions from the Biosphere

The biosphere emits large quantities of gaseous pollutants and particles. The emitted quantities of particles from natural sources at a global level are much larger than the anthropogenic emissions (Seinfeld and Pandis 2006). These include among others resuspended dust, sea salt and primary particles from forest fires. Besides the primary particle emissions there is a secondary particle formation from gaseous precursors as was studied in chap. 5.

Emissions from the biosphere include carbon monoxide, carbon dioxide, nitrogen oxides, volatile organic compounds, methane, hydrocarbons and sulphur

Table 7.1 Anthropogenic gaseous emissions sources

	CO	NH$_3$	NMVOC	NOx	SOx
1. Combustion sources in energy production	√			√	√
2. Non industrial combustion sources	√		√	√	√
3. Combustion sources in the industry	√		√	√	√
4. Production processes	√	√	√	√	√
5. Excavation and disposal of mineral fuels and geothermal energy			√		
6. Use of solvents and other products			√		
7. Transport	√	√	√	√	√
8. Other mobile sources	√		√	√	√
9. Waste management process		√	√		
10. Agriculture		√		√	

7.5 Emissions from the Biosphere 241

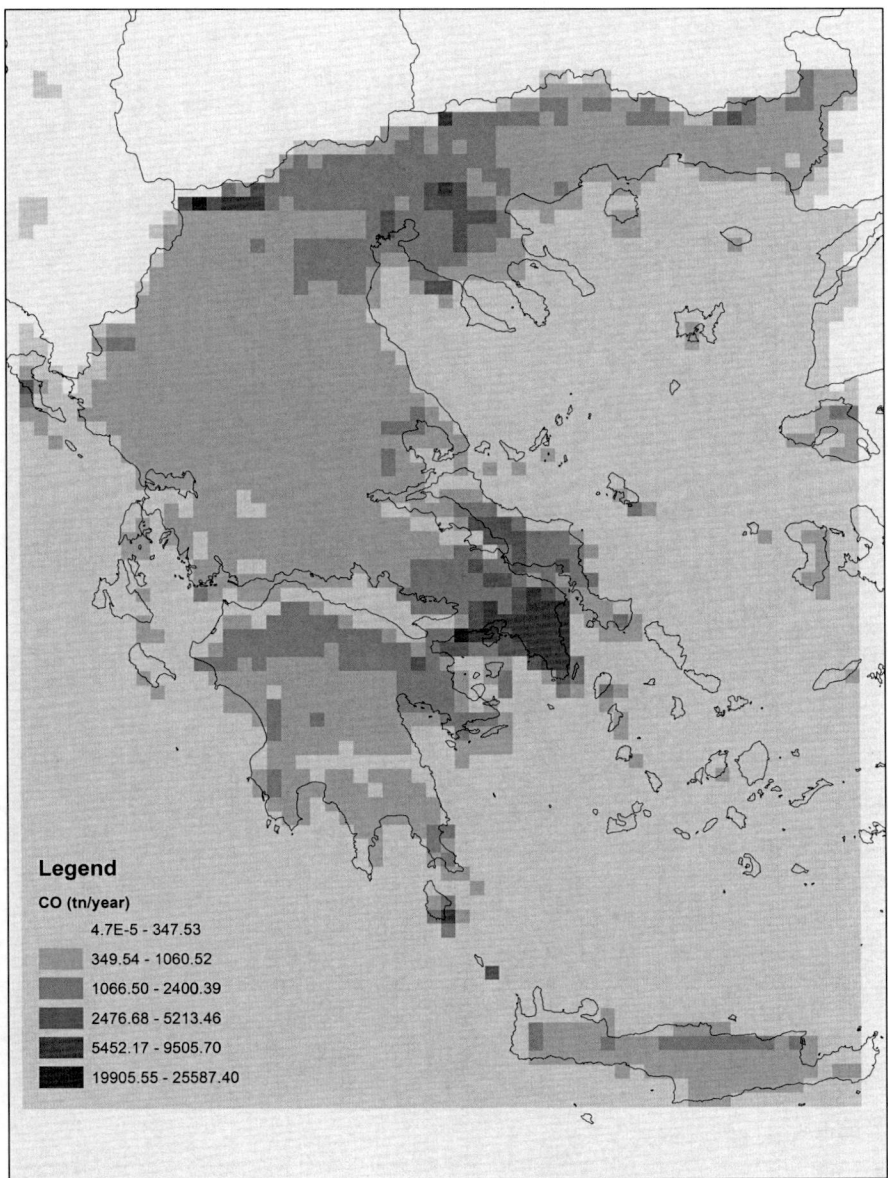

Fig. 7.5 Emissions (annual) of CO during 1999 in Greece

gases. The main sources of gaseous pollutants from the biosphere are vegetation, forest fires, land surface, lightings, wetlands, oceans, volcanos, wild fauna and humans. Emission inventories have been presented for several pollutants (EMEP/CORINAIR 2002) and these emissions contribute significantly to the atmosphere's

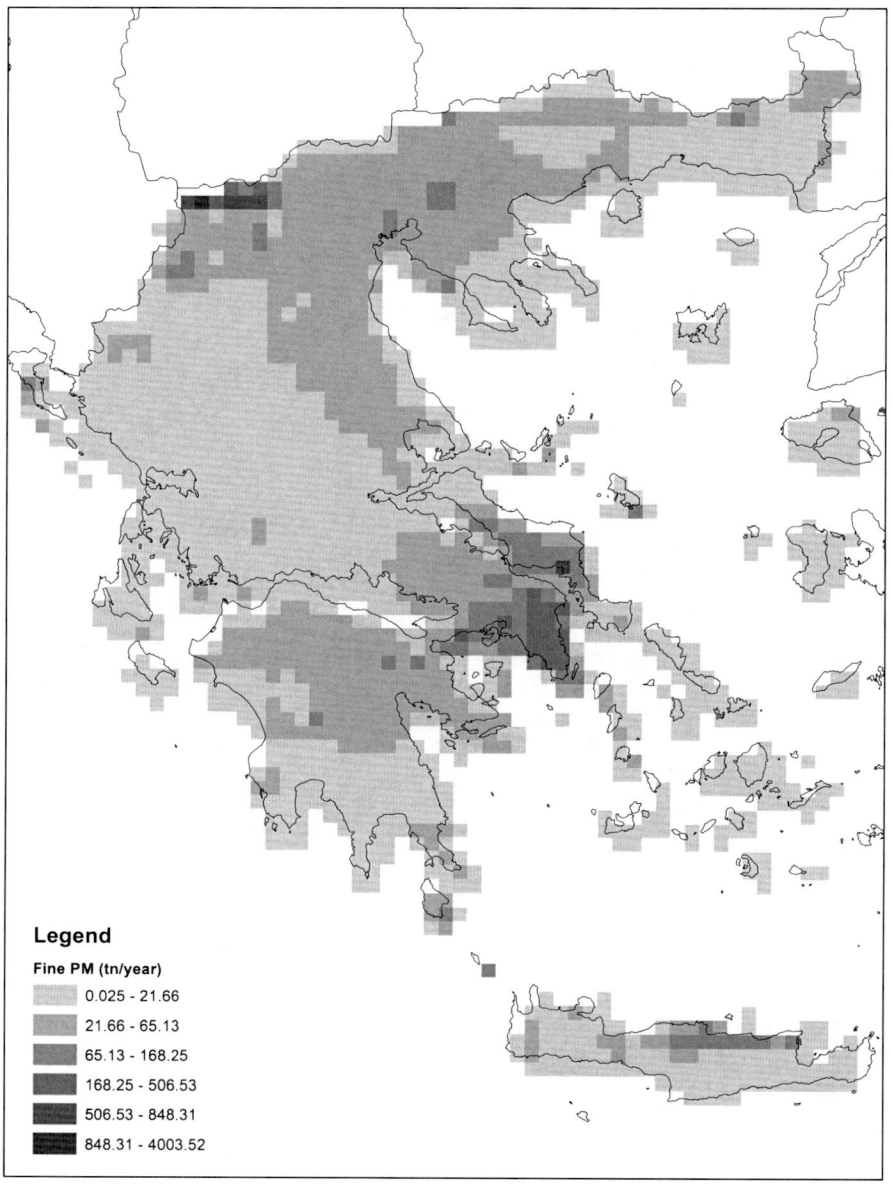

Fig. 7.6 Emissions (annual) of $PM_{2.5}$ during 1999 in Greece

chemistry. For example, emissions of nitrogen oxides from lightings and from wild fauna and the presence of humans. The biosphere is also a sink of pollutants such as methane and carbon dioxide.

7.5.1 Emissions of Volatile Organic Compounds from Vegetation

An important part of secondary aerosol particles in the atmosphere consists of secondarily formed organic matter produced from oxidation of organic compounds. Partitioning of gas-particle organic compounds in the atmosphere is an important task for determining their association with the fine particulate matter. Understanding the mechanisms which control the conversion of organic matter from the vapour phase to particulate matter will provide valuable information for determining future control strategies to reduce the partitioning of organic matter in the particulate phase. However, there is a great complexity of the number of different chemical forms of organic matter and the absence of direct chemical analysis has resulted in the use of experimentally determined fractional aerosol yields, fractional aerosol coefficients and adsorption/absorption methodologies (Lazaridis 2008) to describe the incorporation of organic matter in the aerosol phase.

An important pathway for secondary organic particle formation arises from biogenic hydrocarbons. There are very large quantities of biogenic hydrocarbons that are globally emitted which are also highly reactive. The chemical structures of some biogenic hydrocarbons relevant to SOA formation are shown in Fig. 7.7. SOA formed from the oxidation of VOCs produces highly oxidized compounds which are difficult to measure with current methods. Annual global emissions of biogenic hydrocarbons are estimated to be between 825 and 1,150 Tg C (per year), whereas the anthropogenic emissions are estimated to be less than 100 Tg C (per year). A detailed overview of the formation of organic aerosols from biogenic hydrocarbons is reviewed by Lazaridis (2008)

7.5.2 Calculation of Biogenic Emissions

Of specific importance is the estimation of specific BVOCs due to their amount and role in atmospheric ozone chemistry (isoprene) and to secondary aerosol formation (reaction with OH■ radicals and O_3; monoterpenes). BVOCs hourly emissions were estimated using the methodology presented in the EMEP/CORINAIR (2002) Atmospheric Emission Inventory Guidebook (under SNAP code 11) with modified environmental correction factors for light dependence of isoprene and, additionally, temperature dependence of monoterpenes emissions. In addition other BVOCs (alcohols, aldehydes, oxygenated compounds) emissions were quantified. In particular the emissions E ($\mu g\ C\ m^{-2}\ h^{-1}$) of BVOCs are estimated by the equation:

$$E = \varepsilon D \gamma \quad (7.19)$$

where ε is the emission potential for any species at a standard temperature $T_s = 303$ K and a standard Photosynthetically Active Radiation (PAR) flux equal

Fig. 7.7 Biogenic hydrocarbons influencing SOA formation in the atmosphere. Carbon atom bonds are shown with vertices whereas the hydrogen atoms are not shown

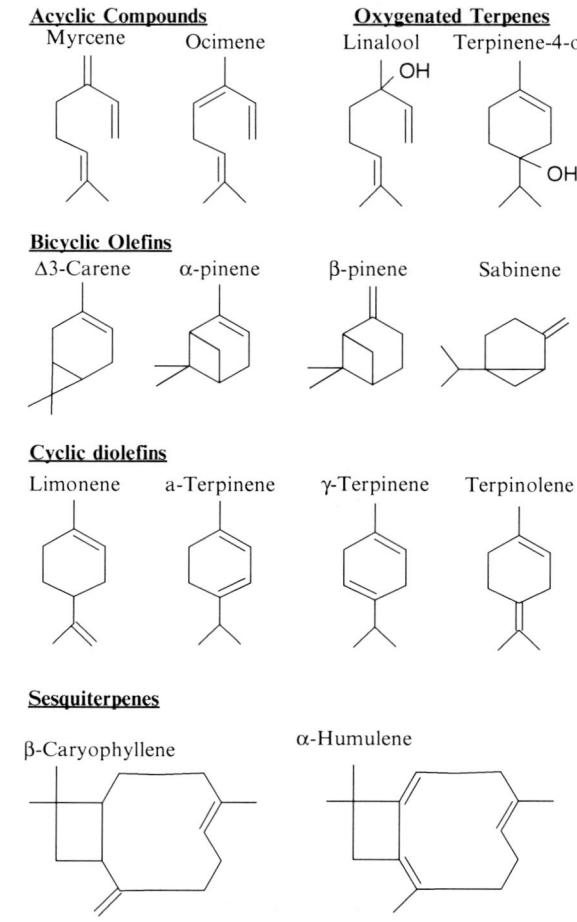

to 1000 µmol photons (400–700 nm) m^{-2} s^{-1}, D is the foliar biomass density (g of dry weight biomass foliage m^{-2}), and γ is a unitless environmental correction factor representing the effects of short-term (e.g. hourly) temperature and solar radiation changes on emissions. The above equation is adjusted for the area covered by vegetation and the actual land area of each cell and gives emissions in µg h^{-1}cell^{-1}.

The emissions of isoprene depend on light and temperature whereas emissions of monoterpenes and OVOCs depend only on temperature. The environmental correction factors for the emissions of the three BVOCs groups were estimated by the equations:

$$\gamma_{iso} = C_T \times C_L$$

$$C_T = \frac{\exp(C_{T1}(T - T_s)/RT_sT)}{1 + \exp(C_{T2}(T - T_M)/RT_sT)}$$

7.5 Emissions from the Biosphere

$$C_L = \frac{F \times L + L_1 + L_2 - \sqrt{(F \times L + L_1 + L_2)^2 - 4F \times L \times L_1}}{2L_1} \quad (7.20)$$

$$\gamma_{mts} = \exp(0.0739 \cdot (T - T_s))$$

$$\gamma_{OVOC} = \gamma_{mts}$$

where C_T and C_L are the environmental correction factors for temperature-dependence and light-dependence of isoprene emissions, respectively, T is the leaf temperature (K), $C_{T1} = 95{,}000$ J mol^{-1}, $C_{T2} = 230{,}000$ J mol^{-1}, $T_M = 314$ K, $T_S = 303$ K, R is the gas constant (8.314 J K^{-1} mol^{-1}), $F = 0.385$, $L_1 = 105.6$, $L_2 = 6.12$

The hourly emission variability (moles/h) is given in Fig. 7.8

In addition Fig. 7.9 shows the mean monthly terpene emissions (Gg/month) during 1997 in Greece.

7.5.3 Sea Salt Emissions

Sea-salt particles are emitted directly from the sea surface as spume from the whitecap cover of waves at high wind speeds (>10 m/s) and in the form of film and jet drops during the breaking of waves on the sea and other surfaces at lower wind speeds. Oceans comprise an important source of primary pollutants such as carbon monoxide and sulphur compounds. In addition, oceans continuously emit into the atmosphere particles of sodium chloride (NaCl) which have sizes between 0.05–10 µm diameter.

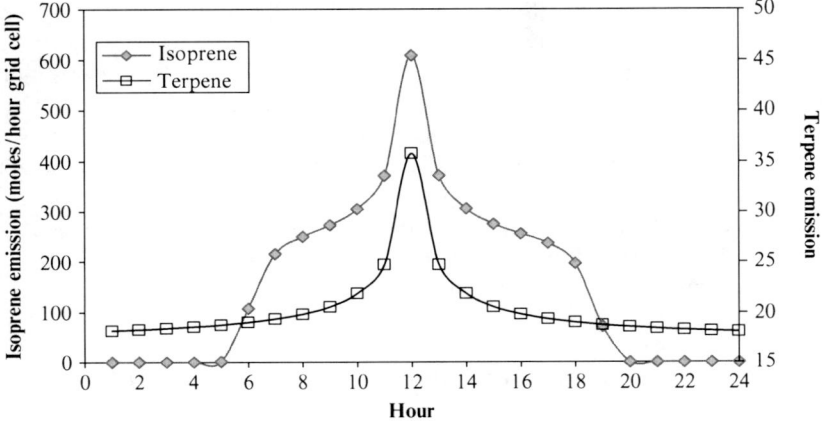

Fig. 7.8 Emissions of isoprene and terpene (moles/hour grid cell) derived from a simulation in Greece (grid cell size: 5 × 5 km^2)

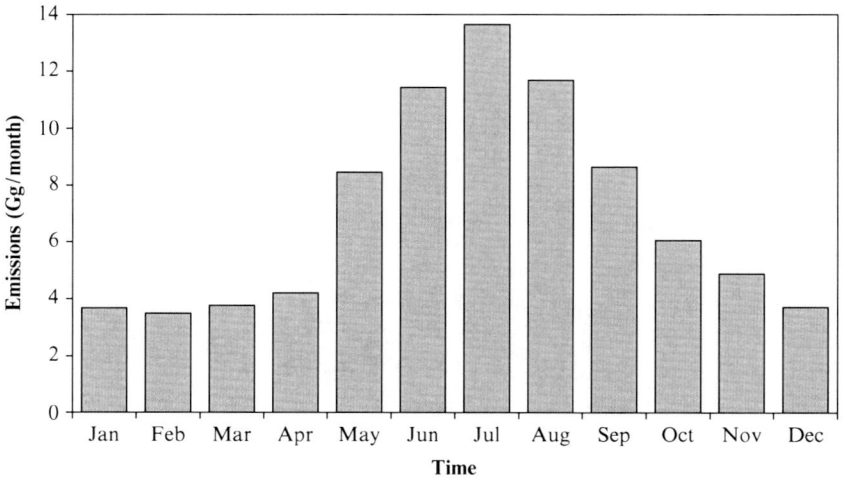

Fig. 7.9 Monthly terpene emissions (Gg/month) in Greece during 1997

Table 7.2 Global emissions of sea salt (Tg) during 2000 (IPCC 2007)

Particle diameter	Northern hemisphere	Southern hemisphere	Globally	Minimum value [a]	Maximum value [a]
$d < 1$ μm	23	331	54	18	100
$d = 1–16$ μm	1,420	1,870	3,290	1,000	6,000
Total	1,440	1,900	3,340	1,000	6,000

[a] Due to uncertainties the real emission values deviate from the values of this table. The presented values give the limits of uncertainty

Table 7.2 shows the emission data (Tg) during the year 2000 as calculated by the Intergovernmental Panel on Climatic Change (IPCC) (IPCC 2007).

The sea-salt particles play an important role in atmospheric chemistry and influence the concentration of gaseous components such as hydrogen chloride (HCl), sulphuric acid (H_2SO_4), nitryl chloride ($ClNO_2$) and ammonia (NH_3). Oceans are also a sink of sulphuric acid, decreasing the potential of sulphuric acid formation inside the marine boundary layer. Specifically, at the coastal zones the sea-salt particles influence the planet's radiation budget directly (with the scattering of the radiation) and indirectly (with the formation of cloud condensation nuclei (CCN)). Determination of the effective particle surface is very important for the specification of their radiation impact.

7.5.4 Emissions of Air Pollutants from the Earth's Surface

Emissions of gaseous and particle pollutants from the Earth's surface can affect the air quality of a region. Specifically, the surface is a huge particle source for the atmosphere due to particle resuspension. Gaseous pollutants (such as nitrogen oxides) are emitted from the surface with rates much smaller than for particles. However, natural emissions of nitrogen oxides (NO_x) from the surface may contribute significantly to the nitrogen oxides level in the atmosphere and specifically in rural areas. Important are also the emissions from inland water surfaces (CH_4, COS, DMS, H_2S, CS_2, VOC, CO_2, N_2O) (EMEP/CORINAIR 2002) and emissions of dimethyl sulfide (CH_3SCH_3, DMS) from the ocean surface.

Furthermore, emissions of nitrogen oxides (NOx) from the Earth's surface comprise 16% of the total concentration in the troposphere at global level. The Earth's surface emits nitrogen oxides through biological processes (bacterial processes).

In addition, resuspended dust emissions present a significant source of coarse particles in the atmosphere. The amount of windblown dust emissions is related to the friction velocity and depends on the surface roughness length, the soil texture and moisture and the threshold friction velocity. Finally, in the next section an analysis of the methodology for the evaluation of forest fire emissions is presented due to the significant role that forest fires play in climate change.

7.5.5 Emissions of Pollutants from Forest Fires

Forest fires emit significant amounts of gaseous and particulate matter pollutants into the atmosphere. The scientific literature has concluded that agricultural fires in Eastern Europe can significantly alter the air pollution levels in the European Arctic. Forest fires can affect the physicochemical properties of the atmosphere, via the release of significant amounts of particulate matter, which interact with solar radiation. Black carbon, for example, absorbs solar radiation strongly, and biomass burning is responsible for as much as 45% of the emissions of black carbon on a global scale. Atmospheric PM also acts as cloud condensation nuclei (CCN), which are important for the radiation balance and the hydrological cycle. Future climate warming may enhance the occurrence and impact of forest fires on regional air quality.

Forest fire emissions can be important for local air pollution levels (IPCC 2007). According to the CORINAIR-2001 inventory, forest fires contribute 0.2% to the emissions of NO_x, 0.5% to the emissions of NMVOC, 0.2% to the emissions of CH_4, 1.9% to the emissions of CO, 1.2% to the emissions N_2O, and 0.1% to the emissions of NH_3 in Europe. As these emissions are constrained to short time periods and limited areas, the impact is more severe for public health (such as respiratory symptoms and illnesses including bronchitis, asthma, pneumonia and upper respiratory infection, impaired lung function and cardiac diseases).

Unlike other anthropogenic sources, forest and agricultural biomass fire emissions are poorly quantified in the literature, due to the difficulties in estimating their temporal and spatial distribution (Lazaridis et al. 2008). Furthermore, the smoke production can vary by an order of magnitude or more from year to year.

The process of burning consists of many stages, producing different compounds at each one of them, while the burnt material is inhomogeneous and difficult to describe in mathematical terms. This fact may cause significant differences between predicted and observed levels of air pollution.

It is complicated to quantify the burnt biomass as it depends on many parameters. The burning material is inhomogeneous, adding complexity. The categorization used in the present study is based on the studies of Lazaridis et al. (2008) and EMEP/CORINAIR (2002). Forest fire fuels include all the materials that can be affected by a fire such as shrubs, trees, leaves, branches, barks and all the organic matter that is present in the upper layers of the ground.

In order to estimate the quantity of a biomass burn, one needs to know the area burnt, the fraction of the average above-ground biomass relative to the total average biomass and the fraction burnt of the above ground biomass. The quantity of dry biomass of a burning species is then estimated as shown in Eq. (7.1) (Seiler and Crutzen 1980:

$$M = a \times b \times A \times B \quad (7.21)$$

where A is the area burnt (m^2), B is the mean biomass quantity per area unit (kg/m^2), a is the fraction of biomass above the surface and b is the burning efficiency of the vegetation which exists above the ground. The coefficients B, a and b depend on the type of the ecosystem. The burnt biomass per area unit (kg/m^2) has a value of 2.81 for Mediterranean forest, 2.40 for scrubland and 0.36 for grassland (Seiler and Crutzen 1980; CORINAIR, 2001).

The main carbon compounds emitted from a forest fire are carbon monoxide, carbon dioxide, methane and hydrocarbons. The quantity of carbon emitted (in kg) is estimated as:

$$C = 0.45 \times M \quad (7.22)$$

where 0.45 is the mean mass fraction of carbon in dry biomass (with mass M) and is considered independent of the type of biomass.

The emissions of carbon compounds are calculated using the following expression:

$$E_j = \varepsilon_j \times \delta_j \times C \quad (7.23)$$

where j is the compound, ε_j is the portion of the total carbon emitted as compound j and δ_j is a conversion factor used to estimate the emissions in tons. The values adopted for the above coefficients were obtained by Andreae (1991).

The nitrogen compounds used in the present study are nitrogen oxides, nitrogen protoxide and ammonia. The emitted nitrogen mass is estimated by

$$N = 0.0045 \times M \quad (7.24)$$

where 0.0045 is the mass fraction of nitrogen in dry biomass and is assumed to be the same for all species. The emissions (in kg) of the nitrogen compounds are:

$$E_j = \varepsilon_j \times N \tag{7.25}$$

where j is the compound, ε_j is the fraction of total nitrogen emitted as compound j. The values of factor ε_j for each species are taken from EMEP/CORINAIR (2002)

The main sulfur compound emitted from forest fires is sulfur dioxide. The mass of SO_2 emitted can be estimated by

$$E_S = 1.6 \cdot 10^{-3} \cdot \times \cdot C \cdot = 0.72 \cdot \times 10^{-3} \cdot \times M \tag{7.26}$$

The total mass of particulate matter emitted from forest fires is found from

$$M_{TSP} = 0.0085 \cdot \times \cdot M \tag{7.27}$$

where 0.0085 is the mass fraction of total suspended particulate matter (TSP) of dry biomass.

7.6 Examples and Applications

Example 1. *Chemically non-reactive compounds have an initial concentration $c_i(0)$ and are emitted with a rate $q_i = 200$ μg m^{-2} h^{-1}. The background concentration is equal to $c_i(0) = 1$ μg m^{-3}. Calculate the steady state concentration above a city which is characterized by an average wind velocity of 3 m s^{-1}. The city has dimensions 100×100 km^2 and a boundary layer height of 1,000 m.*

Since the pollutants are not reactive, the term in the continuity equation which includes chemical reactions is equal to zero ($R_i = 0$). It is also assumed that $\frac{v_{d,i}}{H} c_i << \frac{c_i - c_i^o}{\tau_r}$ since there is no information for particle removal from scavenging. Therefore it is assumed that the variation of the concentration is dependent mainly on advection and the continuity equation can be expressed as:

$$\frac{dc_i}{dt} = \frac{q_i}{H} + R_i - \frac{v_{d,i}}{H} c_i + \frac{u}{\Delta x} (c_i^o - c_i)$$

Consequently the above equation in the specific application can be written as: $\frac{dc_i}{dt} = \frac{q_i}{H} + \frac{(c_i^o - c_i)}{\tau_r}$ *with $c_i = c_i(0)$ for $t = 0$ which can be also written as:*

$$\tau_r \frac{dc_i}{dt} + c_i = \left(\frac{q_i \tau_r}{H} + c_i^o \right)$$

The solution of the above equation is the sum of the general solution of the homogeneous equation and a partial solution of the general equation which can be written as:

$$\frac{dc_i}{dt} + \frac{c_i}{\tau_r} = \left(\frac{q_i}{H} + \frac{c_i^o}{\tau_r}\right)$$

For the homogeneous equation we can write that:

$$\frac{dc_i}{dt} + \frac{c_i}{\tau_r} = 0 \Rightarrow \frac{dc_i}{c_i} = -\frac{dt}{\tau_r} \Rightarrow \int_{c_i(0)}^{c_i} \frac{dc_i}{c_i} = -\int_0^t \frac{dt}{\tau_r}$$

Ordinary differential equations of the form $\frac{dy}{dx} + P(x)y = Q(x)$ have a solution which can be written as $\mu(x)y = \int \mu(x)Q(x)dx$ where $\mu(x) = \exp\left[\int P(x)dx\right]$ and as a result for the partial solution to have:
$\frac{dc_i}{dt} + \frac{c_i}{\tau_r} = \left(\frac{q_i}{H} + \frac{c_i^o}{\tau_r}\right)$ *where $\mu = \exp(t/\tau_r)$.*
Since $\mu(x)y = \int \mu(x)Q(x)dx$ the above equation can be written as:

$$\exp\left(\frac{t}{\tau_r}\right)c_i = \int \exp\left(\frac{t}{\tau_r}\right)\left(\frac{q_i}{H} + \frac{c_i^o}{\tau_r}\right)dt \Rightarrow c_i$$

$$= \left(\frac{q_i}{H} + \frac{c_i^o}{\tau_r}\right)\left[\int \exp\left(\frac{t}{\tau_r}\right)dt\right]\exp\left(-\frac{t}{\tau_r}\right) \Rightarrow$$

$$c_i = \left(\frac{q_i \tau_r}{H} + c_i^o\right)\left(1 + C\exp\left(-\frac{t}{\tau_r}\right)\right)$$

For the general solution at $t = 0$, then $c_i = c_i(0)$ and as a result the constant C is equal to -1. The final solution is:

$$c_i(t) = c_i(0)\exp\left(-\frac{t}{\tau_r}\right) + \left(\frac{q_i \tau_r}{H} + c_i^o\right)\left(1 - \exp\left(-\frac{t}{\tau_r}\right)\right)$$

where τ_r is the time which is needed for the cleaning of the air after cessation of emissions. The contribution to the concentration arises from two terms, the initial concentration (background) which is decreasing exponentially with characteristic time τ_r close to 10 h and the contribution due to emissions and advection.
For $t \gg \tau_r \Rightarrow \exp(-t/\tau_r) \cong 0$ and as a result the steady state concentration can be written as:

$$c_i^{ss} = \frac{q_i \tau_r}{H} + c_i^o \cong \frac{200\,(\mu g\, m^{-2} h^{-1})\,10h}{1000\,m} + 1\,\mu g\,m^{-3} = 3\,\mu g\,m^{-3}$$

Example 2. SO_2 *is emitted in an urban area with rate 2,000 $\mu g\,m^{-2}\,h^{-1}$. The mixing height is 1,000 m, the residence time in the atmosphere is 20 h, and SO_2 reacts with a mean rate 3% h^{-1}. Regions close to the urban area have SO_2 concentration equal to 2 $\mu g\,m^{-3}$. What is the mean SO_2 concentration in the urban area for the above conditions? Assume that the dry SO_2 deposition is equal to 1 cm s^{-1} and the atmosphere is free from clouds and mist.*

7.6 Examples and Applications

The single-cell model (volume model) in the Euler coordinate system can be written as:

$$\frac{dc_s}{dt} = \frac{q_s}{H} - k c_s - \frac{v_s}{H} c_s + \frac{(c_s^o - c_s)}{\tau_r}$$

The solution of the above ordinary differential equation is

$$c_s(t) = \frac{A}{B} + \left(c_s(0) - \frac{A}{B}\right) e^{-Bt}$$

where

$$A = \frac{q_s}{H} + \frac{c_s^o}{\tau_r}$$

and

$$B = k + \frac{v_s}{H} + \frac{1}{\tau_r}$$

with

$$(k = 0.03 h^{-1}), \; (v_s/H = 0.036 \, h^{-1}), \; \left(\frac{1}{\tau_r} = 0.05 \, h^{-1}\right), \; (B = 0.116 h^{-1}),$$

$$\left(\tfrac{q_s}{H} = 2 \, \mu g \, m^{-3} h^{-1}\right), \; \left(\tfrac{c_s^o}{\tau_r} = 0.1 \, \mu g \, m^{-3} \, h^{-1}\right), \; (A = 2.1 \, \mu g \, m^{-3} \, h^{-1})$$

For $t >> 1/B$ the concentration reaches a steady state condition.

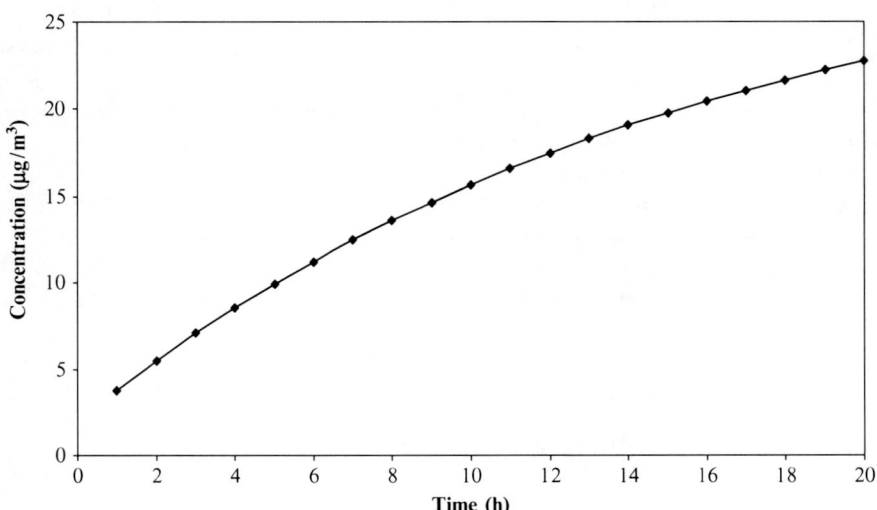

Fig. 7.10 Variation of the SO_2 concentration versus time

The mean concentration of SO_2 which is determined with the Euler method is equal to 18.1 $\mu g\, m^{-3}$ after the replacement of all variables in the general solution. In the case of the Lagrange volume model the result is $\frac{dc_s}{dt} = \frac{q_s}{H} - k c_s - \frac{v_s}{H} c_s$ with solution $c_s(t) = \frac{D}{E} + \left(c_s^o - \frac{D}{E}\right) e^{-Et}$ where $D = \frac{q_s}{H}$, $E = k + \frac{v_d}{H}$ and therefore $D = 2$ $\mu g\, m^{-3}\, h^{-1}$ and $E = 0.066\, h^{-1}$.

In the solution with the method of Lagrange the pollutant concentration reaches a stable state value at times larger than 33.3 h (Fig. 7.10). The mean SO_2 concentration with the Lagrange method is equal to $\mu g\, m^{-3}$

Problems

7.1 Sulphur dioxide is emitted in an urban area with rate 3 mg $m^{-2}\, h^{-1}$. The concentration of the hydroxyl radical changes during the day with a rate

$$[OH] = [OH]_{max} \sin[\pi(t-6)/12]$$

where t is the time (h) having midnight as a starting point. Suppose that the [OH] concentration is zero when $0 \leq t \leq 6$ and also during the period $18 \leq t \leq 24$. The mixing height is variable and is equal to 300 m during night and 1,000 m during day. In the area under study $[OH]_{max} = 10^7$ molecules cm^{-3} and time residence 24 h. The background concentration of sulphur dioxide is equal to 0.2 $\mu g\, m^{-3}$ and the concentration of the sulphur particles is equal to 1 $\mu g\, m^{-3}$. Using the Euler methodology calculate the following:

(a) The mean and maximum concentration of sulphur dioxide and sulphur particles.
(b) The time and the duration that the concentration is maximum. Discuss the results.
(c) Assume that during an air pollution episode the exposure time increases to 36 h. Calculate the maximum concentration of sulphur dioxide and the concentration of sulphur particles during the episode.

7.2 A volume cell has dimensions $\Delta x = 5$ km, $\Delta y = 4$ km and $\Delta z = 0.1$ km. Assume that the concentration of a gaseous pollutant in the volume cell is $N_1 = 10^{11}$ moles/cm^3 and the velocity component at the west end of the cell is equal to $u_1 = +7$ m/s. At the east end of the cell the concentration is equal to $N_2 = 5\, 10^{11}$ moles/cm^3 and the velocity $u_2 = +8$ m/s. Suppose that external sources and sinks as well as dispersion due to turbulence are not present. Calculate the concentration N in the cell after 60 s, when the initial concentration is the average value of the concentration at the two sides of the cell. Calculate also the time which is necessary for the concentration N in the cell to be equal to zero.

References

EMEP/CORINAIR, Atmospheric Emission Inventory Guidebook',3 rd Edition, Prepared by the EMEP Task force on Emission Inventories, EEA Technical Report No 30,(2002)

IPCC Climate Change 2007: *The Scientific Basis. Contribution of Working Group I to the Fourth Assessment Report of the Intergovernmental Panel on Climate Change* (2007)

Lazaridis The Environmental Chemistry of Aerosols. *Edited by I. Colbeck,* (2008).

Lazaridis, M., Latos, V., Aleksandropoulou, Ø., Hov, A., Papayannis, & Tørseth, K. (2008). Contribution of forest fire emissions to atmospheric pollution in Greece. *Air Qual. Atmos. Health, 1,* 143–158.

Seiler, W., & Crutzen, P. J. (1980). Estimates of gross and net fluxes of carbon between the biosphere and the atmosphere from biomass burning. *Clim. Change, 2,* 207–247.

Seinfeld, J. H., & Pandis, S. N. (2006). *Atmospheric chemistry and physics* (2nd ed.). New York: Wiley.

Chapter 8
Indoor Air Pollution

Contents

8.1	Introduction to Indoor Air Quality	256
8.2	Ozone	262
8.3	Nitrogen Oxides	265
8.4	Volatile Organic Compounds	267
8.5	Chemistry of Organic Compounds Indoors	268
8.6	Radon	275
	8.6.1 Radiactive Decay of Radon Isotopes	277
	8.6.2 Exposure and Dose of Radon in Indoor Environment	279
	8.6.3 Examples	282
8.7	Carbon Monoxide	283
8.8	Asbestos	284
8.9	Heavy Metals	285
8.10	Formaldehyde	287
8.11	Pesticides	290
8.12	Polycyclic Aromatic Hydrocarbons (PAH)	291
8.13	Polychloric Biphenyls (Pcbs)	291
8.14	Tobacco Smoke	293
8.15	Bioaerosols	294
8.16	Microenvironmental Models	296
8.17	Air Exchange Rate by Infiltration	299
8.18	Emission Models	300
8.19	Deposition Models	301
	8.19.1 Examples	301
References		303

Abstract People spend about 85% of their time indoors and an additional 3% inside vehicles. Therefore people are exposed to gaseous air pollutants and particulate matter from both outdoor (ambient) sources, through infiltration of outdoor air, and indoor sources. The main question which arises is how safe is the house environment in relation to the air quality indoors. A general overview of indoor air quality is given in Chapter 8. A review of the most common air pollutants and their sources is performed. Air pollutants such as ozone, particulate matter, nitrogen oxides, volatile organic compounds, radon, carbon monoxide, asbestos, heavy metals, formaldehyde, polycyclic aromatic hydrocarbons, polychloric diphenyls and pesticides are examined as well as the chemistry of organic compounds indoors. Furthermore the effect of tobacco smoke indoors is studied, since it is one of the most dangerous and widely found pollutants. We also examine some general aspects of bioaerosols indoors. Finally, we describe the formulation of mass balance models, which are called microenvironmental models.

8.1 Introduction to Indoor Air Quality

People spend about 85% of their time indoors and additional 3% inside vehicles. Therefore people are exposed to gaseous air pollutants and particulate matter from both outdoor (ambient) sources through infiltration of outdoor air and indoor sources (cigarette smoking, cooking, personal activities, dusting of indoor areas, and various other indoor activities), which may have different composition and possibly different toxicities.

The main question which arises is how safe is the house environment in relation to the air quality indoors. The above question is important for children who are spending close to 14.000 h (1.6 years) inside schools from kindergarten to high school. Therefore the study of human exposure from air pollutants in the indoor environment is an important scientific area of study. The health risk to humans from exposure to air pollutants is much higher compared to the exposure in the outdoor air. This is a result of the small volume where the human exposure occurs in the indoor environment which results to high air pollutant concentrations.

Parameters that are studied in relation to the human contribution to indoor air quality are smoking, combustion sources (e.g. heating), food preparation, as well as emissions from humans (e.g. breathing, sweating, hair and skin loss), domestic animals and other living organisms. Other parameters which are studied is the quality of building materials, emissions from detergents and cleaning agents, from synthetic carpets and other surfaces, from deodorants, from candles and other activities. The effect of each source on human health is determined by a number of conditions such as the chemical composition of the emitted pollutants, as well as their concentration. In specific cases, the correct adjustment and maintenance of emission sources are important for determination of the emission source risk. For example, a combustion source (such as an oven that burns gas) emits larger quantities of carbon monoxide if it is not regulated correctly.

There are several indoor sources of air pollutants. These include combustion sources due to heating, cooking and smoking. Additional sources are from building material, carpets, cleaning materials and wooden furniture. The outdoor environment is also a source for air pollutants indoors as well as the soil (radon) (see Table 8.1).

There are several factors that influence the concentration of air pollutants which are infiltrate from the outdoor air or are emitted indoors. The most important of them are the ventilation rate in the house and the meteorological conditions in the area of interest. In particular, the ventilation rate refers to the air exchange rate in the building and consequently to the rate of pollutant influx and outflux of the building. The ventilation rate depends on the building construction and can be achieved

8.1 Introduction to Indoor Air Quality

Table 8.1 Indoor pollutants and sources

Pollutants	Sources
Radon and radioactive daughters (^{222}Rn)	Soil, ground water, building materials
Nitrogen oxides (NO$_x$)	Combustion
Volatile organic compounds (including HCHO)	Building materials, carpets, solvents, paints, personal care products, house cleaning products, room fresheners, pesticides, mothball, humans
Carbon monoxide (CO)	Combustion
Ozone (O$_3$)	Outdoor air, photocopying machines, electrostatic air cleaners
Sulfur dioxide (SO$_2$)	Combustion
Particulate matter	Combustion
Asbestos	Building materials, hair drier
Bioaerosols	Air conditioners, cold water spray humidifiers

mechanically (air conditioning system) or naturally with window opening. In addition, air exchange is also performed also from the small openings and cracks in the building envelope. The meteorological conditions influence the ambient concentration of air pollutants and their indoor levels. Research findings have shown that the air velocity influences the infiltration of air pollutants indoors such as NO$_x$ and SO$_2$.

It is important to understand that air pollutants outdoors infiltrate to the indoor environment with a time delay which is dependent on the rate of air exchange. The fine particles with mass aerodynamic diameter smaller than or equal to 2.5 μm (PM$_{2.5}$) infiltrate indoors easier than do PM$_{10}$ particles. The rate of the particulate matter concentration change indoors is dependent on various parameters such as temperature, relative humidity, air exchange, chemical reactions and several other physico-chemical processes (e.g. condensation, coagulation, nucleation, deposition, resuspension). The equation which can describe all these processes and their effect on the chemical composition of particles (multicomponent aerosol) is the General Dynamic Equation (GDE).

Figure 8.1 shows the main processes which influence the concentration of particles indoors. Particles enter indoors through air infiltration from outdoors and also with the entrance of humans (from shoes and clothes). Chemical compounds may change their phase with the mechanisms of condensation and evaporation. Particles can also coagulate and deposit onto surfaces. Resuspension mechanisms may reintroduce he deposited particles into the air.

The particles which most easily infiltrate indoors are the fine particles (PM$_{2.5}$), which also pose a health risk to humans due to their deposition in the alveolar region. During the infiltration there are losses from the walls and inside the building surfaces. An example is the ozone concentration which is considerably lower indoors compared to the outdoor levels since ozone is deposited on indoor surfaces.

Emissions from indoor sources are not made in similar manner. Specific sources such as building materials, furnitures and carpets emit pollutants in continuous manner. Contrary other activities such as house cleaning and cooking emit pollutants during the activity period. Chemical compounds can be present for long time

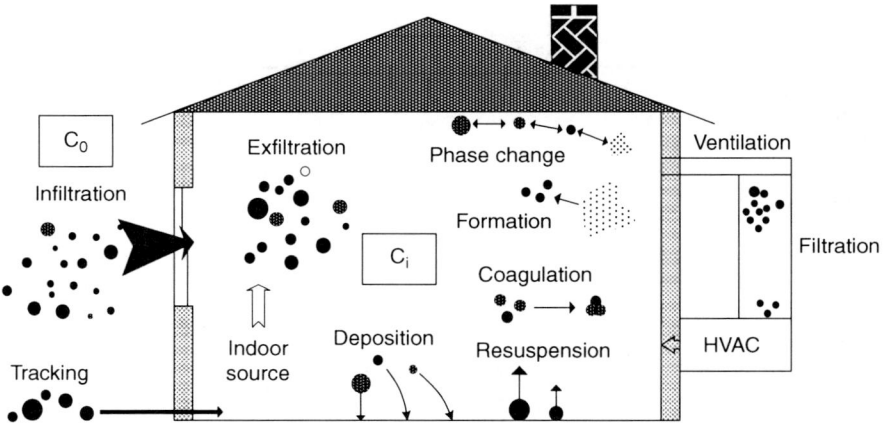

Fig. 8.1 Physico-chemical processes which affect the concentration of particles indoors (C_o, C_i refer to the pollutant concentration outdoors and indoors respectively). (adapted from Morrison 2010)

after their release. An example can be seen in the rugs which are used front of the house entrance for cleaning purposes.

However, the problem of the indoor air quality is complex since there is a multipathway (inhalation, dermal absorption, food intake) and multicomponent (several gaseous components, aerosol number size distribution, mass and chemical composition) human exposure indoors.

Experimental studies have shown that important sources of air pollutants including aerosols are the smoking, frying, cooking, vacuuming. In the case that there are no important indoor sources present, the air pollutant concentration indoors is lower than outdoors. An example is the cigarette smoking indoors which can lead to PM_{10} concentrations of several hundreds of $\mu g/m^3$.

Furthermore, volatile organic compounds (VOC) at elevated concentrations are arising from building materials, cleaning agents, furnitures, carpets and cosmetics. Bioaerosols are in larger quantities in the kitchen and the bathroom. Due to human activities indoors there is elevated concentration of organic matter in the particulate phase. Part of particulate matter chemical components is considered toxic and is consisted among other chemical components from polycyclic aromatic hydrocarbons (PAH) and metals. Increased concentration indoors has also the formaldehyde which is emitted from building materials and furnitures, bioaerosols, insecticides and pesticides.

The understanding of the complex sequence of events starting from the emissions of air pollutants to the atmosphere with the human health effects as a final event is necessary for the prognosis of potential risk to humans from specific chemical compounds and mixtures of them (see Fig. 8.2). Furthermore, the understanding of the chemical composition/size distribution characteristics of particulate matter (PM) and the chemical reactivity of gaseous pollutants together with their

8.1 Introduction to Indoor Air Quality

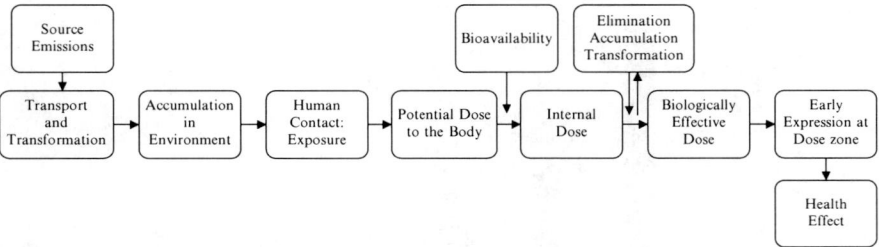

Fig. 8.2 Schematic representation of the complex sequence from emissions of air pollutants to health effects (Adapted from Lioy 1990)

Table 8.2 Comparison of indoor workplace standards with outdoor Federal and California State standards for selected gases

	Indoor			Outdoor	
Gas	8-h PEL[a] and TWA-TLV[b] (ppmv)[+]	15 min STEL[c] (ppmv)[+]	Ceiling (ppmv)[+]	NAAQS[d] (ppmv)	California standard (ppmv)
Carbon monoxide	35	–	200	9.5 (8 h)	9 (8 h)
Nitrogen dioxide	–	1	–	0.053 (annual)	0.25 (1 h)
Ozone	0.1	0.3	–	0.08 (8 h)	0.09 (1 h)
Sulfur dioxide	2	5	–	0.14 (24 h)	0.05 (24 h)

[+] National Institute for Occupational Safety and Health (NIOSH)
[a] *PEL*: Permissible exposure limits
[b] *TWA-TLV*: Time-Weighted average threshold limit values
[c] *STEL*: Short-time exposure limits
[d] *NAAQS*: National Ambient Air Quality Standard

indoor-outdoor characteristics and their relation to human exposure and internal dose are necessary steps for the quantification of human exposure to air pollutants.

Table 8.2 presents a comparison of workplace air quality standards for selected gasses.

Last years several research studies related to indoor air quality have been performed. The interest has been increased after complaints from occupants for indoor environments. Health symptoms were related with eye and nose problems, headaches and tiredness. These symptoms were associated with increased indoor concentration of air pollutants (e.g. formaldehyde), problems in the mechanical ventilation systems and indoor climatic conditions.

The importance of gas and charcoal indoor sources in restaurants is shown in Fig. 8.3. The use of charcoal results to significant increase of the concentration of carbon monoxide (CO) and respirable particles indoors. Both charcoal and gas are also important emission sources of CO_2, NO_2, C_6H_6 and C_7H_8 (toluene).

In this chapter several categories of air pollutants indoors are studied:

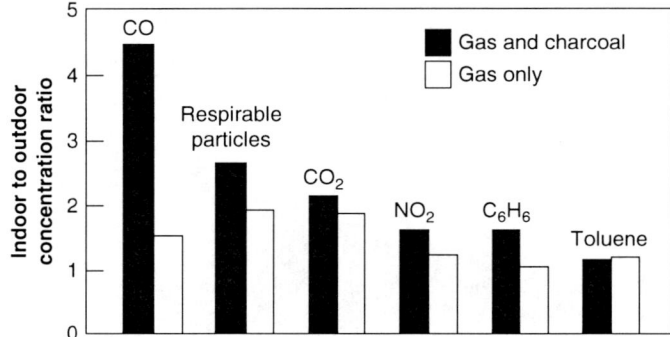

Fig. 8.3 Ratio of indoor/outdoor concentration of CO and several other gasses as measured at restaurants in Korea with the use of gas or gas and charcoal (Adapted from Baek et al. 1997)

- Particulate matter including bioaerosols
- Air pollutants such as carbon monoxide, nitrogen oxides, ozone, volatile organic compounds and formhaldehyde.
- Toxic chemical compounds such as asbestos and lead.
- Radon.
- Polycyclic aromatic hydrocarbons (PAHs) and polychloric diphenyls (PCBs).

Table 8.3 presents the chemical composition of particles ($PM_{2.5}$, PM_{10}) indoors from several sources as well as the processes which control their dynamics. Sources from biomass burning are not included which however are very important in developing countries in Asia and Africa.

Different activities indoors result to an increase of the concentration of airborne particles as described in Fig. 8.4. Even the folding of blankets results to an increase of the particulate matter concentration.

Furthermore, Table 8.4 shows possible chemical indexes (chemical compounds) and tracers which can be used for the identification of pollutant sources indoors.

Table 8.5 shows the importance of the human exposure in indoor environments where chemical compounds which are not reactive have one thousand times higher probability to enter into the human respiratory tract comparing to outdoor exposure.

Prior to the study of the indoor air pollutants it is adequate to remark that the air pollution problems of the indoor air can be classified at five categories:

- Incomplete cleaning and maintance of the indoor environment.
- Insufficient ventilation.
- Emissions of pollutants from indoor sources and activities inside the building.
- Air pollutants from outdoor sources.
- Biological contaminants due to insufficient humidity control indoors.

However, there is a possibility for the improvement and the conservation of the indoor environment with an objective to the minimization of the human exposure to harmful air pollutants. The methods for the management of indoor air pollution include preventive measures at the phase of the building design and construction and daily practices which include:

8.1 Introduction to Indoor Air Quality

Table 8.3 Chemical speciation of airborne particles emitted from indoor sources (Adapted from Ruzer and Harley 2005)

	$PM_{2.5}$	PM_{10}	Processes
Particles from outdoor sources	OC, NH_4^+, SO_4^{2-}, NO_3^-, EC, metals	Components from the Earth's crust such as: Fe, Ca, Si	Infiltration, deposition
Combustion	OC, EC		Pyrolisis, evaporation, combustion, nucleation, condensation
Cooking	Organic acids, aldehydes, ketones		
Wood -fireplace	Levoglucosan, methoxyphenols, resin acids		
Smoking	Organic acids, alkanes, heterocycle compounds		
Candles	Alkanes of high molecular weight, aldehydes, alkanes acids, esters, EC, Pb		Evaporation, condensation, pyrolisis
Aromatics	PAH, aldehydes, $PM_{2.5}$		
Kerosene stoves	$PM_{2.5}$, SO_4^{2-}, HONO		
Biogenic aerosols			
Pets		Allergens on the hair and dead skin of domestic animals (dander), endotoxins	Infiltration
Bacteria		Endotoxins	Infiltration
Household dust mites		Allergens	Infiltration
Human activities			
Use of pesticides		Powder again flea	Infiltration
Air refreshners + O_3	Organic aldehydes and acids		Nucleation, condensation
Walking		Soil elements, faeces from mites	Abrasion, resuspension
Cleaning	Particulate matter of indoor and outdoor places	Particulate matter	Resuspension
Cleaning products, + O_3	Organic aldehydes and acids		Nucleation, condensation
Renovation		Building materials	Abrasion, resuspension

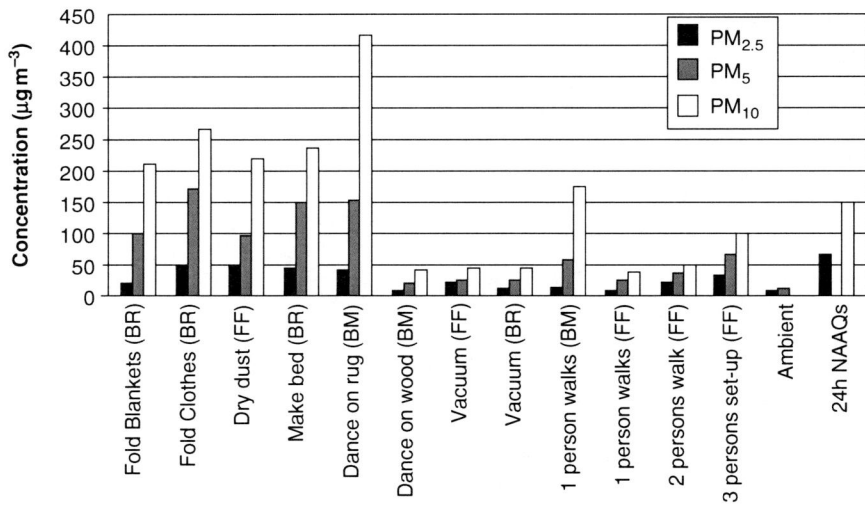

Fig. 8.4 Personal human exposure to particles (PM_{10}, PM_5, $PM_{2.5}$) during specific activities indoors

- Management of air pollution sources (replacement or modification of pollution sources).
- Management of activities indoors.
- Architectural interventions at building stage or a later stage with an objective to reduce air pollution levels.
- Increase of the air exchange rate in case of the existence of important indoor sources and decrease in case of high air pollution levels outdoors
- Cleaning of the indoor environment.

In the following sections the different indoor air pollutants will be examined in more detail. However, before that it is of interest to present some major differences between the indoor and outdoor environment. Table 8.6 shows some major differences between physical parameters indoors and outdoors. An important difference is the existence of large surfaces in the indoor environment (walls, furnitures etc.) which absorb and emit chemical components. Furthermore, many indoor activities emit significant quantities of gaseous species including volatile organic compounds and aerosols. The residence time of the air pollutants indoors is dependent on the air exchange rate with the outdoor environment. In the indoor environment the levels of the ultraviolet radiation are low and the photochemical reactions are not important. However, the heterogeneous chemical reactions are important. Finally, the variability of temperature and relative humidity indoors is small.

8.2 Ozone

The atmospheric ozone is discovered in the middle of the 19th Century (1840–1854) from the C.F. Schonbein. The origin of the word ozone is Greek and

Table 8.4 Possible chemical tracers of particle sources indoors

Particle source	$PM_{2.5}$	PM_{10}
Particles from outdoor sources	SO_4^{2-}, EC (if there are no combustion sources indoors)	Crustal materials such as Fe, Ca, Si
Combustion		
Cooking		
Meat	Cholesterol	
Seed, meat	Triglycerides	
Crop biomass	Levoglucosan	
Wood (from coniferous trees)	Resins, guaiacols, , diterpenoeidis)	
Wood (from deciduous and no coniferous trees, e.g. oak)	Syringols	
Smoking	K-heterocyclic compounds (nicotine), iso-and anti-alkanes, solanesol	
Candles	Wax esters	
Deodorants	–	–
Oil heater	HONO, excess SO_4^{2-}	
Biogenic aerosols		
Pets		Allergens on the hair and dead skin of domestic animals (dander), endotoxins
Strains		Endotoxins
Household dust mites		Allergens
Human activities		
Use of pesticides	Chlorinated compounds	
Deodorants + O_3	Organic acids and aldehydes	
Walking		Soil elements, faeces from mites
Cleaning	Particles	Airborne particles, crustal components, faeces from mites , dander
Cleaning products, + O_3	Organic aldehydes and acids	
Renovation		Cellulose, lignin, $CaSO_4^{2-}$

relates to its smell («όζει » in Greek means something which smells). The role of ozone in the ambient atmosphere is very important and it is one of the main chemical compounds.

The inhalation of ozone from humans can be dangerous. The negative influence of ozone to humans is due to its ability to oxidize the biological tissues. Therefore high concentration levels of ozone result to coughing, pain during deep breath and

Table 8.5 Atmosphere's mass and flux rates to the global atmosphere, the urban atmosphere and indoor places (Adapted from Nazaroff et al. 2003)

Environment	Mass (Kg)	Flux, F (Kg d^{-1})	Inhaled mass, Q^b (Kg d^{-1})	Ratio: Q/F
Global atmosphere	5×10^{18}	–	$\sim 10^{11}$	–
Urban atmosphere ^{c}	$\sim 10^{15}$	$\sim 3 \times 10^{15}$	$\sim 4 \times 10^{10}$	$\sim 10^{-5}$
Indoor atmosphere ^{d}	$\sim 10^{12}$	$\sim 10^{13}$	$\sim 8 \times 10^{10}$	$\sim 10^{-2}$

^{b} Includes air which was inhaled indoors and outdoors
^{c} Sum of all urban environments (globally)
^{d} Sum of all indoor environments (globally)

Table 8.6 Basic physical parameters for the ambient and indoor environment (Adapted from Nazaroff et al. 2003)

Parameter	Urban environment	Indoor atmosphere
Residence time	~ 10 h	~ 1
Light – energy flux	$\sim 1{,}000$ W m^2 (day)	~ 1 W m^{-2}
Ratio of surface – volume	$\sim 0{,}01$ m^2 m^{-3}	~ 3 m^2 m^{-3}
Wet deposition	~ 10–150 cm yr^{-1}	Does not exist

reduction of the human resistance to cold and pneumonia and at the same time favour the occurance of asthma and bronchitis.

High concentrations of ozone which acts as the main photochemical oxidizing agent in the atmosphere may result to the detoriation of materials. For example the ozone affects plastic material affecting the C = C bond which can eventually break. In the past bars of elastic caoutchouc were used for the studies of ozone concentration in the atmosphere. Ozone concentrations above 10 ppb are capable to detoriate synthetic fibers, coloured cloths and other materials. Detoriation of organic materials can be done also with other oxides such as the nitrogen oxides (NO$_x$). Due to the detoriation of materials from elevated ozone levels has been a considerable study in the literature for the determination of ozone levels in museums and historical buildings.

A part of the indoor ozone concentration is originating from the outdoor environment and infiltrates indoors during the air exchange. Indoor sources present the other contribution. The indoor ozone concentration is a function also of its absorption on the different indoor surfaces and gaseous reactions indoors.

The ozone is an undesirable pollutant for the indoor environment. Its effects are not only related to the damage of materials (plastics, organic fibers) but are also related to human health (inhalation and breathing problems).

Emission sources of ozone indoors are the photocopy machines and laser printers which has as a result the occurance of elevated ozone concentrations at office rooms. Also air purification facilities using electrostatic precipitators are also producing ozone.

The variability of ozone indoors is determined from several parameters as discussed above and therefore has significant daily and seasonal variation. The ozone has lower values in the indoor environment since it is absorbed on different materials (clothes, furnitures, human skin etc.) and reacts on their surface.

The concentration half time of ozone inside buildings is close to 7–10 minutes (Weschler 2000). This time is determined mainly from the deposition rate of the pollutant inside the building and the air exchange rate. Secondary importance for the ozone occurance indoors has the reactions in the gaseous phase since ozone deposits quick on the indoor surfaces. The ozone reacts also with organic chemical compounds and eventhough these compounds present a small fraction of all the chemical compounds indoors (close to 10%), the importance of these reactions is high since they are source of free radicals and organic compounds with odour (acids, alcohols, ketones etc.).

Furthermore in the indoor environment have been proposed specific chemical reactions of ozone which lead to the production of NO_3 and HNO_3:

$$O_3 + NO_2 \rightarrow NO_3 + O_2$$
$$NO_3 + NO_2 \leftrightarrow N_2O_5$$
$$N_2O_5 + H_2O \xrightarrow{surfaces} 2\,HNO_3 \quad (8.1)$$
$$NO_3 + RH \rightarrow HNO_3 + R$$

High ozone concentrations during summer, due to elevated Sun's radiation, will result to higher production of nitrogen compounds.

From the above chemical reactions it can be concluded that an indirect method for the determination of indoor sources of the nitrogen compounds can be based on the measurements of the indoor to outdoor ratio of the HNO_3 concentration. The same can be applied for the sulphur sources with the measurement of the indoor to outdoor ratio of SO_2.

8.3 Nitrogen Oxides

Elevated levels of nitrogen oxides have been observed at the indoor environment in case of existence of combustion sources indoors. Sources of nitrogen dioxide indoors include among others heating equipments and cigarette smoking. The combustion sources produce mainly nitrogen monoxide (NO) which is oxidized quickly to nitrogen dioxide (NO_2). The scientific interest is focused on the NO_2 levels mainly due to its effect to public health.

Figure 8.5 shows the concentration ($\mu g/m^3$) of NO_2 at a residential house in London. Higher concentrations are observed outdoors from vehicular traffic, whereas indoors were not obsereved large differences between the kitchen and the sitting room.

In addition, high NO_2 levels have been observed in ice rings where there is use of cars which operate with propane or diesel for the abrasion of ice which results to average concentrations of 200 ppb but have been observed levels close to 3 ppm.

When there are no sources of NO_x indoors then the indoor and outdoor concentrations are similar since the removal on the surfaces indoors is small.

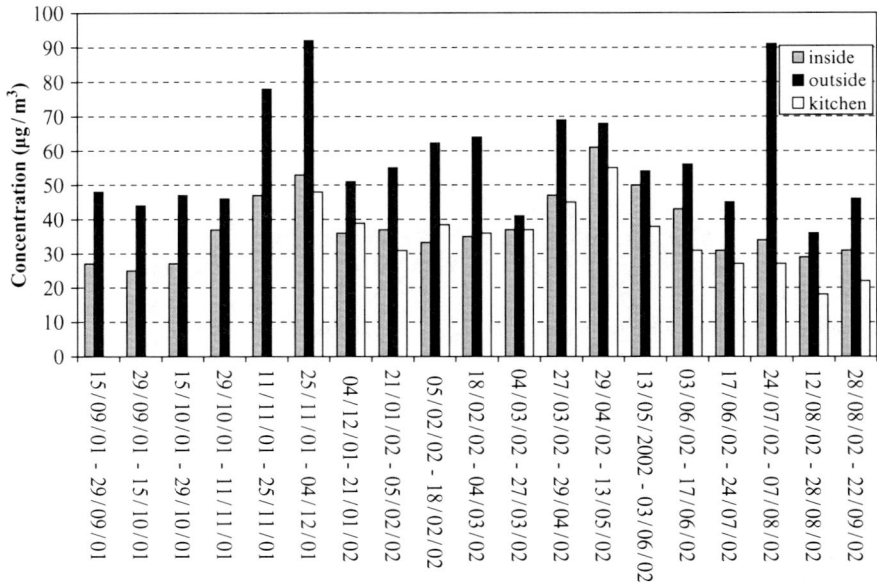

Fig. 8.5 Concentration of NO_2 at the London metropolitan area, indoors and at the kitchen (Source: I. Colbeck, Un. Essex, UK)

The chemical reactions of NO_2 on surfaces lead to the production of nitrous acid (HONO) (Finlayson-Pitts and Pitts 2000):

$$2\,NO_2 + H_2O \xrightarrow{surfaces} HONO + HNO_3 \quad (8.2)$$

In the above chemical reaction the nitrogen dioxide (NO_2) reacts with water at the surface of materials and the reaction rate is different for various surfaces. Higher values of NO_2 and humidity lead to increased concentrations of HONO. Concentrations of HONO at indoor places can reach 8 ppb for average daily concentration and 40 ppb for average concentration value of 6 h. The ratio of the concentrations of HONO and NO_2 outdoors is close to 10^{-2}, whereas for indoor places can reach values close to 0.4 (Finlayson-Pitts and Pitts 2000).

Reactions of NO_2 and HONO have been also observed on surfaces and this leads to the NO production indoors as described by the reaction:

$$NO_2(g) + HONO(g) \rightarrow H^+ + NO_3^- + NO(g) \quad (8.3)$$

The above reaction leads in specific cases to NO concentrations indoors larger than the outdoor levels.

Figure 8.6 shows the deposition rates of NO_2 at different surfaces indoors. Large variability exists for the removal rates of NO_2 at different surfaces which has as a result to differences to the production of NO and HONO.

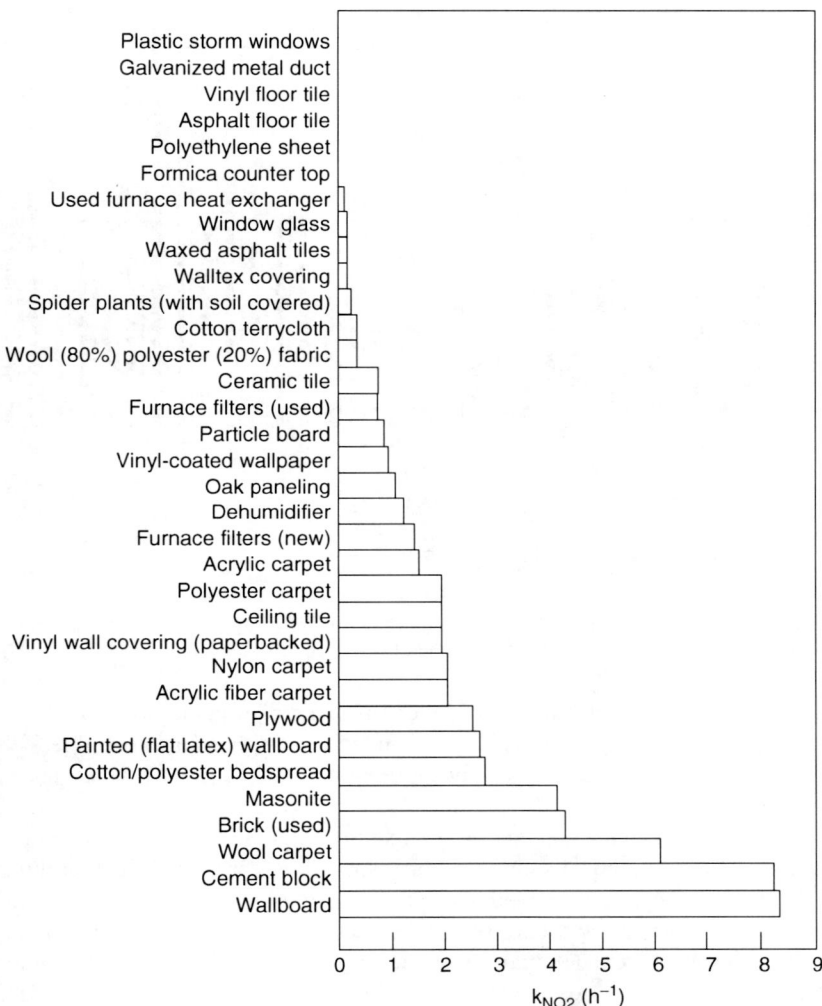

Fig. 8.6 Deposition rates (k_{NO2}) (h^{-1}) at different surfaces (Adapted from Spicer et al. 1989)

8.4 Volatile Organic Compounds

Volatile organic compounds (VOC) indoors are resulted from human activities and building materials. Furthermore, the chemical compounds may arise from the outdoor air with infiltration indoors. Sources of VOC are the paints, the solvents, sprays, air fresheners, floor varnish and wax, insecticides and storaged petrol or other fuels. Fig. 8.7 presents the concentration of several VOCs ($\mu g/m^3$) which are measured indoors in Olso Norway on 16.01.03 and 19.01.03. It is interesting to note the large number of organic chemical compounds present in the indoor environment.

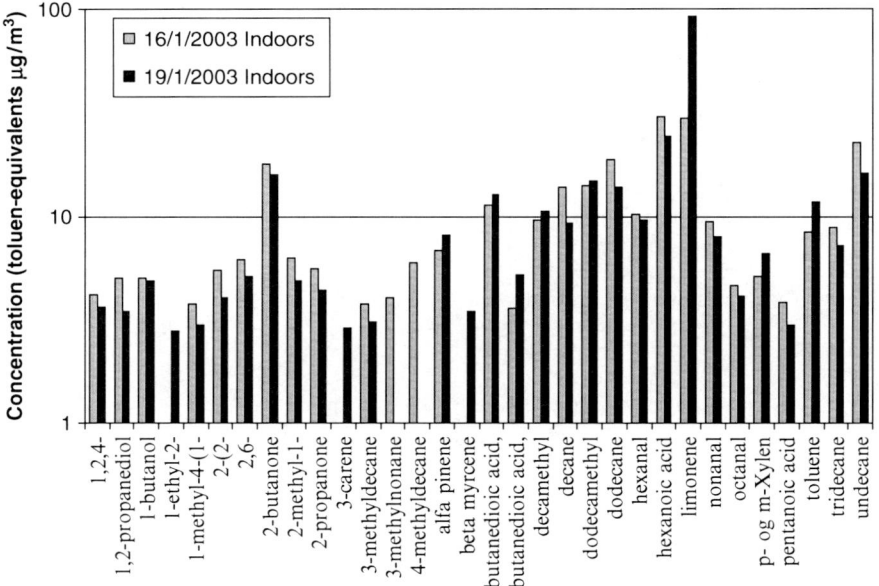

Fig. 8.7 Concentration of specific volatile organic compounds ($\mu g/m^3$) which are measured at Oslo, Norway at 16.01.03 and 19.01.03 (Adapted from Lazaridis and Alexandropoulou 2009)

Since there is a large number of indoor sources of organic compounds, consequenetly there is a large number of VOC indoors as depicted also in Table 8.7. At the absence of indoor sources the concentration of VOC indoors is the same as the outdoor levels.

There are cases where specific outdoor sources may result to elevated concentration of organic compounds indoors. Gasses which are produced from landfills or from fuel leakage can be entrained inside buildings which are located close to these sources through the soil. Furthermore, a significant source of organic compounds is from cars in case there is an indoor garage. Sources of VOCs are also the building materials of the houses, mainly for new built houses. Table 8.7 presents specific VOCs which arise from building materials (Table 8.8).

Furthermore, Table 8.9 shows the indoor/outdoor ratio (I/O) of the concentration of VOCs and typical concentrations in indoor environment. Table 8.10 presents the concentration of specific VOCs which are originating from building materials.

8.5 Chemistry of Organic Compounds Indoors

The chemical reactions which occur indoors have an important effect to the human exposure since there is production of new chemical compounds which come in contact with the human body. A large number of chemical reactions of organic

Table 8.7 Indoor Volatile Organic Compounds (VOCs) and their potential sources (Adapted from Tichenor and Mason (1988))

Functional groups/ Categories	Compounds	Sources
Hydrocarbons	Aliphatic hydrocarbons	Paints and adhesives, benzene, combustion products, floor wax
	Aromatic hydrocarbons (toluene, xylene, ethylbenzene, trimethylbenzene, styrene, benzene)	Insulation, textiles, disinfectants, paints, smoking
	Terpenes (limonene, a-pinene)	Fragrance deodorize, polish, textiles, cloth softener, cigarettes, food, alcoholic beverages
	PAHs	Combustion products (smoking, wood burning, kerosene heaters)
Oxygenated organic compounds	Acrylic acid esters, epichlorohydrin	Monomers can leak/escape from polymers
	Alcohols	Aerosols, glass detergents, paints, paint thinner, cosmetics, adhesives
	Aetones	Varnish, adhesives and varnish removers
	Ethers	Resins, dyes, varnishes, lac, paints, soaps, cosmetics
	Esters	Plastics, resins, plasticizers, lac thinners, spices, fragrances
	Ethylene oxide	Sterilizers (hospital)
Other organic compounds	Toluene-diisocyanate	Polyurethane foam aerosol
	Phthalic anhydride	Epoxy resins
	Sodium dodecyl sulfate (sodium lauryl sulfate)	Carpet shampoo
Chlorinated organic compounds	Benzyl chloride	Vinyl tiles
	Tetrachloroethylene	Dry-cleaned clothes
	Chloroform	Chlorinated water
	1,1,1-trichloroethane	Dry-cleaned clothes, aerosol spray, textile preservatives
	Carbon tetrachloride	Industrial cleaning products
	p-dichlorobenzene	Moth crystals, deodorants

compounds can be observed indoors such as oxidation reactions in the system ozone – terpenes, chemistry of acids on surfaces and hydrolysis of plasticizers. Heterogeneous chemical reactions occur on the surfaces indoors and on the aerosols. Homogeneous chemical reactions occur among molecules in the gaseous phase, inside solids and into the thin water layers on the surfaces of materials. The chemical reactions indoors can be divided in different categories as shown in Fig. 8.8.

Table 8.8 Some Volatile Organic Compounds (VOCs) from building materials (Adapted from Finlayson-Pitts and Pitts (2000 and Tichenor and Mason (1988))

Source	Compounds
Carpet adhesive	Toluene
Floor adhesive (water based)	Nonane, decane, undecane, dimethyloctane, 2-methylnonane, dimethylbenzene
Particleboard	Formaldehyde, acetone, hexanal, hexanol, propanol, butanone, benzaldehyde, benzene
Moth crystals	p-Dichlorobenzene
Floor wax	Nonane, decane, undecane, dimethyloctane, trimethylcyclohexane, ethylmethylbenzene
Wood stain	Nonane, decane, undecane, methyloctane, dimethylnonane, trimethylbenzene
Latex paint	2-propanol, butanone, ethylbenzene, propylbenzene, 1,1'-oxybis[butane], butylpropionate, toluene, formic and acetic acids
Water based acrylic wall paint	1,2-propanediol, isomers of 2,4,4-trimethyl-1,3-pentanediol monoisobutyrate
Furniture polish	Trimethylpentane, dimethylhexane, trimethylhexane, trimethylheptane, ethylbenzene, limonene
Polyurethane floor finish	Nonane, decane, undecane, butanone, ethylbenzene, dimethylbenzene
Room freshener	Nonane, decane, undecane, ethylheptane, limonene, substituted aromatics (fragrances)
Vinyl flooring	Alkyl aromatics, dodecane, 2,2,4-trimethyl-1,3-pentanediol, diisobutylate, 2-ethyl-1-hexanol, phenol, cresol, ethyl hexyl acetate, ammonia
Floor varnish	Butyl acetate, n-methylpyrrolidone
Laminated cork floor tile	Phenol
Carpets	4-Phenylcyclohexene, styrene, 4-ethenylcyclohexene, 2-ethyl-1-hexanol, nonanol, heptanol
Silicon caulk	Methyl ethyl ketone, butyl propionate, 2-butoxyethanol, butanol, benzene, toluene
Paint	Dibutyl phthalate
Acrylic sealant	Hexane, dimethyloctanols
Creosote impregnated timber	Naphthalene, methylnaphthalenes

An important part of chemical reactions indoors are the reactions of ozone with terpenes (Weschler 2004). Terpenes consist an important class of biogenic volatile organic hydrocarbons (VOCs) and have as a common unit the isoprene. Monoterpenes ($C_{10}H_{16}$) are emitted indoors from wood products, air fresheners and cleaners. Many terpenes have a simple or more double bonds and react with ozone. Table 8.11 shows the kinetics of specific terepenes with $_3$, HO and NO_3.

A considerable interest is in relation with specific chemical reactions such as of α-pinene with ozone (Fig. 8.9). Ozone breaks the double bond and results to

8.5 Chemistry of Organic Compounds Indoors

Table 8.9 Indoor/Outdoor concentration ratio (I/O) of VOCs and their typical indoor concentrations (Adapted from Brown et al. (1994))

Compound	I/O ratio (concentration in µg/m^3)	Compound	I/O ratio (concentration in µg/m^3)
n-Alkanes		*Alcohols*	
n-pentane	3	2-propanol	>73
n-hexane	9	*n*-butanol	5 (<1)
n-heptane	4 (1–5)	*Aldehydes*	
n-octane	7 (1–5)	Acetaldehyde	5
n-nonane	14 (1–5)	Butanal	2 (1–5)
n-decane	19 (5–10)	Hexanal	>5
n-undecane	20 (1–5)	Nonanal	5 (5–10)
n-dodecane	20 (1–5)	*Ketons*	
n-tridecane	>6	Acetone	12 (20–50)
n-tetradecane	16 (1–5)	Methyl-ethyl-ketone	4 (1–5)
n-pentadecane	>5 (1–5)	*Esters*	
Cycloalkanes		Ethyl acetate	15 (5–10)
2-methyl-pentane	2	*Aromatic hydrocarbons*	
2-methylhexane	2	Styrene	10 (1–5)
3-methylhexane	3	Benzene	3 (5–10)
Cyclohexane	4 (1–5)	Toluene	6 (20–50)
Halogens		Ethylbenzene	6 (5–10)
Trichlorofluoromethane	10	*m*- and *p*- xylene	6 (10–20)
1,2-dichloroethane	12 (<1)	*o*-xylene	6 (5–10)
Dichloromethane	6 (10–20)	*n*-propylbenzene	4
Chloroform	5 (1–5)	1,3,5-trimethylbenzene	4 (1–5)
Carbon tetrachloride	2 (1–5)	1,2,4- trimethylbenzene	15 (5–10)
1,1,1-trichloroethane	9 (20–50)	(1-methylethyl)benzene	5
1,1-dichloroethane	13 (1–5)	Naphthalene	4 (<1)
Trichloroethylene	6 (1–5)	*Terpenes*	
Tetrachloroethylene	5 (5–10)	Camphene	20 (10–20)
p-dichlorobenzene	5 (5–10)	*a*-pinene	23 (1–5)
m-dichlorobenzene	0.4 (<1)	Limonene	80 (20–50)
Total volatile organic compounds (TVOCs) concentration: 7 µg/m^3			

an ozonide which is further breaks to several products as shown in Fig. 8.9. The production of formaldehyde is occurring with a rate of 0.15 which means that for every 100 ozone molecules whih react with α-pinene there is a production of 12 molecules of formaldehyde. The formation of specific products is dependent from the molecular structure of the molecules which react. As an example β-pinene gives formaldehyde with a rate of 0.7.

Another problem of air pollution indoors is related with the presence of humidity which can result to mold problems. Furthermore, the presence of water helps the hydrolysis of adhesives and plasticizers. An example is presented in Fig. 8.10 which shows the process of hydrolysis of urea-formaldehyde resins which has as a product

Table 8.10 Total weighted median concentration of some indoor VOCs (Adapted from Brown et al. (1994))

Compound	Building type	Number of measurements	Concentration ($\mu g/m^3$)
Hydrocarbons			
n-hexane	Dwelling house	656	5
	Office	26	12
n-nonane	Dwelling house	592	5
n-decane	Dwelling house	1,085	5
Camphene	Trailer house	44	14
Limonene	Dwelling house	584	21
Benzene	Dwelling house	2,171	8
Toluene	Dwelling house	792	37
Ethylbenzene	Dwelling house	1,867	5
o-xylene	Dwelling house	1,518	6
m- and p- xylene	Dwelling house	1,587	18
1,2,4- trimethylbenzene	Dwelling house	619	6
1,2,3- trimethylbenzene	Office	152	9
m-methylbenzene	Office	168	8
Oxygenated organic compounds			
Acetone	Dwelling house	86	32
Methanol	Schoolhouse	11	29
Ethanol	Dwelling house	39	120
Acetic acid	Schoolhouse	5	12
Butyric acid	Schoolhouse	5	25
Methyl-ethyl-ketone	Dwelling house	316	4–21
Diethylketone	Schoolhouse	12	6
Phenol	Schoolhouse	5	9
Nonane	Dwelling house	15	7
Ethyl acetate	Dwelling house	302	8
	Schoolhouse	12	10
Chlorinated organic compounds			
Dichloromethane	Dwelling house	101	17
Chloroform	Office	20	10
1,2-dichloroethane	Dwelling house	35	11
1,1,1-trichloroethane	Dwelling house	1,580	24
Tetrachloroethane	Dwelling house	1,919	7
p-dichlorobenzene	Dwelling house	1,881	8

the formaldehyde. These resins are used extensively to the production of furnitures and wood products. The last years there is considerable efford for the reduction of formaldehyde emissions from wood products with reduction of the formaldehyde percentage in the resins or with the use of specific compounds for the emission reductions.

8.5 Chemistry of Organic Compounds Indoors

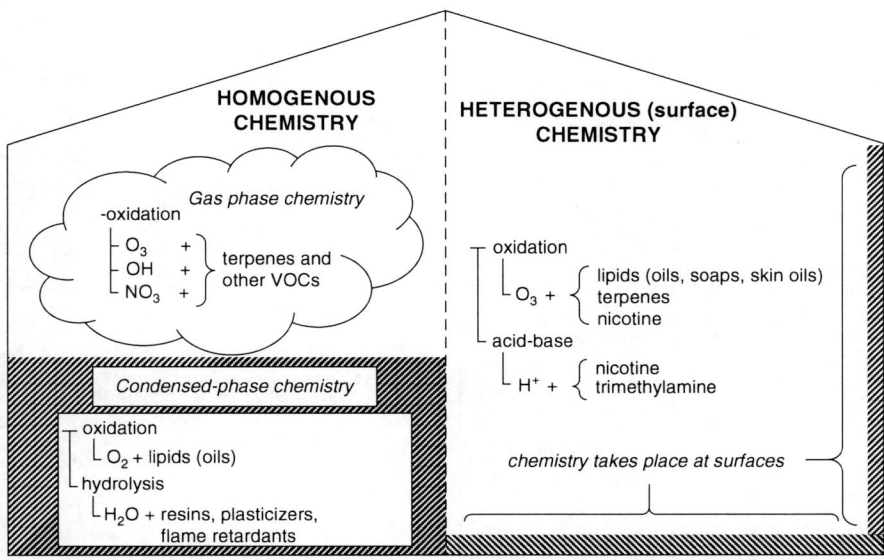

Fig. 8.8 Description of the organic chemical reactions occurring in indoor places (Adapted from Morrison 2010)

Table 8.11 Specific terpenes and their rate constants (h^{-1}) (pseudo-first-order rate) with O_3, •OHO, and NO_3.[a] (Adapted from Morrison 2010)

Component	k', $O_3 \times$ 20 ppb	k', •OHO $\times 10^{-5}$ ppb	k', $NO_3 \times 10^{-3}$ ppb	Sources
Camphene	0.002	0.048	0.016	Cleaners and air fresheners
Styrene	0.03	0.070	0.004	Carpet, adhesives
α-pinene	0.15	0.048	0.55	Cleaners and air fresheners
Limonene	0.37	0.15	1.1	Cleaners, air fresheners and perfumes
Citrnellol	0.43	0.15		Cleaners, insect repellants
α-terpineol[b]	0.54	0.17		Cleaners, pine oil, mold
Linalool	0.76	0.14	1.0	Air fresheners, perfumes

[a] values adapted from Nazaroff και Weschler (2004)
[b] values adapted from Wells (2005)

Another scientific area of research is related with heterogeneous reactions indoors. An example of heterogeneous reactions indoors is related with the ozone chemistry with nicotine as shown in Fig. 8.11. Nicotine and other products from the cigarette smoke are absorbed on the surface of different materials and with the presence of ozone there is formation of several products such as cotton and Teflon (Fig. 8.11). Elevated humidity prevents these chemical reactions on

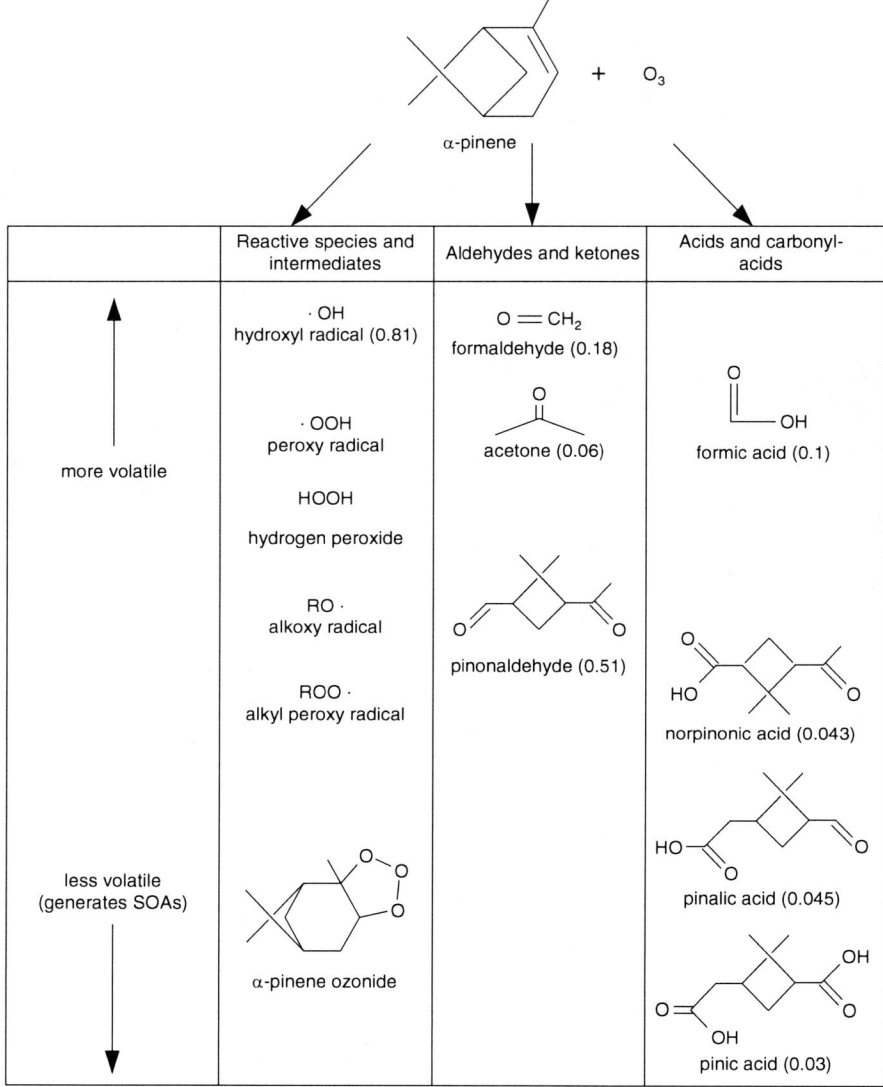

Fig. 8.9 Products of the reaction of α-pinene with ozone and corresponding yields (Adapted from Morrison 2010)

cotton surfaces. Ozone reacts with pyrrolidinic N and produces several volatile organic compounds.

Extended work for the chemistry of organic compounds indoors can be found in the book of Finlayson-Pitts and Pitts (2000).

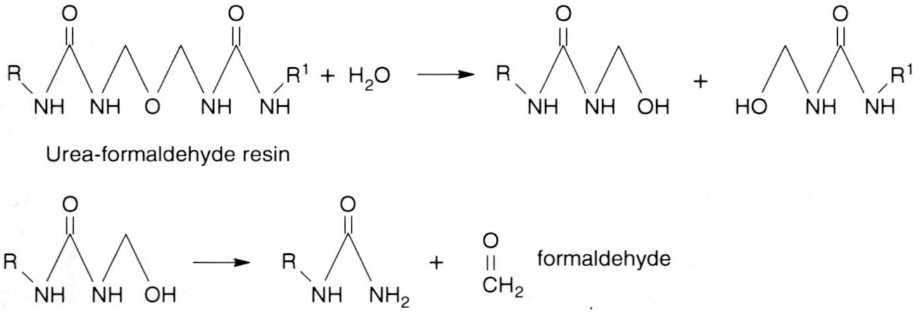

Fig. 8.10 Production of formaldehyde from the hydrolysis of urea-formaldehyde resin

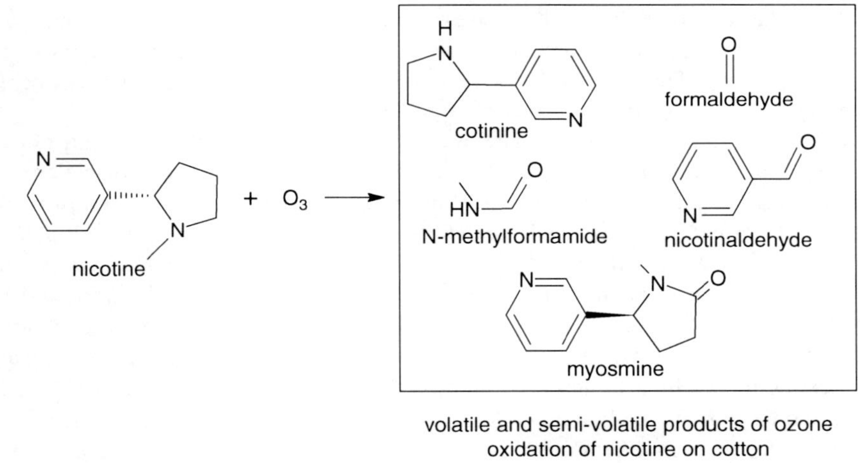

Fig. 8.11 Oxidation of nicotine with ozone on cotton (Adapted from Morrison 2010)

8.6 Radon

Radon is natural radioactive[1] element, tasteless, odorless and colourless gas which is released from the earth's surface to the atmosphere or is diluted in the groundwater. Radon is a product of radioactive elements (Thoron and radium) as a product of their radioactive decay. Specifically from the isotope of radium ^{226}Ra (the exponent is the atomic mass which expresses the sum of protons and neutrons in the nucleus) are produced radon atoms (^{222}Rn) due to radioactive decay with half life time 1,622 years.

---

[1] When an unstable nucleus of an atom decays there is a release of high energy quantity which can be released with several forms which is called radiation. In nature atoms with 84 or less protons are stable. The neutons stabilize the nucleus. Atoms with higher number of protons such as plutonium (Pu, atomic number 92), thorium (Th, atomic number 90), radium (Ra, atomic number 88) and radon (Rn, atomic number 86) decay to lighter atoms.

The radium ^{226}Ra is the decay product of uranium (^{238}U), which is a natural radioactive element in Earth with half life time 4.5 billion years. Furthermore from the isotope of radium ^{224}Ra, which is a decay product of Thoron (^{232}Th), results the radon ^{220}Rn which is known as Thoron.

The radioactive decay of radionuclides uranium, thoron, radium and the radon release occurs with slow rates and usually the soil absorbs important part of the emitted radiation. Consequently radon is considered dangerous to health only for mining workers. One of the reasons which resulted that the radon is considered dangerous for the public health is appeared during December 1984 when an engineer in a nuclear power-plant in Pennsylvania (United States) found to be exposed to large radiation dose. The alarm was set off during his entrance to the facility during the Christmas party. The peculiarity in this situation was that the engineer was entering the facility and therefore the exposure was not happed in the working place. Studies showed that the exposure occurred home which was built on soil with high radiation. The soil originated from uranium-mining operations leftover diggings. The level of the radon concentration in the house was equivalent with the risk of smoking 100 packs of cigarettes every day (Turco 2002).

The inhalation of radon and daughter nuclides indoors is the main exposure source of natural radioactivity in the environment. The inhalation of radon in gaseous form is not dangerous since it is not reaching to decay and is not absorbed in the body (is not dissolved easily in the body fluids) and therefore exhaled easily. The human risk arises with the radioactive decay to other radioactive products which are absorbed in the particulate matter with very small half life time. These nuclides when are inhaled they are deposited in the alveolar epithelium and emit alpha[2] and beta[3] radiation and can make genetic damages (damage or mutation to DNA) and to the destruction of lung tissues. Furthermore part of the nuclides is removed and dispersed through the body fluids to other organs such as stomach and pancreas.

The exposure to radon for many years through inhalation results to the formation of tumors and lung cancer. In the case of radon or daughter products ingestion the greater quantity is transferred to blood and a smaller portion is absorbed from the gastrointestinal tract and is further transported to other organs.

The radon is transferred to the indoor environment with different manners as shown in Fig. 8.12. First radon is released from rocks and enters the indoor

---

[2] The alpha particles are emitted from atom nucleus. They are similar to Helium atoms without electrons. The decay of a radionuclide with emission of an α particle results to the formation of elements with smaller atomic number (by 2 protons). For example: $^{224}Ra decay^{220} Rn + \alpha$ The radiation α does not have large penetration ability and radius of influence. Usually it does not penetrate the skin and the elements which emit α radiation may produce some health problems only if they are introduced inside the body through inhalation or skin wounds. However, the radiation α produces great damage locally to the infected tissues.

[3] The beta particles (β) are electrons which are emitted from the atoms nucleus with high velocity. The radiation decay of a nuclide with emission of β radiation results to the formation of elements isotopes with larger atomic number (e.g. $^{214}Bi decay^{214} Po + \beta$ The radiation β has medium penetrating ability and radius of influence of few meters in air. It penetrates the skin by few millimeters but results to smaller damage than the α radiation.

8.6 Radon

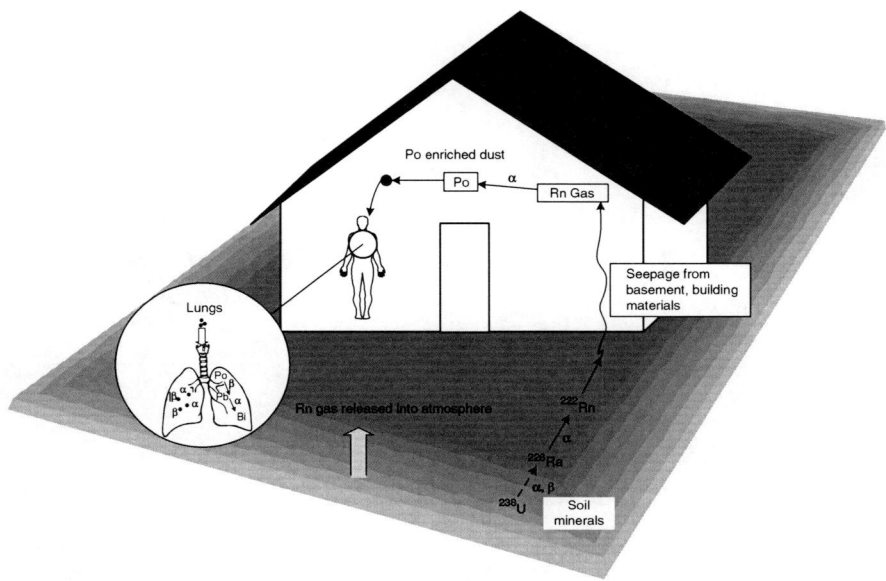

Fig. 8.12 Pathways for the radon transport in the indoor environment

environment. In addition there is the possibility to occur human exposure to radon indoors due to its evaporation from the water tap or/and during the presence in the shower. Water has high concentrations of radon when it is transported through rocks rich in radioactive elements. Study has shown (Osborne 1987) that the concentration of radon increases by two order of magnitudes during the use of water tap and at least 15 minutes after. Finally there is exposure to radon due to the presence of radioactive elements in the building materials.

8.6.1 Radiactive Decay of Radon Isotopes

The half life time of the efficiency[4] of radon and thoron are quite small, 3.825 days and 55.6 s respectively. Thus from a radon atom starts a chain of radioactive decays with products the daughter nuclides of radon and thoron (Fig. 8.13). More specifically ^{222}Rn decays initially to polonium ^{218}Po (half life time 3 min) with the emission of an alpha particle. Furthermore ^{218}Po decays to lead ^{214}Pb with the emission of an alpha particle (half life time of 27 minutes) which furthermore decays to bismuthium ^{214}Bi with the emission of a beta particle. The ^{214}Bi (half life time of 20 minutes) decays to polonium 214 (^{214}Po) which is very unstable with a half life time of 0.00015 s and further decays to lead ^{210}Pb with emission of an alpha

---
[4]The rate of radioactive decays

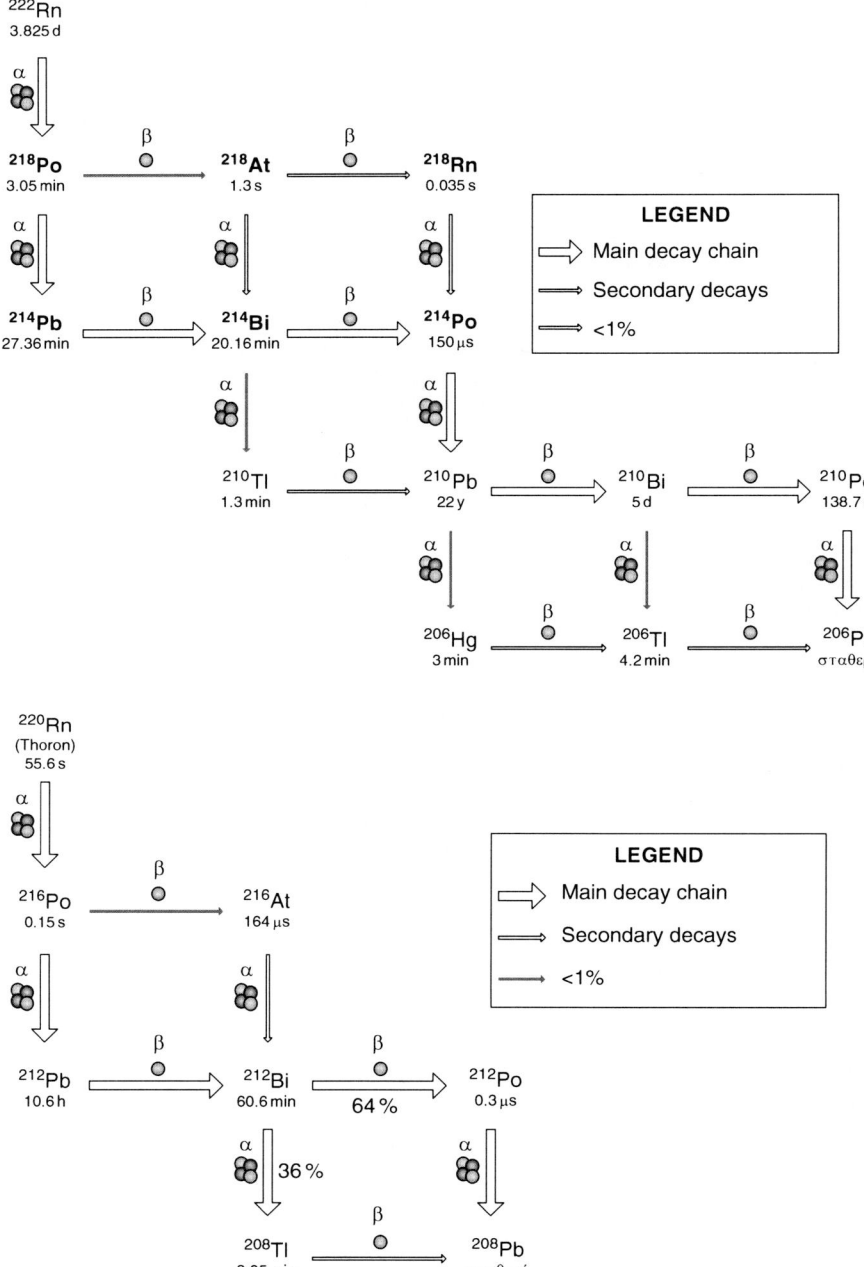

Fig. 8.13 Radioactive decay chains of Radon ^{222}Rn and Thoron ^{220}Rn. (Adapted from Turco 2002)

particle (half life time 22 years). The lead atoms result finally to soil and decay to non radioactive element lead ^{206}Pb.

The first five steps of the radon decay (^{222}Rn) up to its decay to lead ^{210}Pb occur during 50 minutes and the radioactive products present health risk. The ^{218}Po is the most dangerous decay product of radon since it decays quickly and emits α particles. Contrary nuclides of lead ^{214}Pb and bismuthium ^{214}Bi decay with emission of particles β with half life time larger than 20 minutes and therefore can be transferred from the indoor air with adequate ventilation of the area.

Thoron ^{220}Rn decays very fast (half life time 55.6 s) to polonium ^{216}Po (half ^{214}Po life time 0.15 s) with the emission of an alpha particle. Furthermore polonium decays to lead ^{212}Pb with the emission of an alphas particle (half life time 10.6 h) which furthermore decays to bismuthium ^{214}Bi with the emission of a beta particle. The bismuthium ^{212}Bi (half life time 60.6 minutes) decays to polonium ^{212}Po with half life time 0.000003 s and furthermore decays to the non radioactive element lead ^{208}Pb with the emission of an alpha particle. Contrary to radon ^{222}Rn, the decay of thoron is assumed that occurs inside the soil and contributes little to the natural levels of radioactivity in air. Therefore the effects of thoron to human health refer only to part of the population which is on direct contact with soil which has uranium and thorium at high concentrations (e.g. mine workers).

8.6.2 Exposure and Dose of Radon in Indoor Environment

The human exposure assessment of radon can be performed with the aid of a micro environmental model (mass balance model) when are known the sources and scavengers indoors.

The radiation is measured using the unit Curie [Ci][5], which is defined as the number of radioactive decays per unit time and is equal to 37 billion atomic decays per second 1 Ci = 3,7 × 10^{10} decays per second). A practical measurement unit is the 1 pCi which is equal to 10^{-12} Ci. The radon concentration is expressed as pCi per liter (pCi/l). Typical values of radon emission indoors are between 0.01 and 10 pCi/h.

The average radon concentration indoors is close to 1 pCi/l and at outdoor places 0.2 pCi/l. Exposure to high radon values result to an increase of the frequency of occurance of lung cancer. For a population sample of 1,000 people who are not smokers and have a normal life close to 10–20 of them will have lung cancer during their life. This number is resulted partly from radon exposure indoors. Fig. 8.14 shows that 13 deaths from lung cancer for this population sample are due to exposure to radon with concentration 1 pCi/l.

---

[5] In Europe there is the unit Becquerel (Bq) which is a special unit of radiation in the SI system. 1 Bq = 1 radioactive decay s^{-1} and 1 Ci = 3.7 × 10^{10} Bq.

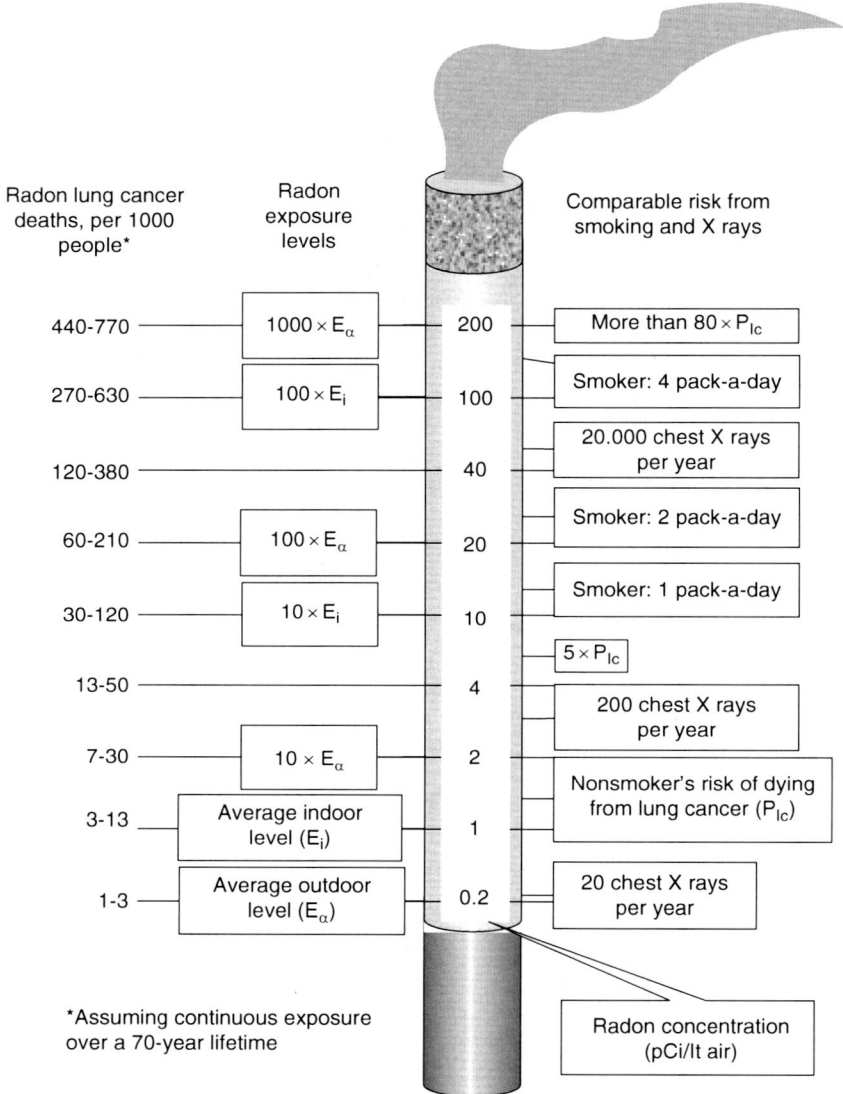

Fig. 8.14 Radon exposure in relation to lung cancer occurance. The radon concentration is expressed as pCi/l. The determined cancer occurance of lung cancer refers to a population of 1,000 persons for a continuous 70-year lifetime exposure. At the right part of the figure the risk is presented for lung cancer occurance from other activities which correspond to equivalent radon exposure (Adapted from Turco 2002)

Acceptable radon concentration indoors for human exposure during a life time (70 years) is adopted the 4 pCi/l from the American Environmental Protection Agency (EPA). Exposure to 4 pCi/l for 70 years results to 13–50 deaths due to lung cancer for a population of 1,000 people and the risk is higher than 200 chest X rays per year. In the United States over 8 million houses have radon levels above 4 pCi/l.

Figure 8.15 presents the radon ^{222}Rn concentration in houses in the United States (Nero et al. 1986). The concentration of ^{222}Rn ranges between values lower than 0.1 pCi L^{-1} and larger than 8 pCi L^{-1}. Furthermore in Fig. 8.16 shows the distribution of the ^{222}Rn concentration in houses in several countries. The higher concentration is observed in the Czech Republic due to natural radiation from the rocks.

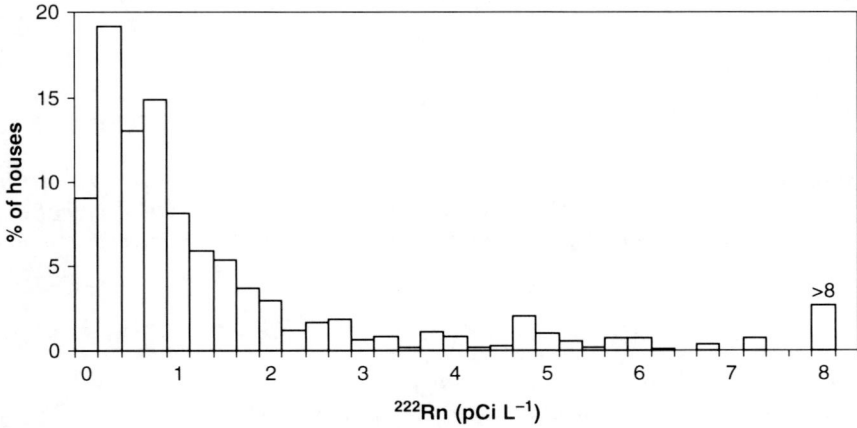

Fig. 8.15 Distribution of the ^{222}Rn concentration in houses in the United States (Adapted from Nero et al. 1986)

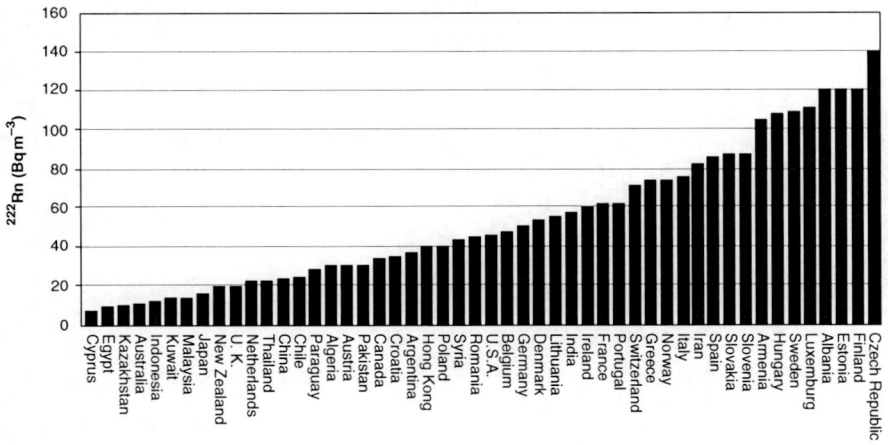

Fig. 8.16 Distribution of the ^{222}Rn concentration in houses at specific countries

The exposure risk at higher radon concentrations can be compared with the frequency of lung cancer occurrence as shown in Fig. 8.14. It is notable to observe that a person who lives in a house with radon levels 100 times larger than the average radon concentration then has the same possibility to die from lung cancer as a smoker who smokes two packets of cigarettes for his whole life (60 – 120 deaths for a population of 1,000 persons).

The dose which a human receives from radon exposure (^{222}Rn) is a result from the daughter radionuclides which have small life time and are adsorbed to particles and are deposited to the respiratory tract with inhalation.

8.6.3 Examples

Example 1 – Half life time. *The half life time ($_{1/2}$) of a nuclide is given as the napier logarithm of 2 divided by the constant of decrease λ. Usually the half life time of a radionuclide and the constant λ can be calculated with the expression:*

$$\lambda = \frac{\ln 2}{T_{1/2}(s)}$$

From the constant λ of a radionuclide and the number of atoms (N) can someone calculate the radiation ς (disintegrations per second ή Becquerel):

$$A\left(dps \; \acute{\eta} \; Bq\right) = N\lambda$$

From the number of atoms per gram of a specific element can be calculated with the help of the Avogadro number ($K = 6.023 \times 10^{23}$) and the atomic number (M). Therefore the specific activity (SA) of an element at Ci/g or Bq/g can be expressed as

$$SA(dps/g \, or \, Bq/g) = \frac{Activity(A)}{SpecificWeight(W)} = \frac{A}{\frac{MN}{K}} = \frac{N\lambda K}{MN} = \frac{K\ln 2}{MT_{1/2}}$$
$$= left(4.17 \times 10^{23})/MT_{1/2}$$

Example 2. *Calculate the specific activity (SA) for the Tc-99 (Technetium) and half life time ($T_{1/2} = 2{,}13 \times 10^5$ years) at Ci/g.*

$$T_{1/2} = 2.13 \times 10^5 \, year$$
$$= 2.13 \times 10^5 \, year \times 365 \, day/year \times 24 \, h/day \times 3600 \, s/h$$
$$= 67171680 \times 10^5 \, s \approx 6.7 \times 10^{12} \, s$$
$$Therefore \quad SA(Tc-99) = \left(4.17 \times 10^{23}\right)/MT_{1/2}(dps/g)$$
$$= \left(4.17 \times 10^{23}\right)/99 \times 6.7 \times 10^{12} \, (dps/g)$$
$$\approx 6{,}29 \times 10^8 \, (dps/g)$$

$$\text{Therefore} \quad SA(Tc-99) = (4.17 \times 10^{23})/MT_{1/2}(dps/g)$$
$$= (4.17 \times 10^{23})/99 \times 6.7 \times 10^{12}(dps/g)$$
$$\approx 6,29 \times 10^{8}(dps/g)$$

It is known that $1\ Ci = 3.7 \times 10^{10}\ Bq = 3.7 \times 10^{10}\ dps$. Therefore for Tc-99 the SA can be expressed as SA (Tc-99) = 6.29×10^8 (dps/g) = 0.017 (Ci/g).

8.7 Carbon Monoxide

Sources of carbon monoxide (CO) are combustion sources indoors. Heating appliances indoors such as gasoline heaters and fireplaces are CO sources indoors. Vehicular ehxaust in garage places as well as cigarette smoke are two important sources of carbon monoxide.

Elevated concentrations of CO have been measured inside vehicle cabins with values ranging between 9 and 56 ppm. In the United States it has been shown that the ratio of CO concentration inside cars to its value outdoors has value close to 4.5 (Chan et al. 1991). Exposure to CO at low concentrations results to tiredness of healthy persons and chest pain to people with heart problems. At higher concentrations there are problems with vision, headache, dizziness, confusion, nausea.

There are cases where the symptoms are similar with the influenza which however are disappeared after moving from the house with the increased concentration values of carbon monoxide. High concentration values may become fatal. Average concentrations at houses without gas stoves range between 0.5 and 5 ppm. Indoor levels of CO close to gas stoves which have adequate regulation can be close to 5–15 ppm and to gas stoves with maladjustment can be 30 ppm or higher.

The global background concentrations of CO range between 0.06 mg/m^3 and 0.14 mg/m^3 (0.05–0.12 ppm). In European towns with high vehicular load the 8-h average CO concentration is lower than 20 mg/m^3 (17 ppm) with possible increases of small duration which can reach 60 mg/m^3 (53 ppm). The CO concentration inside car cabins is greater than the outdoor air. Inside garages, tunnels and other indoor places which have vehichles with no adequate ventilation, the average concentration levels of CO are close to 115 mg/m^3 (100 ppm) for few hours with increases for short time periods. At homes which use gas the CO concentrations are larger than 60–115 mg/m^3. The cigarette smoke in houses, offices, cars and restaurants may lead to an 8-h average CO concentration increase to levels of 23–46 mg/m^3.

The carbon monoxide is diffused quickly to the lungs, to the capillary vessels and to the vital membranes. Close to 80–90% of the absorbed carbon monoxide combines with hemoglobin (HbCO) which represents an exposure index for CO in blood. The attraction of the hemoglobin to the Co is 200–250 times higher than to the oxygen.

The combination of the carbon monoxide with hemoglobin for the production of carbohemoglobin reduces the ability of blood to transfer oxygen and makes more difficult the oxygenation of tissues. The toxic effect of CO is becoming obvious to the organs and tissues which consume elevated quantities of oxygen such as the lung, the heart, skeletal muscles and developing embryos.

The following exposure limits are proposed in order the carbohemoglobin not to exceed the level of 2.5%:

- 100 mg/m^3 for 15 minutes
- 60 mg/m^3 for 30 minutes
- 30 mg/m^3 for 1 minutes
- 10 mg/m^3 for 8 minutes.

8.8 Asbestos

Asbestos is a mineral fiber which is used as a raw material (e.g. $Na_2(Fe^{3+})_2(Fe^{2+})_3$ $(OH)_2Si_8O_{22}$, $Mg_7(OH)_2Si_8O_{22}/Fe_7(OH)_2Si_8O_{22}$, $(Mg,Fe)_7(OH)_2Si_8O_{22}$) for the production of different building insulation materials and other products resistant to high temperatures (fire-retardant). The fibers of the different types of asbestos have differences in their crystal structure and their chemical composition. These differences may result to different health effects which result after exposure to different asbestos forms.

The asbestos fiber properties which present a health risk for humans is their small size and resistant (solid) structure. The asbestos can be found in old houses where has been used as structural and insulation material for pipes and furnaces. In addition has been used for the production of materials for flooring materials, wall and roof materials, materials for thermal insulation, for the gloves production, as well as for electrical applications.

The asbestos fibers consist from thin and long flexible fibers. Their size ranges between 1 and 12 μm (Health Effects Institute 1991). The diameter is between 0. and 15 μm (700 times thinner than the width of human hair). In fact the asbestos fibers consist from a pack of parallel thinner fibers which have diameters close to 20–25 nm. The asbestos fibers tend to be disconnected after mechanical stress or erosion and as a result there is human exposure to asbestos fibers. Fig. 8.17 shows the work for the asbestos removal from pipe covering as well as one asbestos fiber (the picture is 1,250 times enlarged).

Furthermore Table 8.12 presents typical concentrations of asbestos fiber concentrations at different areas.

Since asbestos is carcinogen for the humans there is no a safe threshold of air concentrations. Therefore its concentration in air has to been as low as possible. The Occupational Safety and Health Administration (OSHA) in the United States has possed as a Permissible Exposure Limit (PEL) the concentration of 0.1 fibers/cm^3 for 8-h exposure of an adult person. The corresponding limit adopted from the World Health Organization (WHO) is equal to 2 fibers/cm^3 for 8-h exposure in a working environment. In Europe the limit is 0.2 fibers/cm^3 for workers. Table 8.13 summarizes the adopted human exposure limits to asbestos fibers.

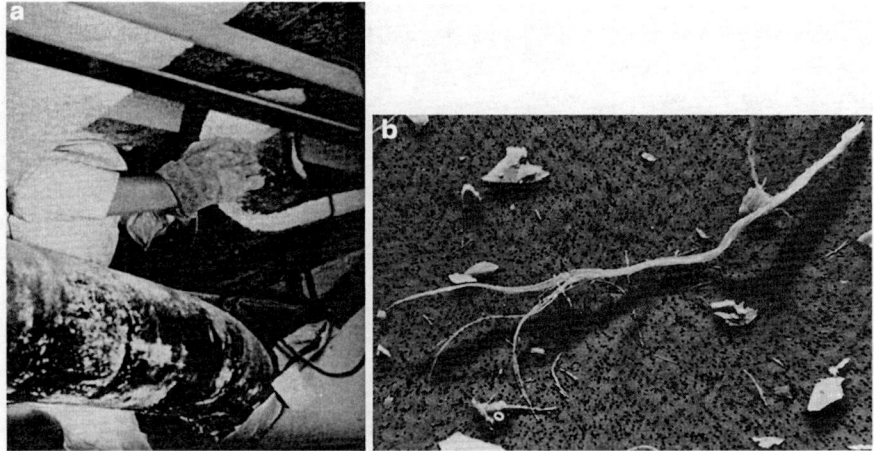

Fig. 8.17 (a) Removal work for asbestos pipe covering. (b) An asbestos fiber magnified 1,250 times (SEM photograph) (Adapted from Hinds 1999)

Table 8.12 Representative asbestos fiber concentrations in air

Area	Conditions	Fiber concentration
Rural areas	At long distance from asbestos sources	<100 fibers/m^3
Urban areas	At main crossroads	900 fibers/m^3
	At distance of 300 m from an industry of asbestos production	2,200 fibers/m^3
	At distance of 700 m from an industry of asbestos production	800 fibers/m^3
	At distance of 1,000 m from an industry of asbestos production	600 fibers/m^3
Indoor areas	Buildings without asbestos materials	~300 fibers/m^3
	Buildings with asbestos materials	700–10.000 fibers/m^3

Table 8.13 Exposure limits of asbestos at working places

	Fiber concentration (fibers/cm^3)
Exposure limit (8-h exposure)	0.2 (European Union)
Limit value	0.1 (OSHA)
World Health Organization (WHO)	2

8.9 Heavy Metals

Metals are incorporated in primary particles which are emitted in the atmosphere during combustion, evaporation or sublimation of materials. Metals can be found both in fine and coarse particles in the atmosphere. Inside fine particles are incorporated more often compounds of Pb, Cd, V, Ni, Cu, Zn, Mn, Fe etc., whereas in coarse particles can be found oxides of elements which exist in soil (e.g. Si, Al, Fe, Ca). Fine particles which include heavy metals present higher health risk than the coarse particles since they are suspended in air for large time intervals which in the range from few hours to weeks.

Figure 8.18 shows the emission sources of heavy metals in the environment and their transport, as well as, the chemical processes which occur in the atmosphere.

Metals include lead (Pb), cadmium (Cd), mercury (Hg) (Fig. 8.19), chromium (Cr), iron (Fe), copper (Cu), manganese (Mn), calcium (Ca), arsenic (As), nickel (Ni), aluminium (Al) or in general every metallic chemical compound with high density and it is toxic at low concentrations.

In general the heavy metals produce toxicity through the formation of complex organometallic compounds. These affect the biological molecules which can lead even to their necrosis. When molecules of metals attached to molecules of oxygen, nitrogen or sulphur it is possible to be disabled important enzyme systems or to be affected the form of proteins.

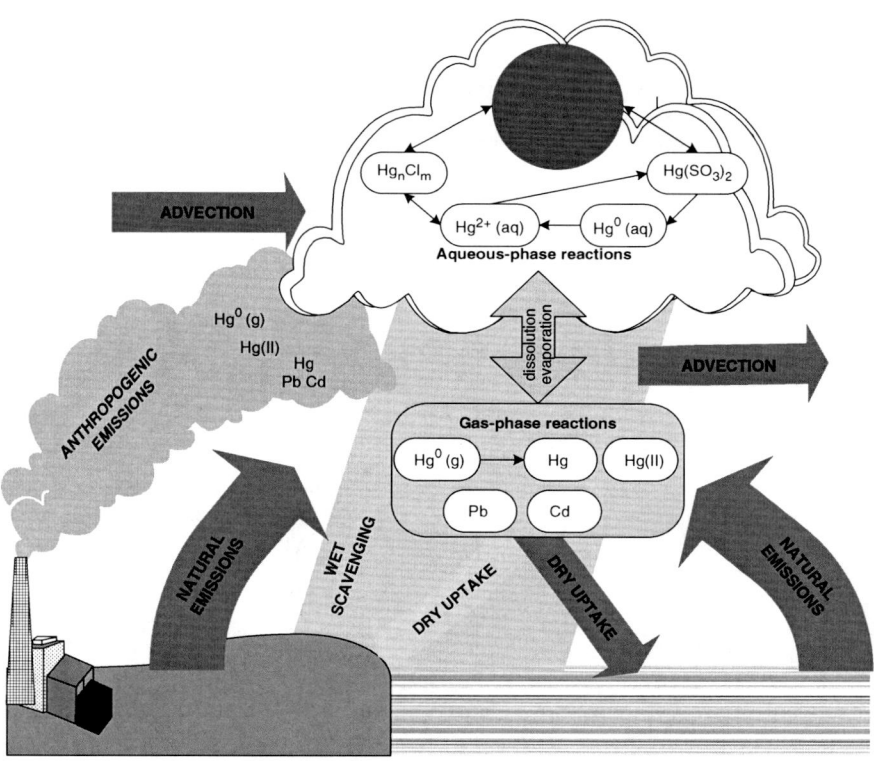

Fig. 8.18 Emission sources of metals and their transport in the environment

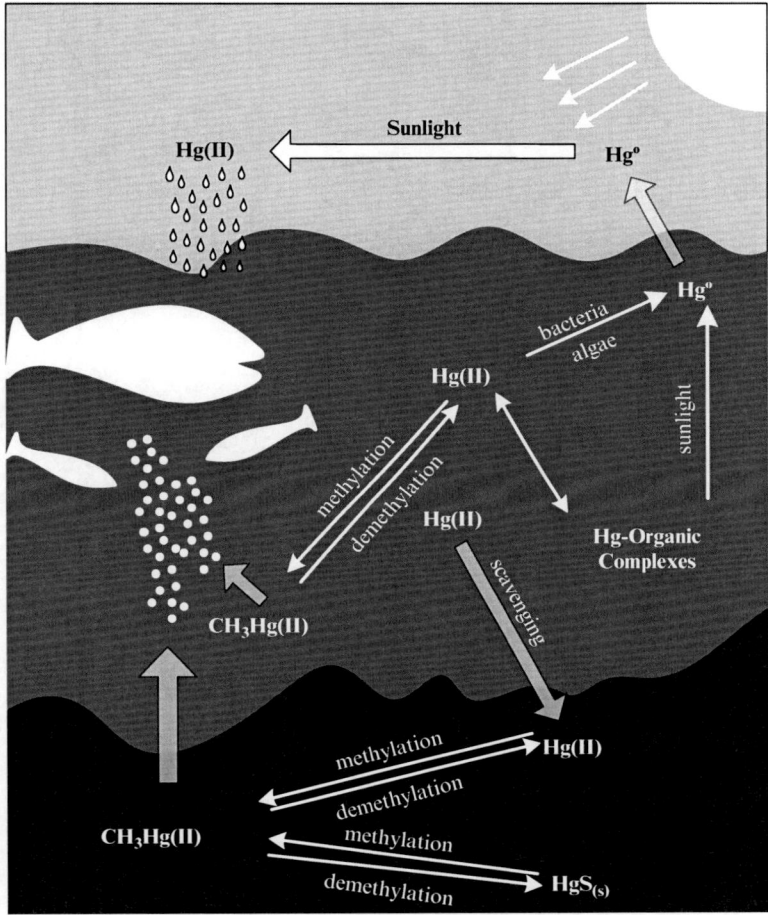

Fig. 8.19 Chemical forms of Hg in the ocean environment

8.10 Formaldehyde

Formaldehyde is a volatile organic compound (HCHO) which is used widely as insulation, adhesive and as protective material for wood. Table 8.14 shows the emission rate of formaldehyde from different materials. The emission range is dependent from the material temperature, its age, the covering method (e.g. painting) and other parameters.

Exposure to formaldehyde occurs when it is infiltrated indoors. The infiltration can occur from the connections of the walls with the floor, the windows, the doors and the electrical system. Smoking is an important source of formaldehyde due to combustion. The cigarette smoke can have 40 ppm formaldehyde concentration and

Table 8.14 Emission sources of formaldehyde

Product	Emission rate range (10^{-6} g/g material/day)
Board	0.4–8.1
Plywood	0.03–9.2
Imitation panel	0.8–2.1
Insulation with fiber glass	0.3–2.3
Clothing	0.2–4.9
Carpeting	0–0.06
Paper products	0.03–0.4

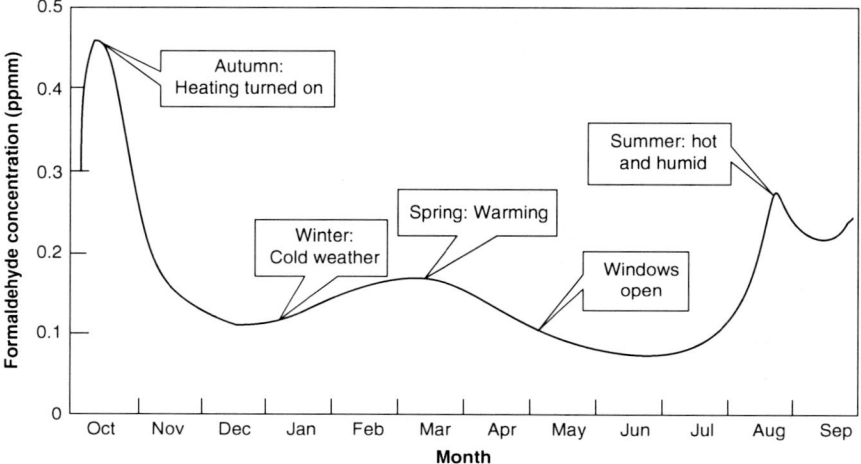

Fig. 8.20 Formaldehyde concentration inside a home during a year. The concentrations are average monthly values. The specific house insulated with material containing formaldehyde. (Adapted from Gammage, R. B. and K. C. Gupta, "Formaldehyde", included in the book *Indoor Air Quality*, Walsh, P. J., C. S. Dudney and E. D. Copenhaver, eds. Boca Raton, Fla: CRC Press, 1984)

one cigarette can produce 1–2 mg smoke. In a smokers room the formaldehyde concentration can reach levels of several ppm which corresponds to high exposure.

The formaldehyde concentration indoors changes at different seasons and is dependent from construction home materials and the life style of people. Fig. 8.20 shows that temperature rise at the insulation walls of a home increase the formaldehyde emissions, whereas lower temperatures during winter result to lower emissions. During the Spring period the formaldehyde concentration increases since the temperature is not high enough to open the windows. When the air exchange rate of the house increases the formaldehyde concentration is reduced.

The formaldehyde concentration at different environments and the exposure limits are presented in Table 8.15. The outdoor air has formaldehyde concentrations between 0.4 and 80 ppb where the higher concentrations refer to environments with elevated anthropogenic emissions. At the photochemical smog which occurs often

8.10 Formaldehyde

Table 8.15 Concentrations of formaldehyde (adapted from Turco 2002)

Area	Concentration range (ppmv)
Outside air	0.0004–0.08 (0.01 mean value)
House without insulation of UFFI[a]	0,01–0,08 (0.03 mean value)
House with UFFI insulation	0.01–3.4 (0.12 mean value)
Mobile home	0.01–2.9 (0.38 mean value)
Textile plant	<0.1–1.4
Fertilizer plant	0.2–1.9
Bronze foundry	0.12–8.0
Iron foundry	<0.02–18.3
Plywood industry	1.0–2.5
Hospital autopsy room	2.2–7.9
Exposure limit	
Indoor air, United States	0.1 (maximum)
Indoor air, Sweden, old structures	0.7 (maximum)
Occupational air, NIOSH[b]	1.0 (maximum 30 minutes)
Occupational air, OSHA[c]	3.0 (maximum 8 h)
Occupational air, OSHA	5.0 (maximum)

[a] Urea-Formaldehyde-Foam Insulation
[b] National Institute for Occupational Safety and Health (United States)
[c] Occupation Safety and health Administration (United States)

Table 8.16 Health effects of formaldehyde

Concentration (ppmv)	Exposure duration (min)	Health effects
0.01	5	Eye irritation
0.05	1	Odor threshold
0.08	1	Cerebral cortex affected
0.2	60	Eye, nose, and throat irritation
0.8	10	Brain alpha-wave rhythm and autonomous nervous system changes
4.0	1	Unbearable without respiratory protection

at urban areas there is secondary formation of formaldehyde (concentrations between 10 and 80 ppb) from the presence of several volatile hydrocarbons in air. However, the formaldehyde concentration is in general higher indoors.

Exposure to formaldehyde occurs through inhalation. The formaldehyde is water soluble and therefore during inhalation can be absorbed at the upper respiratory tract. In the case of nose breathing the formaldehyde remains at the nose walls. In the case that formaldehyde is absorbed from airborne particles then it can penetrate at the lower parts of the respiratory tract such as lungs.

Table 8.16 presents health effects related to human exposure to formaldehyde. For the population protection is proposed the air quality limit level of 0.1 mg/m^3 for

30 minuted exposure. The minimum concentration, which is responsible for the irritation of human throat and nose cavities after a short exposure period, is of the order of 0.1 mg/m^3.

8.11 Pesticides

Pesticides consist a pollutant in indoor places since they are used to control and exterminate the population of household pests. The majority of pesticides are synthetic chemical compounds and in few cases biological components such as bacteria or viruses with specific toxic properties against specific parasites. Their chemical composition varies. They can be products of inorganic composition (products of chlorine, sulfur, potassium bromide, etc.), organometallic compounds (e.g. organic derivatives of zinc), volatile organic compounds (e.g. Methyl bromide, naphthalene) and semivolatile compounds (e.g. diazinon, pentachlorophenol) or products of non volatile organic compounds (e.g. dimethylamine salt).

In the past it was widely used for agricultural crops the p,p$'$-dichlorodiphenyltrichloroethane (DDT) and other organochlorinated compounds (Fig. 8.21). At indoor environments the main pesticides (such as o-Phenylphenol, Diazinon, Heptachlor, Malathion, Carbanyl) have as target the disinfection and perfuming of different surfaces and areas.

The vapours and pesticide particles are deposited and absorbed at carpets, walls, clothes, furnitures and other materials indoors. Many household pests have seasonal use. Therefore the use of pesticides is higher during spring and summer months and therefore their concentration is higher indoors up to 100%. This is also may due to their evaporation due to higher temperatures.

The health effects due to pesticide exposure include irritation to eyes, nose and throat, damage to central nervous system and kidney and increased cancer risk. For example the exposure to high pesticide concentrations which include cyclohexadiene, may result to headache, dizziness, muscle spasm, weakness and nausea. The recommended average concentration limit from the United States Industrial Hygiene Association (ACGIH) for exposure to pesticides is between 0.5 and 5 mg/m^3 for exposure of 8–10 h for private houses, offices and public enterprises.

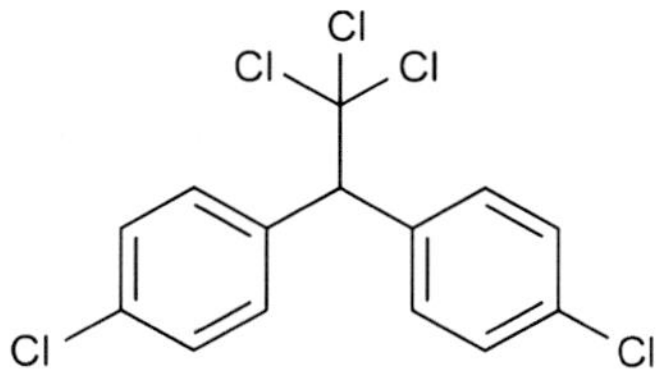

Fig. 8.21 The molecule of the widely known pesticide DDT

Table 8.17 PAH concentrations in the tobacco smoke

PAH	Concentration in the cigarette smoke (ng/cigarette)
Benz(a)anthracene	20–70
Benzo(b)fluoranthene	4–22
Benzo(j)fluoranthene	6–21
Benzo[k] fluoranthene	6–12
Benzo[a]pyrene	20–40
Chrysene	40–60
Dibenz[a,h] anthracene	4
Dibenz [a,i]pyrene	1.7–3.2
Indeno[1,2,3-cd]pyrene (IP)	4–20
5-methyl-chrysene	0.6

8.12 Polycyclic Aromatic Hydrocarbons (PAH)

Polycyclic Aromatic Hydrocarbons (PAH) are produced during the incomplete combustion of organic matter as for example wood burning, gasoline and coal fuel combustion and tobacco smoke (Table 8.17). PAHs consist of two or more benzene rings and contain only carbon and hydrogen (Table 8.18). PAHs are suspected to be carcinogenic and mutagenic compounds and present a major constituent of atmospheric aerosols.

Table 8.18 shows PAHs which are possible to be found in the cigarette smoke and are responsible for cancer occurance in humans.

Combustion tracers such as PAHs have been used in source apportionment studies. More than 100 PAH compounds have been identified in urban areas and their composition ranges from species with two rings (such as naphthalene which is mainly in the gas phase) to species with seven and more rings (such as coronene which is mainly in the aerosol phase) (Lazaridis 2008). Anthropogenic combustion sources contribute to PAHs from naphthalene to coronene whereas natural biomass burning contributes mainly with salicylic compounds.

8.13 Polychloric Biphenyls (Pcbs)

The polychloric biphenyls compose a category of chlorinated hydrocarbons which consist of a diphenyl in which the chloride atoms can substitute hydrogen atoms (Fig. 8.22).

The polychloric biphenyls have the tendency to be transported and remain at soil, sediments and water and less in air. However, they can be detected in air including indoor places and degrade reacting with hydroxylic radical or due to photolysis. The degradation is a slow process and the time required increases as less chlorinated is the compound.

Table 8.18 Structure and properties of PAH species

Structure	Nomenclature	Molecular Weight	MP[a] (Melting point)	BP[b] (Boiling point)	Log [p (torr)] 20 °C	Solubility µg/L[d]
	Fluorene (Fl)	166.23	116	295	−2.72	31,700
	Phenantherene (Phe)	178.2	101	339	−.3.5	1,290
	Athracene (An)	178.2	216.2	340	−3.53	73
	Pyrene (Py)	202.3	156	360	−4.73	135
	Fluorathene (Flu)	202.3	111	375	−4.54	260
	Benz(a)anthracene (BaA)	228.3	160	435	−6.02	14
	Chrysene (Chr)	228.3	255	448	−6.06	2
	Benzo(b) fluoranthene (BbF)	252.33	168	481	−5.22	14[e]
	Benzo(k) fluoranthene (BkF)	252.33	217	481	−7.13	4.3[e]
	Benzo(a)pyrene (BaP)	252.33	175	495	−7.33	0.05
	Benzo(e)pyrene (BeP)	252.31	178.7	493	−7.37	3.8
	Indeno[1,2,3-cd] pyrene (IP)	276.34	163	530[e]	−10[e]	0.5[e]
	Benzo(ghi) perylene (BghiP)	276.34	277	525	−9.35	0.3
	Dibenz(a,h) anthracene (DBA)	278.35	267	524	−10[e]	0.5
	Coronene (Cor)	300.36	439	590	−12.43	0.1

[a,b,d] Finlayson-Pitts and Pitts 2000
[c] Pankow and Bidleman 1992, [e]ATSDR 1997

Fig. 8.22 Polychlorinated biphenyls (PCBs)

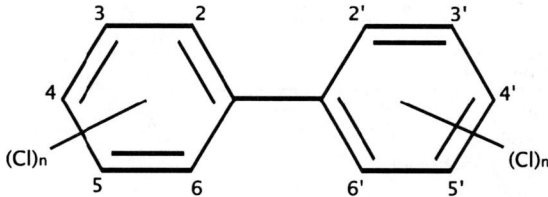

Studies have shown that the concentration of PCBs indoors is higher than outdoor levels. High concentrations are observed when there are sources of such materials, as example capacitors, transformers or plastics materials. For example in schools and laboratories in Germany have been measured concentrations of PCBs from $6 - 7,500$ ng/m^3, whereas the mean measured value for the 5% of the most polluted schools was higher than 3,000 ng/m^3. Analogous measurements have been performed in houses with mean concentrations to range between 5 and 580 ng/m^3. It has to be noted that the destruction of these components is very slow and in case that a building has PCBs it is very difficult to be totally cleaned.

8.14 Tobacco Smoke

The tobacco smoke is one of the most dangerous and widely found pollutants. It is composed from a mixture of gaseous and particulate matter. The gases are products of the cigarette combustion whereas the particles are composed mainly from tar, nicotine and organic chemical compounds which have low volatility and are deposited at pre-existing particles in the atmosphere.

Table 8.19 presents the main chemical compounds which can be found in the tobacco smoke. Tar is the black colloid compound which remains in the cigarette filter but enters also inside the human lungs. Nicotine is also an important compound which results to addiction such as cocaine and morphine. Cigarette smoking emits also carbon monoxide. The carbon monoxide as cigarette combustion product at high concentrations can result even to death. The particles from the tobacco smoke have small diameter and can deposit in the human respiratory tract. Dangerous chemical gaseous components which are emitted from cigarette smoking are acrolein, hydrogen chloride and formaldehyde. In the cigarette smoke exist also radioactive polonium (Po-210).

In table 8.19 there is a distinction between the mainstream and sidestream smoke. The mainstream smoke is passed through the unburned tobacco and the filter which captures part of the chemical components of smoke. Furthermore, with the burning of the whole cigarette there is release of these chemical components in air. With the entrance of air through the cigarette then it is burned at high temperature and fast. The sidestream smoke is produced when the cigarette is burned slowly

Table 8.19 Emissions of specific chemical components from cigarette smoking

Component	Emission (mg/cigarette)	
	Mainstream smoke	Sidestream smoke
Tar	10.2–20.8[a]	34.5–44.1
Nicotine	0.46–0.92	1.27–1.69
Carbon monoxide	18.3	86.3
Ammonia	0.16	7.4
Hydrogen cyanide	0.24	0.16
Acetone	0.58	1.45
Phenols	0.23	0.60
Formaldehyde	–	1.44
Toluene	0.11	0.60
Acrolein	0.084	0.825
Nitrogen oxides(NO_x)	0.014	0.051
Polonium-210 (pCi)	0.07	0.13
Fluoranthenes	7.7×10^{-4}	1.6×10^{-3}
Benzofluorenes	2.5×10^{-4}	1.0×10^{-3}
Pyrenes	2.7×10^{-4}	1.5×10^{-3}
Chrysene	1.9×10^{-4}	1.2×10^{-3}
Cadmium	1.3×10^{-4}	4.5×10^{-4}
Perylenes	4.8×10^{-5}	1.4×10^{-4}
Dibenzanthracenes	4.2×10^{-5}	1.4×10^{-4}
Anthanthrene	2.2×10^{-5}	3.9×10^{-5}

[a] Range between filtered and unfiltered cigarettes

and at lower temperature. The slow cigarette burning produces a different mixture of gasses and in general is more toxic. The sidestream smoke is not passing through the cigarette filter and therefore its chemical composition is not changing considerably after the emission. Furthermore, the mainstream smoke passes directly inside the human respiratory tract whereas the sidestream smoke disperses first to the air and after is introduced in the human respiratory tract through inhalation .

Figure 8.23 shows the human exposure of total particles in relation to the active smoker density at different indoor environments. The measurements were performed in the United States. At the left end of the figure are presented the particle concentrations at houses of non smokers which are below the ambient air quality standards. The concentration of particles increases in relation to the number of smokers. At places with a large number of smokers the concentration of particles is considerably higher than the ambient air quality levels.

8.15 Bioaerosols

Bioaerosols may originate from almost all the natural and anthropogenic activity and every source has different emission bioaerosol characteristics. The bioaerosol concentration varies from relatively low levels ($\leq 10^2$ cfu/m^3) (e.g. at clean rooms

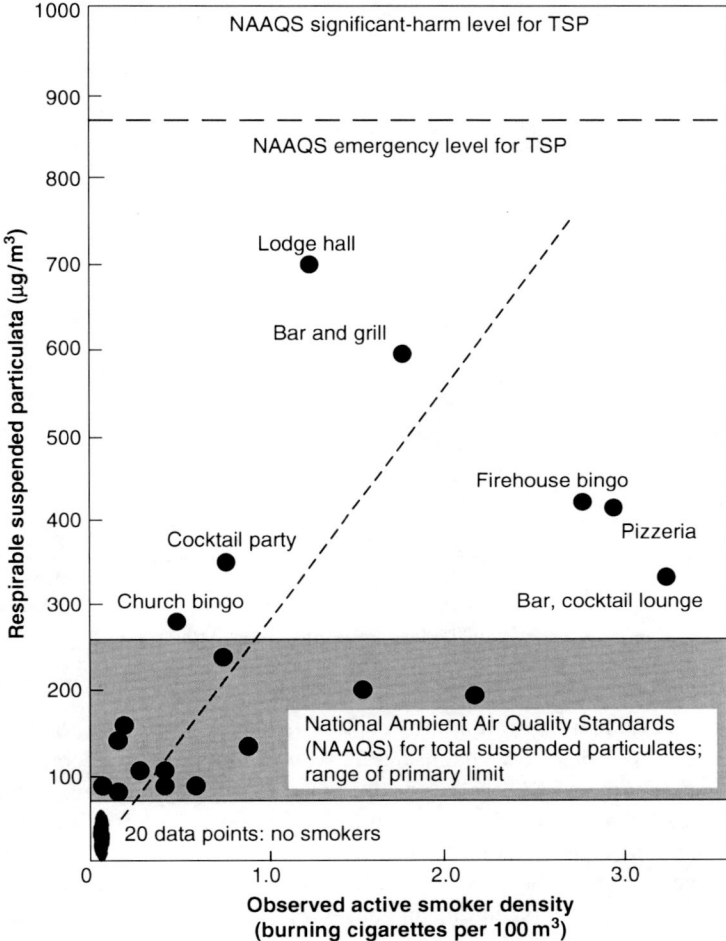

Fig. 8.23 Concentration of suspended particles at several microenvironments in relation to the density of active smokers (Adapted from Turco 2002)

and surgery rooms in hospitals) until very high (10^5–10^{10} cfu/m^3) during specific operations in the industry and farming. The bioaerosol concentration is expressed in colony forming units per unit of volume (m^3) (cfu/m^3). Table 8.20 shows some examples of bioaerosol emission sources and their ambient concentration.

The size of bioaerosols varies and their diameter ranges between 0.5 and 30 μm (other scientists define the range bewteen 0.01 and 100 μm). The bioaerosol size and the size of the cells of microorganisms inside them is an important parameter which defines the microbial contamination and airborne diseases. The size of the three categories of microorganisms (a cellular, prokaryotes and eukaryotic microorganisms) which exist ranges from 0.025 to 40 μm, as shown in Table 8.21.

Table 8.20 Bioaerosol concentration at specific environments

Category	Activity	Bacteria (cfu/m^3)	Fungi (cfu/m^3)
Farming	Livestock	10^3–10^5	10^2–10^8
	Composting	10^3–10^6	10^2–10^7
	Harvest	10^2–10^3	10^3–10^9
	Storage	10–10^3	10^2–10^7
Air conditioning	Heating, ventilation and air conditioning	10–10^4	10–10^3
Indoor places	Walls	10–10^3	10–10^4
	Carpets	10^3–10^6	10^2–10^5
	Plants	10–10^4	10^2–10^5
	Working room	10–10^2	10–10^2
Industry	Foodv	10–10^3	10^2–10^4
	Construction	10^2–10^6	10^2–10^6
Waste water treatment	Aeration tank	10^2–10^3	10–10^2
	Activated sludge	10^2–10^6	10–10^3

Table 8.21 Size of microorganisms (μm)

Category	Microorganisms	Size of cells (μm)
Accellular	Viruses	0.025–0.25
Prokaryotes	Bacteria	0.25–2
Eukaryotic	Fungi, algae, protozoa	1–40

During the human expiration are released to the atmosphere droplets of size between 0.3 and 8 μm, most of them smaller than 2 μm. Furthermore, bioaerols with diameter smaller than 5 μm are released during talking and other human activities such as sneezing, blowing, whistling, singing or wiping nose. These bioaerosols can be transported through inhalation to the lungs or through ingestion to the pharynx of other people. Scientists have been observed that during coughing or loud speech there is release of droplets with mean concentration close to 10^5 per m^3 air (mean diameter smaller than 1 μm) and concentration close to 10^4 per m^3 air (mean diameter larger than 1 μm) respectively.

8.16 Microenvironmental Models

The study of the air quality indoors can be performed with the use of mathematical models which simulate the physico-chemical processes indoors and the air exchange rate with the outdoor environment. For this reason mass balance models are used which are called micro environmental models. These models assume that the indoor air is well mixed. In the scientific literature there is also use of computational fluid mechanic models (CFD models) which examine the effect of local in homogeneities (e.g. temperature, wind velocity) to the air pollutant mixing.

8.16 Microenvironmental Models

According to the simple mass balance model the indoor aerosol concentration depends on the strength of internal sources, strength of the sinks, air exchange rate and outdoor aerosol concentrations. Assuming a room with perfect mixing, the rate of change in indoor aerosol concentration can be expressed as:

$$V \times \frac{dC_i}{d\tau} = Q(\tau) \times P \times C_o(\tau) - Q(\tau) \times C_i(\tau) - D \times V \times C_i(\tau) + S(\tau) \quad (8.4)$$

where

V is volume of room [m^3]
$C_i(\tau)$, $C_o(\tau)$ are indoor and outdoor aerosol concentrations at time τ [μg m^{-3}]
$dC_i/d\tau$ is time rate of indoor concentration change [μg m^{-3} s^{-1}]
$Q(\tau)$ is flow rate of outdoor air entering room at time τ [m^3 s^{-1}]
P is penetration coefficient [-]
D is deposition rate [s^{-1}]
$S(\tau)$ is production rate of indoor sources at time τ [μg s^{-1}]
τ is time [s]

With indoor particle sources the temporal indoor concentrations can exceed the outdoor ones. Without the indoor particle sources, indoor particle concentrations are often lower than outdoor concentrations. This is a consequence of particle removal processes as particle deposition on indoor surfaces or particle removal by filtration. The particle deposition is usually governed by Brownian and turbulent diffusion and gravitational settling. As a result the particle deposition is size dependent with higher deposition velocities observed for ultrafine particles due to diffusion and coarse particles due to sedimentation. Thus larger differences between indoor and outdoor aerosol particle concentration are usually observed for ultrafine and coarse particles.

Figure 8.24 shows a schematic description of air pollutant mass balance in an indoor environment. The rate of increase of an air pollutant concentration in the indoor air volume ($V \frac{dc}{dt}$) ($V_{building}$ is the volume V of the indoor environment) is equal to the rate with which the pollutant enters inside the volume from outdoors ($Q\,c_a$), plus the emission rate from indoor sources (N), minus the rate by which the pollutants exit from the indoor environment ($Q\,c$) and minus the rate by which the pollutants are lost due to the chemical reactions (kcV). The flux towards indoors of the air (Q_{in}) equals the air flux from the indoor environment outdoors (Q_{out}) which is also equal to the product of the air volume with the number of air exchanges per hour. The coefficient for the chemical reaction rates is denoted by κ. Table 8.22 shows the infiltration rate of specific pollutants. In Table 8.23 are given the infiltration rates of specific chemical reaction rates indoors. In addition, Table 8.24 presents the coefficients of average chemical reaction rates.

It is supposed that there is well mixing for the chemical components indoors and the equation of the mass balance in relation to the infiltration rate from the outdoor air, the indoor sources and the scavenging parameters can be written as:

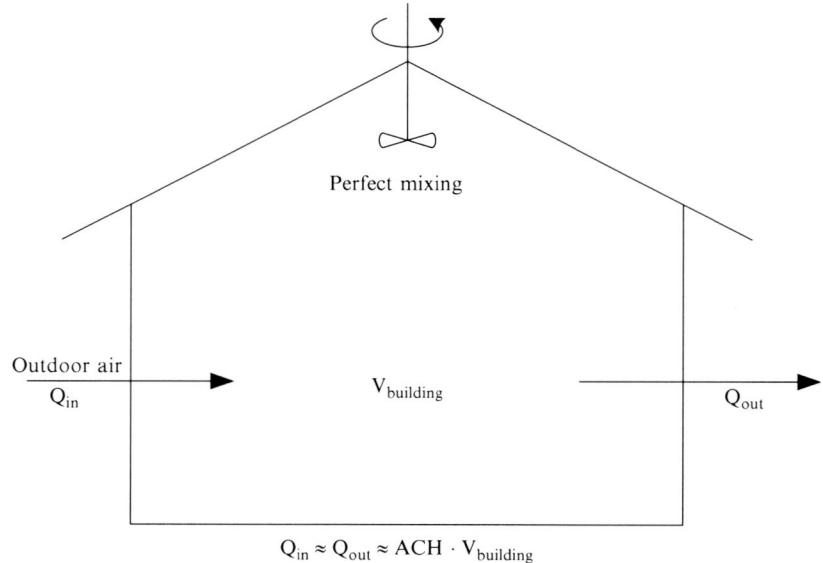

Fig. 8.24 Description of a microenvironmental model for the mass balance of an air pollutant

Table 8.22 Infiltration rate values of specific pollutants

Pollutant	Infiltration rate
VOCs	1
Aldehydes	1
PM_{10}	0.5
$PM_{2.5}$	0.8

Table 8.23 Coefficients of average chemical reaction rates

Pollutant	$k\ (s^{-1})$
CO	0.0
HCHO	1.11×10^{-4}
NO	0.0
NO_x	4.17×10^{-5}
Radon	2.11×10^{-6}
SO_2	6.39×10^{-5}

Table 8.24 Emission rates of total volatile organic compounds (VOC) for specific pollution sources

Source	$R_o\ (mg/h/m^2)$	$k\ (h^{-1})$
Wood painting	17,000	0.4
Polyurethane	20,000	0.25
Varnish of wooden floors	20,000	6.0
Old glassware	14,000	0
Dry cleaning	1.6	0.03
Liquid nail painting	10,000	1.0

$$V\frac{dc}{dt} = Qc_a + E - Qc - kcV \tag{8.5}$$

where Q (m³/s) is the infiltration arte from the outdoor air and c_a (g/m³) is the pollutant concentration outdoors.

The general solution of the Eq. (8.5) can be expressed as:

$$c = \frac{\frac{E}{V} + c_a \frac{Q}{V}}{\frac{Q}{V} + k}\left[1 - \exp\left(-\left(\frac{Q}{V} + k\right)t\right)\right] + c_o \exp\left[-\left(\frac{Q}{V} + k\right)t\right] \tag{8.6}$$

where c_o is the initial concentration of pollutant indoors. The steady state solution of the above equation can be written as:

$$c = \frac{Qc_a + E}{Q + kV} \tag{8.7}$$

In the following it is examined the case of a gaseous pollutant which has a constant indoor source without important removal process and with very low outdoor concentration. These pollutants include chemical components which are emitted from house construction materials, furnitures and carpets. For the concentration calculation indoors it is used the Eq. (8.7) without the chemicl reactions term and putting the outdoor concentration equal to zero. With these assumptions the solution is:

$$c = \frac{E}{Q} = \frac{E}{ACH \times V} \tag{8.8}$$

In the presented methodology it is assumed also that there is uniform mixing which is valid often for individual rooms and specific houses. However, the pollutant concentration at different rooms is different. In the simulation of houses with several rooms the equation of mass balance can be used for each room and to connect the different rooms through the process of infiltration.

8.17 Air Exchange Rate by Infiltration

Infiltration is the uncontrolled flow of air through unintentional openings driven by wind, temperature difference, and/or appliance-induced pressures across the building envelope. Unintentional openings can be gaps in openable components (windows, doors), gaps in construction joints, cracks in walls and ceilings and porous building materials.

Several procedures exist to measure infiltration rates, usually involving so called tracer gas technique when an inert gas is used to mark the indoor air and the tracer concentration is monitored and related to the room's infiltration rate.

Infiltration rates can be also obtained using building simulation software packages, which are capable of calculating pressure differences between outdoors and indoors at given height and orientation of building's exterior surface. Suitable weather databases containing hourly data about outdoor air temperature, solar radiation, wind speed and wind direction are available worldwide as these are used for building energy and air flow calculations. In buildings without mechanical ventilation the pressure difference across the building envelope consists of two components – stack pressure and wind pressure. While modelling and simulation of stack effects (caused by indoor-outdoor temperature differences) are already mastered, some difficulties still prevail in wind pressure calculations. Substantial uncertainties exist in determining surface pressure induced by wind particularly in the case of low-rise buildings (e.g. family houses or middle-sized blocks of flats) in urban areas. Infiltration rates should be studied on the local basis, taking into account the weather conditions, building's type and geometry, its surroundings and the size and position of openings (gaps) in the building envelope.

8.18 Emission Models

An important parameter for the simulation of pollutant dynamics indoors is the quantification of emissions and their variation with time. The most widely used methodology is based on the use of empirical relations which are based on experimental data from smog chambers. The model which is mainly used for the emission rates of pollutants R(t) (mg/h/m^2) is in the form:

$$R(t) = R_o e^{-kt} \tag{8.9}$$

where t is the time (h), R_o is the initial emission rate (mg/h/m^2) and κ is a constant (h^{-1}). The total mass which is emitted per surface unit is given as:

$$M = \int_0^\infty R(t)\,dt = \int_0^\infty R_o e^{-kt} dt = \frac{R_o}{k} \tag{8.10}$$

A large number of sources can be described with the Eq. (8.11). In the case that the source is constant the coefficient κ is equal to zero. In the literature sources with coefficient κ larger than 0.2/h are dependent from the mass transfer inside the air volume. Contrary, for values lower than 0.01/h the transport is dependent from the mass transfer from the source. An example of emissions which are dependent from the mass transfer from the source to air indoors (gas-phase-limited sources) is the liquid nail painting and wood painting (Table 8.24). Contrary the dry cleaning is mainly dependent from the mass transfer from the source to the air close to the source (source-phase-limited sources). Furthermore, emission rates for specific chemical compounds from wood painting are given in Table 8.25.

Table 8.25 Emission rates for specific chemical compounds from wood painting

Source	R_o (mg/h/m^2)	k (h^{-1})
Nonane	1,973	0.89
Decane	1,887	0.39
Hendecane	181	0.11

Table 8.26 Deposition (k_a) and resuspension (k_d) rates for specific chemical compounds

Material	Pollutant	k_a (m/h)	k_d (h^{-1})
Carpet	TVOC	0.1	0.008
Painted walls	TVOC	0.1	0.1
Wooded roof	TVOC	0.1	0.1
All surfaces	Ethylene glycol	3.2	0.0001
All surfaces	Propylene glycol	3.2	0.0
All surfaces	Texanol Ester Alcohol(2,2,4-Trimethyl-1,3-pentanediol Monoisobutyrate)	1	0.0002
All surfaces	p- dichlorobenzene	0.35	0.01

8.19 Deposition Models

Deposition of gaseous pollutants indoors occurs on various surfaces which retain them and furthermore re-emits (resuspends) part of them back to air. This behaviour is common for all gaseous components but it is more complex process for the airborne particles. Here the deposition of volatile organic compounds (VOC) will be examined since it has been studied extensively in the literature. A simple deposition model is based on the Langmuir isotherm as following:

$$R_{\sin k} = k_a C A - k_d M A \quad (8.11)$$

where R_{sink} is the net rate of scavenging, k_a is the deposition constant at surfaces and k_d is the resuspension rate of the chemical compound in air. In addition, A is the surface available for deposition, C is the concentration of the chemical compound in air and M is the mass of the chemical compound per unit surface. The above model works satisfactory for the deposition of volatile organic compounds but underestimates the resuspension rates. Table 8.26 shows the deposition and resuspension rates for specific chemical compounds.

8.19.1 Examples

Example 1. *A heating appliance which uses kerosene operates for one hour in an apartment wth volume of 200 m^3. The appliance emits SO$_2$ with a rate of 50 μg/s.*

The outdoor concentration of SO_2 and the initial concentration indoors is $100\ \mu g/m^3$. If the penetration rate is 50 l/s and the air is well mixed, which is the concentration of SO_2 indoors after one hour?

An application of Eq. (8.6) results to:

$$c = \frac{\frac{50\ \mu g/s}{(200\ m^3)} + 100\ \mu g/m^3\ \frac{0.050\ m^3/s}{200\ m^3}}{\frac{0.050\ m^3/s}{200\ m^3} + 6.39 \times 10^{-5}\ s^{-1}} \times \left[1 - \exp\left(-\left(\frac{0.050\ m^3/s}{200\ m^3}\right)(3600\ s)\right)\right]$$
$$+ (100\ \mu g/m^3)\exp\left[-\left(\frac{0.050\ m^3/s}{200\ m^3}\right)(3600\ s)\right] = 625.39 \mu g/m^3$$

Example 2. *Analytical solution of the continuity equation of the pollutant concentration indoors*

It has been shown in this chapter that the equation which determines the concentration balance of pollutants indoors is given by the equation:

$$V\frac{dc}{dt} = Qc_a + E - Qc - kcV$$

The above equation can be written as:

$$\frac{dc}{dt} + \left(\frac{Q}{V} + k\right)c = \frac{Qc_a + E}{V}$$

The solution of the above equation is the sum of the general solution of the homogeneous and the partial solution of the full equation. For the homogenous solution it can be written $\frac{dc}{dt} + \left(\frac{Q}{V} + k\right)c = 0 \Rightarrow lnc = -\left(\frac{Q}{V} + k\right)t + c_o$ where c_o is a constant and as a result it can be written $c = c_o \exp\left[-\left(\frac{Q}{V} + k\right)t\right]$.

For the spatial solution of the full equation can be $c = \frac{1}{\mu(t)}\int \mu(t)Q(t)\,dt$ where $P(t) = \frac{Q}{V} + k$, $\mu(t) = \exp\left[\int P(t)dt\right]$ and $Q(t) = \frac{Qc_a+E}{V}$. The solution can be written as $c = \left(\frac{Q}{V}c_a + \frac{E}{V}\right)\left\{\frac{1}{\frac{Q}{V}+k} + A\exp\left[-\left(\frac{Q}{V} + k\right)t\right]\right\}$ where ς is a constant.

The final solution is the sum of the above two solutions and can be written as

$$c = c_o \exp\left[-\left(\frac{Q}{V} + k\right)t\right] + \left(\frac{Q}{V}c_a + \frac{E}{V}\right)\left\{\frac{1}{\frac{Q}{V}+k} + A\exp\left[-\left(\frac{Q}{V} + k\right)t\right]\right\}$$

Using the initial conditions that for $t = 0$ the concentration is given as $c = c_o$ which also results that $A = -\frac{1}{\frac{Q}{V}+k}$.

Problems

8.1. In a new office complex a synthetic carpet has been placed. Formaldehyde (HCHO) is emitted from the carpet with a rate of $10\ \mu g/h$ in an office of $13\ m^3$. A worker is complained for nausea due to the formaldehyde emissions from the carpet. The formaldehyde concentration in the air indoors reached a level of 5 ppb. Which must be the ventilation rate in the office in order to have a concentration decrease of formaldehyde to 0.1 ppb ? It is assumed that the density of air at room temperature is 1.2 kg/m^3.

8.2. Gas stove operates in a closed room of 35 m^3 volume. The gas stove produces 12.000 Btu/h with the use of gas (mainly propane (C_3H_8) which has 20.000 Btu/h). If the infiltration rate is 1.3 h^{-1} calculate (a) the steady state CO_2 concentration (b) the CO_2 concentration after half an hour operation of the gas stove.

References

Agency for Toxic Substances and Disease Registry (ATSDR). (1997). *Toxicological profile for polychlorinated biphenyls*. Atlanta, GA: U.S. Department of Health and Human Services, U.S. Public Health Service.
Baek, S.-O., Kim, Y.-S., & Perry, R. (1997). Indoor air quality in homes, offices and restaurants in Korean Urban areas – Indoor/Outdoor relationships. *Atmospheric Environment, 31*, 529–544.
Brown, S. K., Sim, M. R., Abramson, M. J., & Gray, C. N. (1994). Concentrations of volatile organic compounds in indoor air – A review. *Indoor Air, 4*, 123–134.
Chan, C. C., Ozkaynak, H., Spengler, J. D., & Sheldon, L. (1991). Driver exposure to volatile organic compounds, CO, Ozone and NO_2 under different driving conditions. *Environmental Science & Technology, 25*, 964–972.
Finlayson-Pitts, B. J., & και Pitts, J. M. (2000). *Chemistry of the upper and lower atmosphere*. CA: Academic.
Health Effects Institute – Asbestos Research. (1991). *Asbestos in public and commercial buildings: A literature review and synthesis of current knowledge*. Cambridge: Health Effects Institute.
Hinds, W. C. (1999). *Aerosol technology*. New York: Wiley.
Lazaridis, M. (2008). The environmental chemistry of aerosols. In *In: I. Colbeck organic aerosols* (pp. 91–115). UK: Blalckwells Publication.
Lazaridis, M., & Alexandropoulou, V. (2009). Variability of indoor and outdoor gaseous aerosol precursors. *Water, Air, & Soil Pollution: Focus, 9*, 3–13.
Lioy, P. J. (1990). Assessing total human exposure to contaminants. *Environmental Science & Technology, 24*, 938–945.
Morrison, G. C. (2010). Indoor organic chemistry. In Colbeck Ian & Lazaridis Mihalis (Eds.), *Human exposure to pollutants via dermal absorption and inhalation*. New York: Springer.
Nazaroff, W. W., & Weschler, C. J. (2004). Cleaning products and air fresheners; exposure to primary and secondary air pollutants. *Atmospheric Environment, 38*, 2841–2865.
Nazaroff, W. W., Weschler, C. J., & Corsi, R. L. (2003). Indoor air chemistry and physics. *Atmospheric Environment, 37*, 5431–5453.
Nero, A. V., Schwehr, M. B., Nazaroff, W. W., & Revzan, K. L. (1986). Distribution of airborne Radon-222 concentrations in US homes. *Science, 234*, 992–997.
Osborne, M. C. (1987). Four common diagnostic problems that inhibit radon mitigation. *JAPCA, 37*, 604–606.
Pankow, J. F., & Bidleman, T. F. (1992). Interdependence of the slopes and intercepts from log-log correlations of measured gas-particle partitioning and vapor pressure 1. Theory and analysis of available data. *Atmospheric Environment, 26A*, 1071–1080.
Ruzer, L. S. and Harley, N. H. (Editors) (2005). *Aerosols Handbook – Measurement, Dosimetry and Health Effects*. CRC Press.

Spicer, C. W., Coutant, R. W., Ward, G. F., Joseph, D. W., Gaynor, A. J., & Billick, I. H. (1989). Rates and mechanisms of NO_2 removal from indoor air residential materials. *Environment International, 15,* 643–654.

Tichenor, B. A., & Mason, M. A. (1988). Organic emissions from consumer products and building materials to the indoor environment. *JAPCA, 38,* 264–268.

Turco, R. P. (2002). *Earth under siege: From air pollution to global change.* New York: Oxford University Press.

Wells, J. R. (2005). Gas-phase chemistry of alpha-terpineol with ozone and OH radical: Rate constants and products. *Environmental Science & Technology, 39,* 6937–6943.

Weschler, C. J. (2004). Chemical reactions among indoor pollutants: what we've learned in the new millennium. *Indoor Air, 14,* 184–194.

Weschler, C. J. (2000). Ozone in indoor environments: concentrations and chemistry. *Indoor Air, 10*(4), 269–288.

Chapter 9
Human Exposure and Health Risk from Air Pollutants

Contents

9.1	Human Exposure and Doses from Air Pollutants	306
9.2	Exposure Pathways	309
	9.2.1 Dermal Absorption	309
	9.2.2 Inhalation Exposure	310
9.3	Calculation of Dose–Response Functions	315
	9.3.1 Dose Calculation Through Intake	315
	9.3.2 Internal Dose Calculation Through Dermal Absorption	316
	9.3.3 Internal Dose Calculation Through Inhalation and Food Intake	318
	9.3.4 Functions of Dose–Response	319
9.4	Particulate Matter Dose Through Inhalation	322
	9.4.1 Deposition of Particles in the Respiratory Tract	322
	9.4.2 Classification of Particles Based on Their Ability to Penetrate the Respiratory Tract	328
	9.4.3 Calculation of Particle Deposition in the Respiratory Tract	331
	9.4.4 Particle Clearance in the Human Respiratory Tract	333
	9.4.5 Particle Deposition Measurements	338
9.5	Application: Internal Dose from Radon Inhalation	341
9.6	Health Effects from Air Pollutants	346
9.7	Health Effects from Exposure to Particulate Matter	350
References		354

Abstract Human activities continuously modify the environment, on a variety of scales, from indoor to global, and the resulting changes of environmental properties act as stressors that can affect human and ecological health. Assessing the health risks associated with environmental factors requires one to understand and quantify the complete sequence of events and processes involved in the "environmental health sequence" from the release or the formation of a stressor to the development of an environmentally caused disease. Chapter 9 studies the human exposure and health risk from air pollutants through dermal absorption and inhalation. Calculations of dose–response functions are derived. In particular calculations of particle deposition and clearance in the human respiratory tract through inhalation are performed. Finally health effects from exposure to gaseous and particulate matter air pollutants are examined.

9.1 Human Exposure and Doses from Air Pollutants

Human activities continuously modify the environment, on a variety of scales, from indoor to global, and the resulting changes of environmental properties act as stressors that can affect human and ecological health. Assessing the health risks associated with environmental factors requires one to understand and quantify the complete sequence of events and processes involved in the "environmental health sequence" from "source" (e.g. the release or the formation of a stressor) to "outcome" (e.g. development of an environmentally caused disease).

The procedure by which chemical compounds enter into the human body can be divided in two steps. At the beginning the person comes into contact with the chemical compound (exposure) and then follows the intake of the compound into the body (transport through the contact boundary[1]). Specifically, exposure is defined as the pollutant concentration that comes in contact with the human body for a specific time period. The contact occurs in the body area that is close to the breathing paths (mouth and nose). Exposure occurs through inhalation or with dermal contact. In addition, exposure from air pollutants can occur indirectly through the consumption of food and water if air pollutants have been deposited on them from the atmosphere or through pesticide use.

The human response is related mainly to the dose and not to its initial exposure. The dose is defined as the pollutant amount which is uptaken or absorbed at different parts of the human body for a specific time interval. The dose of a specific chemical is dependent on different factors such as the time of contact of the chemical with the human body and its clearance rate. The dose that is delivered to organs or tissues of the human body can lead to its injury or even to changes of its performance. In particular, the available quantity of the chemical which can be absorbed at the contact boundary (e.g. skin) is called a delivered dose. The chemical which is absorbed at a specific contact boundary is called an absorbed dose.

The way of penetration of chemicals into the human body through a contact boundary is difficult to be described. One can however make a distinction between two processes that occur during this phenomenon:

Intake: the introduction of a chemical compound inside the human body with the introduction of a related medium (air, food or water). The introduction of the chemical occurs through the external contact boundaries of the body (nose, mouth). The quantity of the chemical compound entering the body is calculated from the product of its quantity per intake frequency (e.g. m^3 air/day, kg food/day, lt water/day) with the concentration of the chemical in the medium. It is expressed as the rate of intake of the chemical compound which is the product of the quantity of the chemical per unit of time.

Uptake: the introduction of a chemical compound inside the human body through dermal absorption or other exposed organs such as eyes. Uptake also occurs

---

[1]The contact boundary refers to specific human organs, for example skin, lungs, gastrointestinal system, through which the chemical compound enters into the human body.

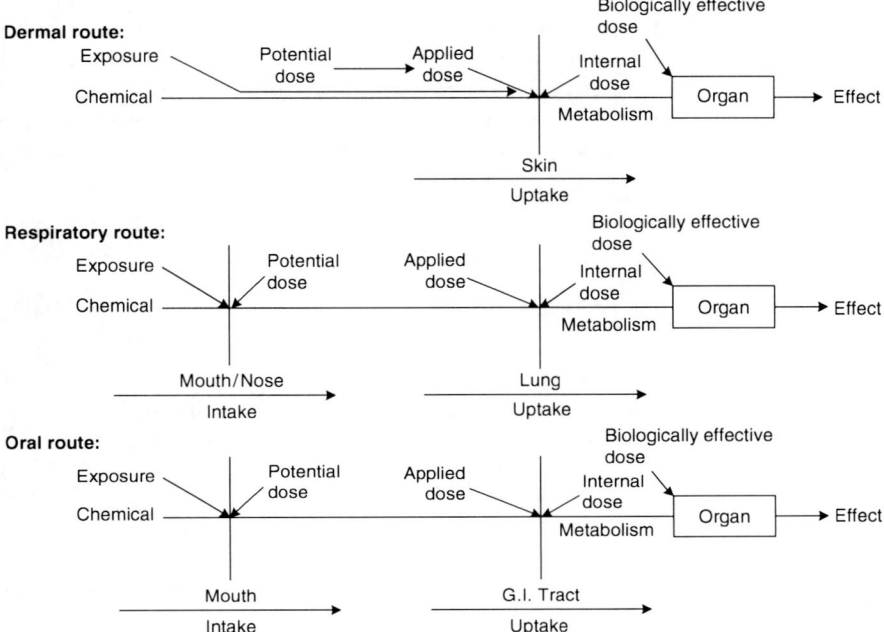

Fig. 9.1 Exposure and dose through different pathways (Adapted from USEPA 1992)

internally in the lungs, the gastrointestinal system and other internal organs after introduction of the chemical compound through inhalation or through food intake. The uptake depends on the penetration to the limit of contact and the concentration gradient of the chemical compound between the two ends of the contact limit. Therefore the uptake rate can be expressed as a function of the concentration exposed, the penetration coefficient and the exposed surface.

Usually part of the chemical compound concentration is removed (with urine, through the bowels, with sweating) or exhaled. The quantity of the chemical compound which is introduced into the human body is the potential dose (Fig. 9.1). The concept of a potential dose is useful in cases where there is exposure to a specific chemical compound but it is difficult to calculate the actual dose received (e.g. swimming in a pool). The delivered dose is usually lower than the applied dose due to the bioavailability of the chemical compounds. Finally, an internal dose is defined as the concentration of the chemical compound which is absorbed into the body and is available to act on tissues of high biological significance inside the body. Of course the route of the chemical compound continues and its concentration can be further metabolized, transported to the whole body, absorbed or removed. Finally, the quantity of the chemical compound which will reach a tissue or an organ is the delivered dose and the part of it that will reach tissues of an area in the body and will interact with them resulting in health effects is the biologically active dose.

The time of exposure to the respective dose can be long or short term. Long term exposure to a specific chemical can occur from occupational exposure (e.g. solvents, benzene) or house exposure (e.g. construction materials, cleaning agents, smoking). In this case the concentration of the chemical in the air is relatively small and for the calculation of the dose several simplifications can be made. On the other hand, short term exposure occurs after an accident or an extreme air pollution episode.

The total exposure (between times t_o and t_1) is given by the expression:

$$\bar{\varepsilon}_i(t_o, t_1) = \int_{t_o}^{t_1} c_i(t)\, dt \qquad (9.1)$$

where, $c_i(t)$ is the concentration of the chemical compound which is in contact with a person i (from a population) at time t. However, for practical reasons the exposure is expressed as:

$$\Delta \bar{\varepsilon}_i = \Delta t \times c_i(\Delta t) \qquad (9.2)$$

where Δ expresses the limited interval in which the exposure is calculated.

Figure 9.2 shows graphically the representation of total, acute, maximum and average exposure.

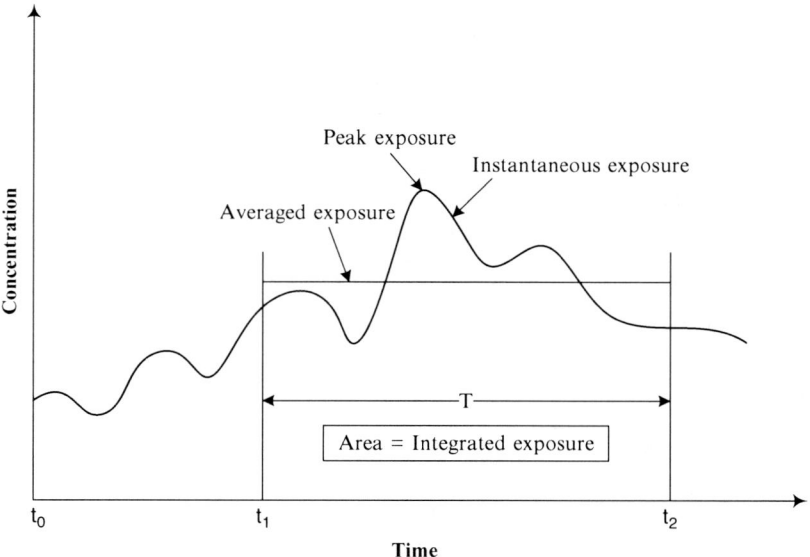

Fig. 9.2 Graphical representation of an average, maximum, accute and total exposure

9.2 Exposure Pathways

The exposure pathway expresses the routes by which a chemical compound will be introduced inside the human body (through food, inhalation or dermal absorption). It is possible for a chemical compound to be introduced directly into the blood circulation through open injuries of the skin and further transported to the whole body.

An analysis of the exposure pathways and description of the human physiology and especially of the human respiratory tract is presented in the following subsections, since the human anatomy affects the intake and uptake of the chemical compounds inside the human body.

9.2.1 Dermal Absorption

Dermal absorption of environmental contaminants has been shown to contribute significantly to an internal dose in many exposure scenarios. Percutaneous absorption is highly influenced by the microstructure and biochemical composition of the skin. A brief review of the skin's properties is necessary for the interpretation and better understanding of specific dermal absorption/transport rate data during exposure/risk assessment.

The skin is a highly organized, heterogeneous, and multilayered organ system. It is not only the largest organ in the body, but also perhaps the most complex, with at least five different cell types contributing to its structure, and other cell types from circulatory and immune systems being transient residents of the skin. In terms of the number of functions performed, the skin simply outweighs any other organ. Primarily, it constitutes a living envelope surrounding the body and offering protection, which covers physical, chemical, immune, pathogen, UV radiation and free radical defenses. In addition, skin is a sensory organ, also contributing to thermoregulation and endocrine functions. In addition to producing mediators of inflammatory and immune responses, the skin produces factors that regulate growth and differentiation. These factors make the skin more than an inert barrier and it should be viewed as a dynamic, living tissue whose permeability characteristics are susceptible to change. Therefore, there are abundant biochemical and physiological factors that remain to be systematically investigated.

The skin consists of two distinct layers (Fig. 9.3): the epidermis, a nonvascular layer about 100 μm thick, and the dermis (forming the bulk of skin), a highly vascularized hydrous tissue about 500–3,000 μm thick, made up of connective tissue elements.

The dermis is largely acellular, but is rich in blood vessels, lymphatic vessels and nerve endings. An extensive network of dermal capillaries connects to the systemic circulation, with considerable horizontal branching from the arterioles and venules in the papillary dermis to form plexuses and to supply capillaries to hair follicles and glands. Dermal lymphatic vessels help drain excess extracellular fluid and clear antigenic materials. The elasticity of the dermis is attributed to a network of protein

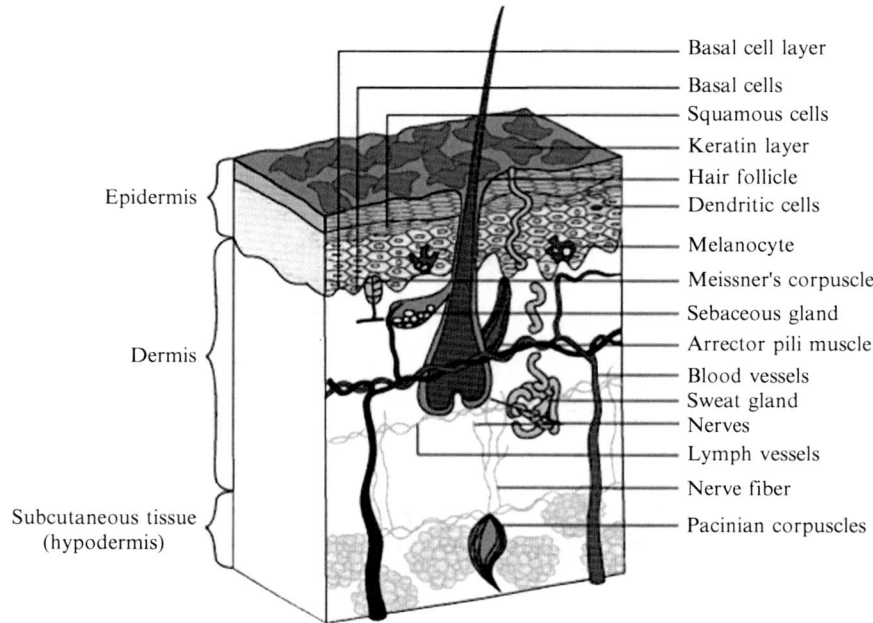

Fig. 9.3 Skin structure

fibres, including collagen (type I and III) and elastin, which are embedded in an amorphous glycosaminoglycan ground substance. The dermis also contains scattered fibroblasts, macrophages, mast cells and leukocytes.

The hair follicles and sweat ducts (skin appendages) originate deep within the dermis and terminate at the external surface of the epidermis (Fig. 9.3). These occupy only about 1% of the total skin surface, and therefore their role as transport channels for the passage of substances from the external environment to the capillary bed is thought to be negligible for most chemicals.

The rate-limiting diffusion barrier for most compounds deposited on skin surface has been located in the superficial epidermal layer, the stratum corneum. Because of the importance of this layer in determining the rate and extent of dermal absorption, the following discussion will focus on its structure and function.

9.2.2 Inhalation Exposure

The understanding of the particle deposition in the human respiratory tract (RT) requires the definition of a detailed morphometric model for the respiratory system. The basic aspects of the respiratory system are known so a simple inhalation model has been constructed (e.g. ICRP 1994). However, there is limited information available for the detailed geometry of the respiratory tract such as the dimensions of its numerous parts. Besides, there are differences in the RT due to human

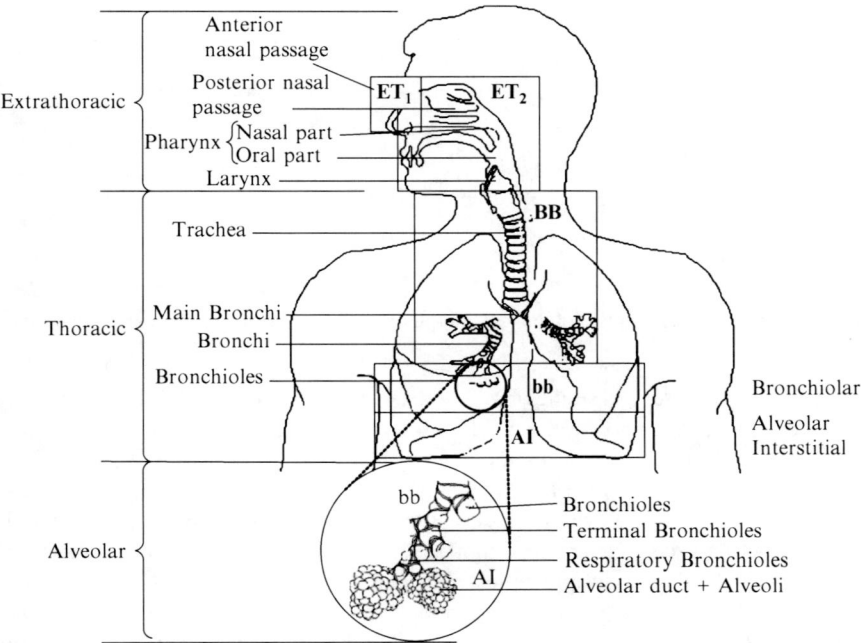

Fig. 9.4 The anatomy of the human respiratory system

variability and due to time dependent changes of the respiratory system arising from the breathing cycle.

The basic regions of the respiratory system are shown in Fig. 9.4 (ICRP 1994). The human respiratory system can be viewed as three anatomical regions. In each region the deposition characteristics for particles, as well as their structure, are different. The first region includes the *extrathoracic* airways ET_1 (anterior nose) and ET_2 (posterior nasal passages, larynx, pharynx and mouth). This region serves for warming and humidification of the inhaled air. The *extrathoracic* part is known also as *head airways region* or *nasopharyngeal region*.

The second region is the *tracheo-bronchial* region or *lower airways* which includes the bronchial (BB) and the bronchiolar (bb) regions. This region includes the airways which conduct air from the extrathoracic region to the gas exchange regions of the lung. The BB region consists of the trachea and bronchi and the bb region includes the bronchioles and terminal bronchioles. The trachea branches into two airways which are called main bronchi. This branching continues until the terminal bronchioles and the structure of the division is similar to an inverted tree. Both the extrathoracic and tracheo-bronchial airways have also the name of *conducting airways* since they conduct air to the gas exchange regions of the lung.

Finally, the third region is the alveolar-interstitial region (AI), where gas exchange is performed. The AI region is called also the *alveolar region*, or the *pulmonary region*, or the *acinar region*. This region consists of the respiratory bronchioles, the alveolar ducts, the alveoli and the interstitial connective tissue.

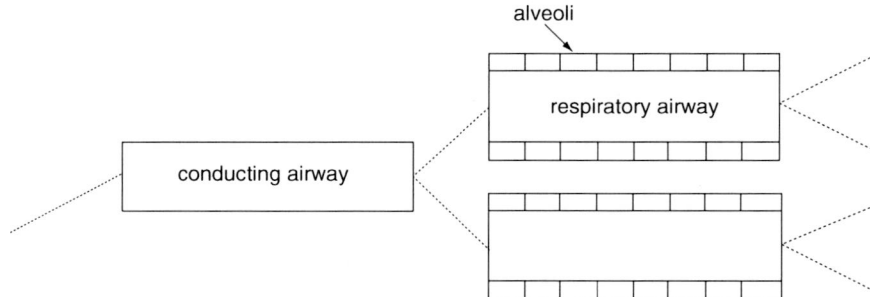

Fig. 9.5 Schematic presentation of the lung morphology as used in lung models with dichotomous branching

There have been several studies to measure the dimensions of the different parts of the respiratory tract, which however is a difficult task since the airways have three- dimensional structure. In the scientific literature many researchers have used measurements of casts of normal lungs to define the dimensions of the airways (e.g. Weibel 1963). Since many airways resemble a cylindrical structure, then the definition of the length and the diameter can be used for their definition (see Fig. 9.5). It is apparent that the above description of the RT is a simplification of its real structure.

The most extensively used lung model is the symmetric model of Weibel (Weibel 1963) which is referred to as the Weibel A model (Table 9.1). In this model there are 24 airway generations (23 levels of bifurcations), which branch symmetrically (dichotomously), with trachea being generation zero and alveoli generation 23. Since every generation divides into two branches the number of branches in each generation, z, is $N_z = 2^z$. Table 9.1 shows the characteristics of the Weibel A model for the structure of the human respiratory tract (ICRP 1994). The generations 0–15 correspond to the conducting airways and the remainder to the gas exchange region. A healthy adult has a number of alveoli between 400–1,200 million with a surface for the air exchange close to 80 m^2.

Since the total cross-sectional area usually increases with penetration distance, the average air velocity in distal airways decreases. Airway gravity angles, which in single pathway lung models are uniform within any generation, vary in accordance with the specific body posture considered (upright position, sleeping position etc.).

There are many simplifications made in the Weibel A model such as symmetric branching and underprediction of the tracheobronchial airways. Besides, the diameters and lengths of the alveolated airways are too small and also start at a high generation number (Finlay 2005). Therefore Weibel's picture is a simplification of the human bronchial tree, where there is irregular dichotomous branching, not well defined, and invariant over time with respect to geometric characteristics of the different bronchial branches. In addition, the tracheobronchial airway surface is lined with ciliated epithelium, but a constant amount of mucus at the ciliated epithelium does not exist all the time, making time dependent changes in the geometry of the human airways, which are not considered in Weibel's model. The main difficulty in the modeling of the RT is the description of a dynamic

Table 9.1 Architecture of the human lung according to Weibel A symmetric model for adults with lung volume of 4,800 ml and approximately ¾ increase due to inhalation (International Commission on Radiological Protection ICRP 1994)

		Generation z	Number, n (z)	Diameter d (z)(cm)	Length l(z) (cm)	Area S(z) (cm²)	Volume V(z) (cm³)	Cummulative Volume (cm³)	Flow rate 1 l s⁻¹ Speed (cm s⁻¹)	Reynolds No.	Residence time (ms)
BB	Trachea	0	1	1.8	12	2.54	30.5	30.5	393	4.350	30
	Main bronchus	1	2	1.22	4.76	2.33	11.25	41.8	427	3.210	11
	Lobar bronchus	2	4	0.83	1.9	2.13	3.97	45.8	462	2.390	4.1
		3	8	0.56	0.76	2	1.52	47.2	507	1.720	
	Segmental bronchus	4	16	0.45	1.27	2.48	3.46	50.7	392	1.110	3.2
	Bronchi with cartilage	5	32	0.35	1.07	3.11	3.3	54	325	690	4.4
	in wall	6	64	0.28	0.9	3.96	3.53	57.5	254	434	
		7	128	0.23	0.76	5.1	3.85	61.4	188	277	
		8	256	0.186	0.64	6.95	4.45	65.8	144	164	
bb		9	512	0.154	0.54	9.65	5.17	71	105	99	
		10	1.024	0.13	0.46	13.4	6.31	77.2	73.6	60	
	Terminal bronchus	11	2.048	0.109	0.39	19.6	7.56	84.8	52.3	34	7.4
	Bronchioles with	12	4.096	0.095	0.33	28.8	9.82	94.6	34.4	20	16
	muscle in wall	13	8.192	0.082	0.27	44.5	12.45	106	23.1	11	
		14	16.384	0.074	0.23	69.4	16.4	123.4	14.1	6.5	
	Terminal bronchiole	15	32.768	0.066	0.2	113	21.7	145.1	8.92	3.6	31
AI	Respiratory	16	65.536	0.06	0.165	180	29.7	174.8	5.40	2.0	60
	brronchiole	17	131.9 × 10³	0.054	0.141	300	41.8	216.6	3.33	1.1	
		18	262 × 10³	0.05	0.117	534	61.1	277.7	1.94	0.57	
	Alveolar duct	19	524 × 10³	0.047	0.099	944	93.2	370.9	1.10	0.31	210
		20	1.05 × 10⁶	0.045	0.083	1.600	139.5	510.4	0.60	0.17	
		21	2.10 × 10⁶	0.043	0.07	3.220	224.3	734.7	0.32	0.08	
		22	4.19 × 10⁶	0.041	0.059	5.880	350	1.0847	0.18	0.04	
	Alveolar sac	23	8.39 × 10⁶	0.041	0.05	11.800	591	1675	0.09	–	550
	Alveoli, 21 per duct		300 × 10⁶	0.028	0.023		3.200	4.800			

system as the human respiratory tree, with time dependent changes and variability characteristics between different people. However, Weibel's model is the most well-known model with very extensive applications.

Table 9.1 shows that the majority of the lung volume is contained in the alveolar region. The extrathoracic airways in an adult have a volume of approximately 50 ml, the tracheo-bronchial region a volume of approximately 100 ml, whereas, the remainder of the lung volume (between 2,000 and 4,000 ml during tidal breathing and 6,000 ml when fully inflated for an adult male) is in the alveolar region (Finlay 2005).

The dimensions of the airways change due to age, gender, race and weight. The Weibel A model can be viewed as a sequence of filters and the deposition fractions for each region are calculated after accounting previously for the filtering effect of the airways. At each region of the lung the deposition is calculated using empirical functions.

The anatomy and physiology of the RT determine the deposition of particles. Of particular importance are the parameters which determine the rhythm and process of the breathing cycle (physiology). The breathing mode (nose breathers, nasal augmenters) intensity depends upon activity. Inhalability depends also on the breathing mode. The lung physiology is based on several parameters such as (see also Fig. 9.6):

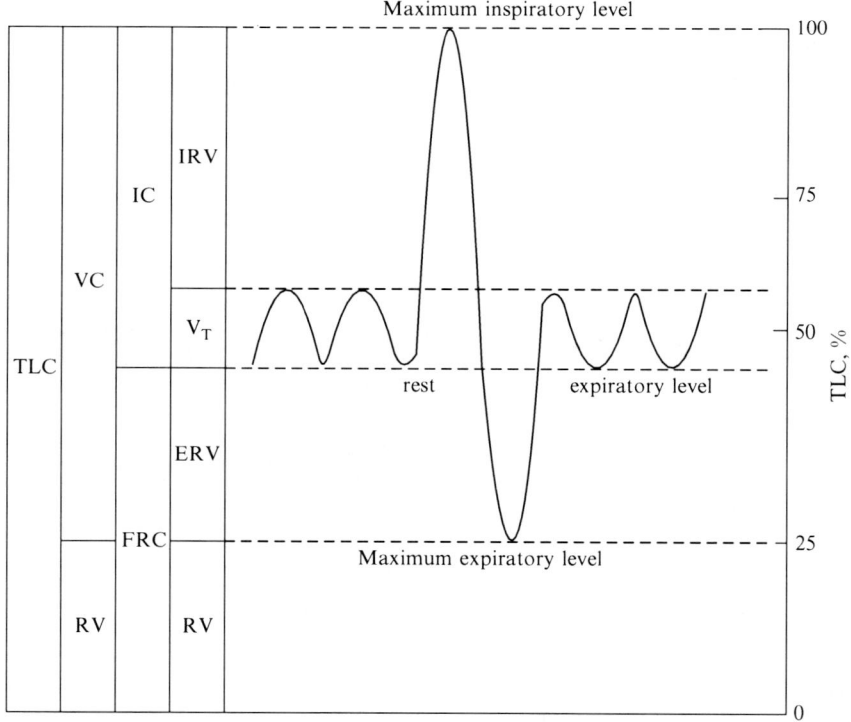

Fig. 9.6 Diagram of the different air volume parameters (respiratory volumes) used in the *RP* for the description of the breathing cycle. (*TLC* = (Total Lung Capacity), *VC* = (Vital Capacity), *RV* = (Residual Volume), *FRC* = (Functional Residual Capacity), *IRV* = (Inspiratory Reverse Volume), *ERV* = (Expiratory Reverse Volume), V_T = (Tidal volume), *IC* = (Inspiratory Capacity)

- The tidal volume V_T, which is the air volume inhaled and exhaled during a periodic breathing (500–3,800 ml depending on exercise level)
- The breathing frequency f, which is defined as the number of tidal breaths per minute (12–18 min^{-1} for normal activity)
- The total lung capacity (TLC), which is the total volume of the airspaces in the lung when it is totally inflated (maximum breath) (about 6,000–7,000 ml)
- The functional residual capacity FRC, which is the lung volume at the end of normal exhalation or at the start of normal inhalation (close to 3,000 ml for adults)
- The residual volume (RV), which is the remaining volume of the airspaces when someone exhales the maximum volume of air
- The vital capacity (VC), which is maximum air volume that someone can inhale (close to 4,000 ml for adults)
- The forced expiratory volume in one second (FEV_1), which is the maximum volume of air that a typical individual can exhale during one second when there is the maximum volume in the airspaces (TLC).

9.3 Calculation of Dose–Response Functions

9.3.1 Dose Calculation Through Intake

The calculation of the potential dose D_{pot}, after the intake of a chemical compound is given by the integral of the intake rate (with inhalation or ingestion) and the exposure time:

$$D_{pot} = \int_{t_1}^{t_2} C(t)IR(t)dt \qquad (9.3)$$

The intake rate is the product of the concentration of the chemical compound inside the medium, C(t), with the uptake rate of the medium, IR(t). The exposure duration is expressed with the term $t_2 - t_1$. The dose can be expressed as a summation of doses in specific time intervals and conditions of exposure i:

$$D_{pot} = \sum_i C_i \times IR_i \times ED_i \qquad (9.4)$$

If the concentration of the chemical compound and the medium uptake rate are constants, then the above expression can be written as:

$$D_{pot} = \overline{C} \times \overline{IR} \times ED \qquad (9.5)$$

where, ED is the total duration of the exposure. The above hypothesis is valid when the exposure duration is small.

In cases when the dose calculation is performed within the scope of a calculation of the human risk due to exposure from a specific chemical compound, then the dose (expected or internal) is expressed as the average rate in the specific time interval. Specifically, in cases when the exposure to the chemical compound is continuous and smooth, the commonly used term is the Average Daily Dose (ADD). The ADD is calculated as the dose integral at specific time interval (days) weighted with the human weight. However, it is practically impossible to collect data which refer to the time profile of the exposure and the intake of chemical compounds for long time intervals (chronic exposure). Therefore for the calculation of the ADD, average values are used for the concentration and the rate of uptake:

$$ADD_{pot} = [\overline{C} \times \overline{IR} \times ED]/[BW \times AT] \tag{9.6}$$

where, BW is the human weight and AT is the duration of exposure expressed in days. When the calculations refer to the determination of the probability of cancer appearance through dose–response functions, then one commonly uses the average daily dose that the human receives during life (Lifetime Average Daily Dose-LADD) which can be calculated with the expression:

$$ADD_{pot} = [\overline{C} \times \overline{IR} \times ED]/[BW \times LT] \tag{9.7}$$

where, LT is the life span of a human expressed in days.

9.3.2 Internal Dose Calculation Through Dermal Absorption

The dose calculation through uptake with dermal contact of a human with a medium that contains a chemical compound can be achieved in two ways. The first way is applied in cases of partial or total immersion in liquid (e.g. a swimmer) from which the chemical compounds are absorbed due to dermal contact. In this case the expected dose is expressed as the integral of the rate of chemical intake at the specific time interval t_2-t_1:

$$D_{pot} = \int_{t_1}^{t_2} C(t) \times K_p \times SA(t) \times dt \tag{9.8}$$

In Eq. (9.8) the term C is the concentration of the chemical compound in the medium, SA is the exposed surface and K_p is the permeability coefficient. The concentration and the exposed surface can be varied with time, whereas the permeability coefficient is changed in relation to the area of the body (analogous to

9.3 Calculation of Dose–Response Functions

the structure and the dermis width). In the case of the dermal absorption the medium in which the chemical compound is dissolved does not pass the contact limit and therefore there is a flux of the chemical compound through the contact limit which is dependent on the chemical structure, the physical and chemical characteristics of the contact limit and the concentration gradient of the chemical compound at the two sides of the contact limit.

The flux of the chemical components (J) using Fick's law is given with the expression $J = K_p C$ under constant concentration C. The permeability coefficient remains constant for a large range of concentrations and can be measured experimentally. The internal dose can be calculated from the relationship:

$$D_{int} = \overline{C} \times K_p \times \overline{SA} \times ED \tag{9.9}$$

and the average daily internal dose can be expressed as:

$$ADD_{int} = [\overline{C} \times K_p \times \overline{SA} \times ED]/[BW \times AT] \tag{9.10}$$

In case the concentration and the surface change significantly during the exposure period, Eq. (9.10) does not give very realistic results.

A second approach for calculation of the dose through uptake and dermal contact is based on empirical observations and can be applied in the case that the quantity of the substance which is in contact with the skin is low or known. The expected dose can be calculated when the quantity of the medium (M_{medium}) in which the chemical compound is dissolved and the concentration of the chemical compound (C) is known. For example the dose arising from dermal contact with soil can be calculated with the expression:

$$D_{pot} = \overline{C} \cdot M_{median} = \overline{C} \cdot F_{adh} \cdot \overline{SA} \cdot ED \tag{9.11}$$

where, F_{adh} is the sticking coefficient of soil on the skin which express the soil quantity which is applied and sticks on the skin per unit surface and time. In case of dermal exposure, the difference between the applied and expected dose is related to the quantity of the chemical compound, which for the determination of dose is the quantity that is dissolved in the medium that is applied to the skin (applied dose), whereas for the expected dose it is only the quantity of the chemical compound which is in direct contact with the skin.

Theoretically the expression between the internal and applied dose can be expressed with the equation:

$$D_{int} = D_{app} \int_{t_1}^{t_2} f(t)dt \tag{9.12}$$

where, f(t) is a complex non-linear absorption function having units of absorbed mass of the chemical compound per unit mass of the medium which is applied on the skin per unit time. The above function is dependent on the concentration slope of the chemical compound, the skin type, the humidity and the properties of the skin, the exposed surface and the properties of the medium (e.g. in the case that the medium is soil it is the organic content, the granulus form and the humidity).

The integral of the function f during the exposure period is the absorbed ratio AF, which can increase with time until the value 1 (100%) or to reach a stable position at lower value. Therefore the expression between the internal and the expressed dose is given by:

$$D_{int} = D_{app}AF \tag{9.13}$$

where the absorbed ratio AF is expressed as an absorbed/expressed mass and is dimensionless.

Assuming that the total quantity of the chemical component which exists in the medium is in direct contact with the skin, the internal dose can be expressed as a function of the applied dose through the expression:

$$D_{int} = D_{pot}AF \tag{9.14}$$

and the average daily internal dose is expressed as:

$$ADD_{int} = [\overline{C} \times M_{median} \times AF]/[BW \times AT] \tag{9.15}$$

The use of the above equation requires caution since it gives realistic results only in the case that the quantity of the chemical component inside the medium is in direct contact with the skin. However, often the quantity of the chemical component is not in direct contact with the skin. Especially if the quantity is large, then the time which is required for the chemical species to be transported inside the medium in order to be absorbed from the skin is large and can exceed the exposure period. It is also possible for the exposure period to be smaller than the time which is required for the absorption function to attain steady state, especially for chemical components which are absorbed with a medium rate.

9.3.3 Internal Dose Calculation Through Inhalation and Food Intake

The quantity of a chemical component which is introduced into the human body through ingestion and inhalation is accumulated in places with increased biological importance (e.g. lung, gastrointestinal tract). Furthermore it is moved to specific organs. The internal dose through the intake of chemical components through the surface of the lungs or through the gastrointestinal tract can be calculated. In the same manner the internal dose through dermal absorption can be calculated. However, even though details for the pharmacokinetics of the chemical components can be known, it is not possible to know several other data concerning the surface of the lungs, the gastrointestinal system and the permeability coefficient. These data are not directly measurable and besides are affected by the physiology, anatomical characteristics of each individual and health condition (e.g. diseases of the alveolar region). If it is assumed that the whole quantity of a chemical component which is introduced into the body is absorbed (the applied and expressed dose are equal), then the internal dose can be calculated from the equation:

$$D_{int} = D_{app}AF = D_{pot}AF = \overline{C} \times \overline{IR} \times ED \times AF \qquad (9.16)$$

Therefore the average daily internal dose of chemical components after exposure through inhalation or ingestion is given by the expression

$$ADD_{int} = ADD_{pot}AF = \left[\overline{C} \times \overline{IR} \times ED \times AF\right]/[BW \times AT] \qquad (9.17)$$

where the use of average intake rate and average concentration involves their smooth variation versus time.

9.3.4 Functions of Dose–Response

Adverse effects due to environmental contamination occurs as a result of a sequence of events, beginning with release of a contaminant from a source, and including environmental fate and transport, accumulation in microenvironments (relatively uniform environments in the immediate vicinity of receptors), uptake by human and biological transport to sensitive tissues/site within organisms, and finally a dose–response expressing the adverse effect.

Empirical observations have generally revealed that as the dosage of a chemical is increased, the toxic response (in terms of severity and/or incidence of effect) also increases. Hence, for a specific pollutant (chemical agent in general) a dose–response relationship may quantify the effect of a measurable exposure (dose) on the proportion of subjects demonstrating specific, biological changes (response). In other words, it estimates how different levels of exposure to a pollutant change the likelihood and severity of health effects. The dose–response relationship concept is widely used in the theory and practice of toxicology, pharmacology and epidemiology, yet, the discipline of exposure/internal dose assessment is young and rapidly evolving and much fundamental knowledge on the mechanisms involved is still lacking.

For most chemicals and pollutants, there is a lack of appropriate information on effects in humans. In such cases, the principal information is drawn from experiments conducted on nonhuman mammals, most often the rat, mouse, rabbit, guinea pig, hamster, dog, or monkey. Human data are occasionally used in qualitatively establishing the presence of an adverse effect in exposed human populations. When there is information on the exposure level associated with an appropriate endpoint, epidemiological studies can then provide a basis for a quantitative dose–response assessment. The presence of such data makes obvious the necessity of extrapolating from animals to humans; therefore, human studies, when available, will be given first priority, with animal toxicity studies serving to complement them.

In the literature the terms "dose" and "response" do not follow a standard definition. Depending on the nature of the study, "dose" may well be the amount of a chemical administered orally, applied externally (e.g. environmental concentration), or absorbed internally. Similarly, "response" may be the measured or observed incidence, the percent response in groups of subjects (or populations), or the probability of occurrence within a population. Generally, in toxicological studies the dose is expressed in terms of administered amount per kilogram of body weight of the subject per day. In epidemiological cohort studies the dose is usually expressed in terms of external exposure, e.g. concentration in air, drinking water etc. Also, there are recent works where the dose is expressed using internal intake parameters, for instance, the concentration of the chemical in the serum.

Hence, in developing the dose–response relationships, certain choices and assumptions must be made in arriving at an appropriate measure of the health effects from the anticipated multiroute environmental exposure. The most important aspects that will be considered are:

1. Selection of appropriate mathematical functions to describe the low-dose response, in conjunction with using a correct measure of dose (i.e. external dose, absorbed dose, or target organ dose). The proposed tool is well suited to perform such refined assessments because it integrates multi-route exposure modeling (inhalation, skin) together with (PBPK) modeling.
2. Interspecies extrapolation, as required to convert data obtained on animals to equivalent human doses.
3. Effect of dose patterns and timing, as needed to assess long-term, chronic exposures from short-term acute impacts.
4. Joint response due to simultaneous intake of multiple chemicals.

From the physiological point of view, there is a clear distinction between the various measures of "dose", which, according to USEPA (1992) can be classified as external dose (or exposure), applied dose, internal dose and delivered dose. "External dose" (or "exposure") means contact of an agent with the outer boundary of an organism, "applied dose" refers to the amount of an agent presented to an absorption barrier and available for absorption (e.g. deposited mass in the respiratory tract), "internal dose" refers to the amount crossing an absorption barrier (e.g. the exchange boundaries of skin, lung, and digestive tract), whereas "delivered dose" for an organ or cell means the amount available for interaction with that organ or

9.3 Calculation of Dose–Response Functions

cell. However, the available dose–response data generally do not include such refined differentiations. For the purposes of the proposed work, and in the absence of appropriate information, the algebraic dose–response functions that will be utilized, will be used to quantify the response to all forms of dose, be it an external dose, applied dose, internal dose, or delivered dose.

For cancer-causing pollutants a linear relationship will be assumed, as suggested by the United States Environmental Protection Agency (USEPA). The underlying assumption is that even an infinitesimal dose does have some risk, even if only by a small amount. Obviously, such a dose–response relationship is completely determined if either the slope, or a single non-zero point is specified (for example the ED_{10} effective dose- the dose associated with a 10% increased response).

For non-carcinogenic pollutants a dose level may exist below which no adverse health effects are expected to occur (NOAEL: No-Observed-Adverse-Effect Level). In this case the frequency or severity of biological response does not vary proportionately with the amount of dose, and consequently the dose–response relationship will exhibit a nonlinear behaviour. There is no response up to a dose level corresponding to the so-called Reference Dose (RfD). The RfD point is derived from the NOAEL point by consistent application of generic order-of-magnitude uncertainty factors. Then, the response curve evolves according to the characteristic S-shape, consistent with many biological experiments.

From the diagrams of dose – response one can resolve the Threshold dose (Th) which is required for observable responses of organisms. It is also possible to obtain from these diagrams the toxicity of different chemical compounds and their doses as well as determined responses at selected time intervals. A dose – response curve has typically a canonical distribution, in vivo and in vitro studies, versus dose. However, cumulative curves are used usually since they offer more direct information and are easy to determine toxicity differences between different chemical components.

Figure 9.7 shows three curves of dose–response. The curves 1 and 3 correspond to the dose of the same chemical component A but at a different response, whereas the curve 2 corresponds to the same response as the curve 1 but for a component B. In the figure are shown the threshold value (Th) at which are observed adverse effects to the bodies under study. This limited value is known as NOAEL (No Observed Adverse Effect Level), LOAEL (Lowest Observed Adverse Effect Level) or TLV (Threshold Limit Value). It is also shown that the cumulative curve of dose–response reaches a plateau and at higher doses the response is not changing. The two chemical components A and B are compared (curves 1 and 2) and it is observed that even though the mean response results from the same dose (ED_{50} – Effective dose), the doses at different percentages of the cumulative response are different (e.g., ED_{99}, ED_1). The mean dose (ED_{50}) is defined as the quantity of the chemical component (toxic component) which causes the specific response (sign of illness) for 50% of the exposed population. In the case that the response is death for 50% of the population under study, then the dose is know as lethal and is denoted with LD_{50}. The margin of safety is also shown which is the difference of the maximum limited dose which elicits a specific response (e.g.

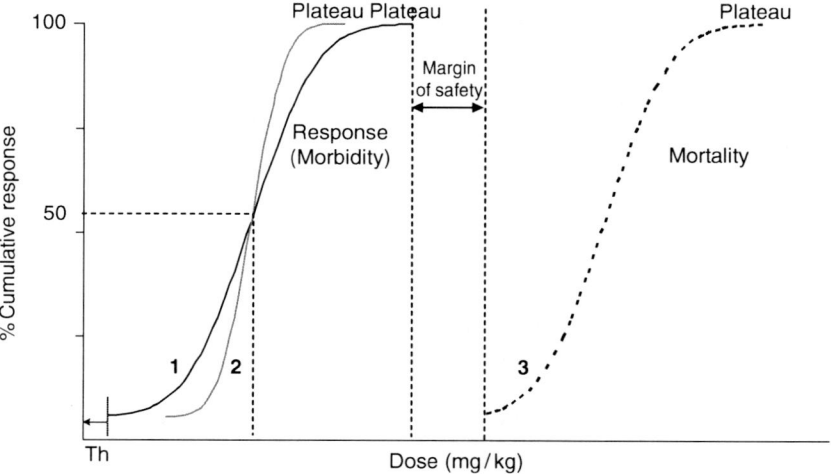

Fig. 9.7 Cumulative curves for the dose – response relationships and their basic characteristics. Curves 1 and 2 refer to different components A and B which have the same mean response but there are differences in relation to their effects at other dose levels. Curve 3 refers to the mortality versus dose of component A

morbidity) and the minimum dose of the same chemical component which results in death (e.g., LD_{01}/ED_{99} or LD_{50}/ED_{50}). One toxic substance is more drastic than another if for the same percentage of cumulative response the dose is smaller than another substance, whereas it is characterized as more effective if the same dose corresponds to a larger range of responses.

9.4 Particulate Matter Dose Through Inhalation

9.4.1 Deposition of Particles in the Respiratory Tract

The deposition characteristics of particles in the RT depend on their size, their physico-chemical properties and the physiology of the person. The main deposition mechanisms in the RT are inertial impaction, settling and Brownian diffusion (see Fig. 9.8). The deposition mechanisms of interception and electrostatic deposition are important in specific cases (Hinds 1999; Finlay 2005). The total deposition is calculated as a superposition of independent deposition efficiencies of the different mechanisms.

Regional deposition profiles in the lungs are usually given in terms of deposition fractions in specific functional compartments: the nasopharyngeal region (the extrathoracic compartment - ET), the tracheobronchial region (TB), and the pulmonary–alveolar region (P). On average, there is a clear relationship between

9.4 Particulate Matter Dose Through Inhalation

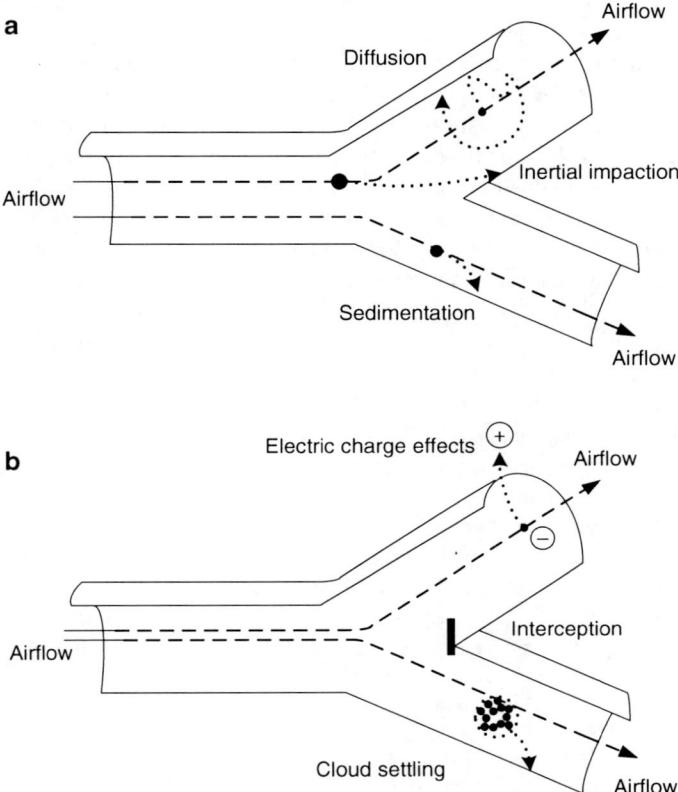

Fig. 9.8 Schematic description of deposition mechanisms of particles in the RT. (Adapted from Ruzer and Harley (2005))

particle penetrability class (i.e., inhalable, thoracic, respirable) and the compartment in which most of them tend to deposit, since particle size is the main attribute of both processes. Apart from size and concentration, the deposition profile is also affected by the particle chemical properties. Processes such as gas-to-particle conversion, particle–particle interactions, and growth are likely to occur in the respiratory tract, and once materialized have a considerable effect on the deposition pattern. Gas-to-particle conversion includes the following processes: reaction (neutralization), nucleation, and vapor condensation (hygroscopic growth). The latter is the most important growth mechanism in the RT under normal conditions, since prevalent atmospheric and anthropogenic aerosols tend to show affinity to water vapor. The very nearly saturation conditions in the human RT promotes hygroscopic growth of unstable aerosols, even if they have already attained equilibrium at the ambient conditions. Nucleation within the framework of inhalation dosimetry is unlikely even in extreme occupational environments, but coagulation of post-combustion (cigarette smoke particles) may be of some importance.

In addition to health effects caused by particular properties of individually deposited particles, inhalation of pollutant mixtures may introduce additional pathways of exposure, resulting from biochemically-induced events and dose-modulating effects. It is well known that exposure to a highly reactive oxidation agent, such as ozone, has multiple effects on the lung, ranging from alterations of the pulmonary mechanical function and ventilation to modulating the susceptibility of the epithelium layer, sensitizing the lungs to other irritants, and promoting tissue injury. Further effects include alteration of mucus secretion and therefore impairing the lungs' defense and clearance mechanisms. Simultaneous exposure to PM and ozone, both are products of atmospheric photochemistry of post-combustion precursors, is thought to synergistically affect the adverse health effects associated with exposure to each of these chemicals separately.

The deposition characteristics of particles in the RT depend on their size, their physico-chemical properties and the physiology of the person. The main deposition mechanisms in the RT are inertial impaction, settling and Brownian diffusion (see Fig. 9.8). The deposition mechanisms of interception and electrostatic deposition are important in specific cases (Hinds 1999; Finlay 2005). The total deposition is calculated as a superposition of independent deposition efficiencies of the different mechanisms.

The dose of particles in each part of the RT is calculated using the following expression:

$$H = n_0 c_A B \left(a \sum_i n_{fine,i} + (1-a) \sum_i n_{coarse,i} \right) \quad (9.18)$$

where H is the dose-rate (μg/h), n_0 is the inhalability ratio, c_A is the aerosol concentration in air, B is the ventilation rate, α is the fine mode mass fraction, $n_{\text{fine},i}$ is the retention in region i of lungs for fine particles and $n_{\text{coarse},i}$ is the retention in region i of lungs for coarse particles (ICRP 1994).

The *inertial impaction* mechanism is a result of the inadequacy of particles to follow the streamlines at places that air changes abruptly direction inside the RT. As a result the particles continue their original direction for a short distance due to their inertia. This mechanism is important for particles having an aerodynamic diameter larger than 0.5 μm mainly in the upper RT (ICRP 2002). The probability of particle deposition due to impaction is proportional to the airflow velocity, the inhalation frequency and the particle size and density. Inertial impaction is important in areas of sharp curvatures in the streamlines, such as in the nasal turbinates and airway bifurcations. Inertial deposition is proportional to the Stokes number ($St = \frac{\tau \bar{u}}{l}$) of the particles that enter in the RT (τ is the particle relaxation time and the ratio $l/\bar{u}$ is the average residence time of the particle in the airway, where l is the length of the airway and $\bar{u}$ is the airflow velocity). The highest probability of particles to deposit by impaction is in the bronchial region (Hinds 1999).

9.4 Particulate Matter Dose Through Inhalation

Gravitational settling refers to the deposition of particles in the RT due to gravity and together with the mechanism *inertial impaction* are the most important mechanisms for deposition of particles with aerodynamic diameter greater than 0.5 μm. The mechanism of *gravitational settling* is important for deposition in small airways and the alveolar region. The probability of particle deposition due to gravity is proportional to the residence time in the airways, their size and density and inversely proportional to the breathing frequency. Gravitational settling and inertial impaction are competing deposition mechanisms in the RT (Hinds 1999). The inertial impaction is most important for increased flow rates. The relative contributions of these two mechanisms to the particle deposition at the bifurcations of the first generations of the alveolar region are about the same. However, their difference is at the deposition location with the major fraction of deposited particles due to impaction at the airway bifurcations, whereas, the particles deposited due to gravitation are located along the whole airway (Balashazy et al. 1999; Housiadas and Lazaridis 2010).

Particles smaller than 0.2 μm have a Brownian motion and may deposit at the airway walls especially in smaller airways (alveolated airways). *Diffusional deposition* is the main deposition mechanism for particles in the bronchioli and the alveolar region where the flow velocity is low and the residence times long. In addition, the deposition of ultrafine particles (smaller than 0.01 μm) due to diffusion is significant at the head airways due to high diffusion coefficients (ICRP 2002).

Electrically charged particles are attracted to airway walls in the RT due to an electrostatic image charge induced in the airway surfaces. The *electrostatic attraction* is inversely proportional to the size of the particles and the velocity air flow. However, the concentration of electrical charged particles in the atmosphere is not high since the particles are neutralized from the atmospheric ions. Therefore the importance of the mechanism of electrostatic attraction is relatively small compared to the other deposition mechanisms. Experimental data have shown that the deposition mechanism of electrostatic attraction is a main deposition mechanism for ultrafine particles in the lower tracheobronchial area (USEPA 2004).

Another deposition mechanism is the particle *interception* at which particles make contact with the airway surfaces due to their size. The interception is dependent on the particle size and its morphology. This mechanism is important for fibrous particles of large aspect ratio, and for highly elongated fibers (ICRP 1994).

There are several dosimetry models which were developed to predict the total and regional deposition of particles in the RT. The most used models are those of the International Commission on Radiological Protection (ICRP 1994) and the National Council on Radiation Protection and Measurement (NCRP 1994). Deposition differences predicted by the two models are small compared to the subject variability differences and are both recommended for particle deposition calculations.

The ICRP model has been used here to study the particle deposition at different parts of the RT versus their diameter for a man, woman and a child (3 months old) (see Figs. 9.9–9.11). The results show that at the upper respiratory tract (ET1 and ET2 regions) the deposition of particles with a diameter smaller than 0.2 μm is higher for men compared to women due to higher volumetric flow rates. For larger particles the deposition percentage for males and females is almost equal.

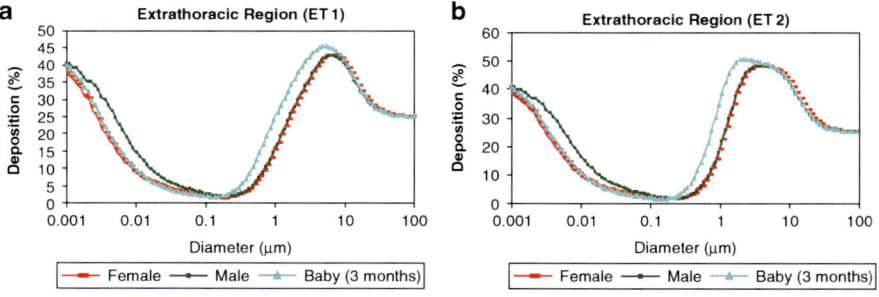

Fig. 9.9 Mass particle percentage which deposits in the (**a**) ET1 and (**b**) ET2 thoracic regions of the RT for a man, woman and a child (3 months) in relation to the particle diameter

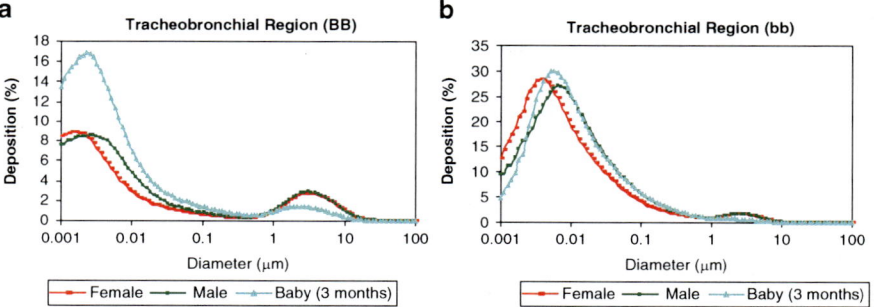

Fig. 9.10 Mass particle percentage which deposits in the (**a**) BB and (**b**) bb tracheobronchial regions of the *RT* for a man, woman and a child (3 months) in relation to the particle diameter

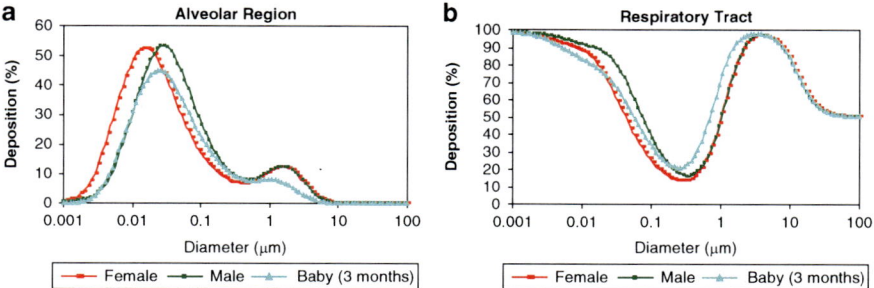

Fig. 9.11 Mass particle percentage which deposits in the (**a**) alveolar region of the RT and (**b**) total deposition of particles in the *RT* for a man, woman and a child (3 months) in relation to the particle diameter

9.4 Particulate Matter Dose Through Inhalation

In the BB part of the tracheobronchial region (bronchi) coarse particles have the same deposition characteristics for males and females. However, particles with a diameter smaller than 0.002 μm deposit with higher probability in the female's RT, whereas, particles with diameter in the range between 0.002– 0.2 μm deposit with higher probability in the men's RT. In the bb part (bronchioli) and in the alveolar region there is the same behaviour as the BB region. Particles with diameters smaller than 0.006 and 0.021 μm deposit more in the RT of women for the regions bb and AL (alveolar) respectively. Particles with diameter in the ranges between 0.006– 0.85 μm and 0.021–0.7 μm deposit with higher probability in the men's RT for the regions bb and AL respectively.

Smaller particles deposit easier in the lower parts of the female's RT. This is due to anatomical differences between women and men. The total deposition fraction is larger for men for the small particles and almost equal between men and women for the coarse fraction.

A 3 months old child has higher deposition percentage of larger particles in the upper respiratory and extrathoracic regions compared to an adult, whereas, for the smaller particles the deposition profile is the same as that of an adult. A smaller percentage of particles deposits in the alveolar region of a child since the dimensions of this region are smaller than that of an adult. However, the deposition particle mass per surface area in the alveolar region of a child is large (USEPA 2004). Finally, the total particle deposition percentage of a child compared to an adult is lower for small particles and higher for larger particles. In addition, due to a child's activities during a day the volumetric flow rate is high and the particle dose in the RT is also higher compared to the average activity conditions applied to adults.

Elderly and individuals with chronic obstructive pulmonary disease (COPD) breath with higher frequency and have higher deposition percentage in the tracheobronchial region compared to healthy individuals (USEPA 2004). In addition, individuals with COPD have also a higher percentage of particle deposition in the alveolar region compared to healthy people. However, the volumetric flow rate for individuals with COPD is not evenly distributed due to obstruction of parts of the RT. Therefore, whereas some parts of the alveolar region are not reachable some other parts have locally intense deposition.

Age plays an important role for the dimensions of the different volume parameters in the RT. For an elderly person the residual volume and the functional residual capacity are increased, whereas, the vital capacity, the inspiratory reverse volume and the expiratory reserve volume are decreased. In addition, the dead volumes at each part of the RT are changed since the alveolar region losses its elasticity and the entrained air is not divided uniformly at the different airways. Therefore a number of alveoli do not receive air continuously and as a result the air volume that is not involved actively in the breathing cycle is increased.

The physiology parameters which are used for the description of the breathing cycle are dependent on the race and the age of the human subject. For example an adult person inhales more air than a child, whereas, the breathing frequency is decreased. Table 9.2 presents representative values for physiology parameters of

Table 9.2 Representative values of anatomy and physiology parameters which are used for the calculation of particle deposition in the RT (International Commission on Radiological Protection ICRP 1994)

Parameter	Exposed person		
	Male	Female	Child (3 months)
FRC: Functional residual capacity (ml)	3,301	2,681	148
$V_D(ET)$: Anatomical dead space of the Extrathoracic region (mL)	50	40	2,6
$V_D(BB)$: Anatomical dead space of the trachea and bronchi (mL)	49	40	4,5
$V_D(bb)$: Anatomical dead space of bronchioli (ml)	47	44	6,8
d_0:Diameter of Trachea (cm)	1.65	1.53	0.616
d_9:Diameter of the first bronchioles (cm)	0.165	0.159	0.099
d_{16} : Diameter of the terminal bronchiole (cm)	0.051	0.048	0.020
Mean values during sleeping			
B: Volumetric flow rate during inhalation (m^3/h)	0.45	0.32	0.99
V_T: Air volume during inhalation (ml)	625	444	39
V: Volumetric flow rate (ml/s)	250	178	50
f: Breathing frequency (breaths/min)	12	12	38
Mean values at rest			
B: Volumetric flow rate during inhalation (m^3/h)	0.54	0.39	–
V_T: Air volume during inhalation (mL)	750	464	–
V: Volumetric flow rate (mL/s)	300	217	–
f: Breathing frequency (breaths////min)	12	14	–
Mean values for light exercise			
B: Volumetric flow rate during inhalation (m^3/h)	1.5	1.25	0.19
V_T: Air volume during inhalation (mL)	1,250	992	66
V: Volumetric flow rate (mL/s)	833	694	106
f: Breathing frequency (breaths////min)	20	21	48
Mean values for heavy exercise			
B: Volumetric flow rate during inhalation (m^3/h)	3	2.7	–
V_T: Air volume during inhalation (mL)	1,920	1,364	–
V: Volumetric flow rate (mL/s)	1,670	1,500	–
f: Breathing frequency (breaths/min)	26	33	–

Caucasian persons under different activities. In addition, the main anatomical parameters which are used for the calculations of particle deposition in the RT are given for male, female and a child.

9.4.2 *Classification of Particles Based on Their Ability to Penetrate the Respiratory Tract*

The entrance of particles inside the human respiratory tract (RT) depends mainly on their size and the air flow characteristics close to the mouth and nose. The entry of particles into the mouth and nose is characterized by the inhalability (I). Inhalability is

9.4 Particulate Matter Dose Through Inhalation

the intake efficiency at which ambient aerosols enter the mouth and nose. It is defined as the ratio of the particle mass concentration inspired through the nose or the mouth (M_o) for a specific size to the ambient number concentration ($M_{o,amb}$):

$$I = \frac{M_o}{M_{o,amb}} \tag{9.19}$$

Series of experiments have been performed for the determination of inhalability in low-velocity wind tunnels with the use of a full-size, full-torso mannequin connected to a mechanical breathing machine (Hinds 1999). These experiments have been done using all possible orientations of the mannequin in respect to the wind flow. Orientation averaged data from mouth inhalation are shown in Fig. 9.12. The particles are distinguished at respirable, thoracic and inhalable depending on their ability to enter specific areas of the human respiratory tract. Specifically, inhalable are the particles that can enter and deposit to the upper respiratory system. Thoracic are the particles which can penetrate the trachea and bronchi. Finally, respirable are the fraction of thoracic that enter the alveoli region. These definitions are designed for practical reasons and their correspondence to the equivalent particle diameter has been proposed by the International Standards Organization (ISO 7708) and the American Conference of Governmental Industrial Hygienists (ACGIH 1997). The corresponding curve for PM_{10} (particle mass aerodynamic diameter equal to or smaller than 10 μm) is almost identical with the thoracic fraction curve, whereas, the curve for $PM_{2.5}$ (particle mass aerodynamic diameter equal to or smaller than 2.5 μm) is found to the left of the respirable fraction (USEPA 2004).

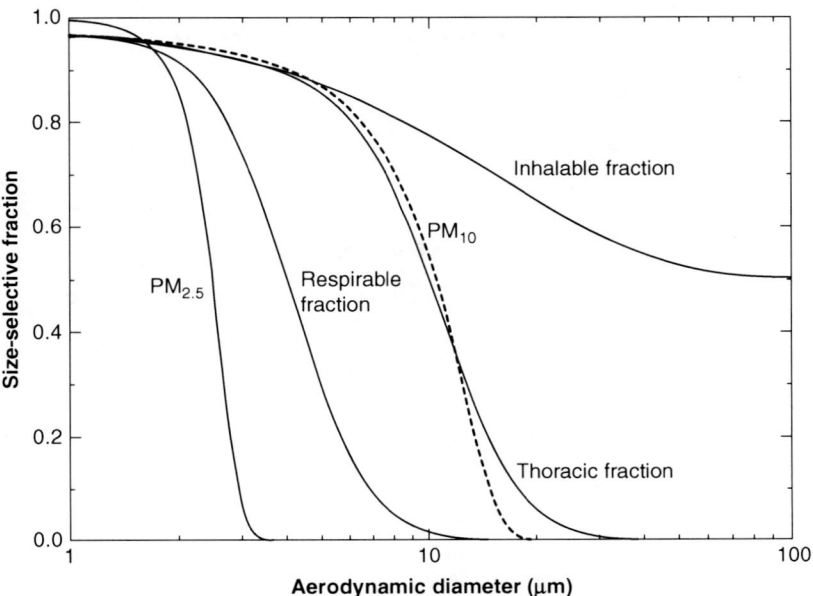

Fig. 9.12 Sampling criteria for respirable, thoracic and inhalable particle fractions (ACGIH 1997) together with the PM_{10} and $PM_{2.5}$ curves

The equation that defines inhalability for ambient conditions, where the wind velocity is lower than 4 m s^{-1} is given as:

$$I = 0.5\,(1 + \exp(-0.06\,d_{ae})) \tag{9.20}$$

where d_{ae} is the particle diameter (μm).

At wind velocities higher than 4 m s^{-1},,,, the inhalability can be described by an expression which depends not only on the particle diameter but also on the wind speed U (m s^{-1}) (Vincent et al. 1990):

$$I = 0.5\,(1 + \exp(-0.06\,d_{ae})) + 10^{-5} U^{2.75} \exp(0.055 d_{ae}) \tag{9.21}$$

The expression for inhalability used in the ICRP66 inhalation dosimetry model is (ICRP 1994):

$$I = 1 - 0.5\left[1 - \left(7.6 \times 10^{-4} d_{ae}^{2.8} + 1\right)^{-1}\right] + 10^{-5} u^{2.75} \exp\{0.055 d_{ae}\} \tag{9.22}$$

The inhalability depends also on whether the subject inhales from the mouth or nose. A mathematical expression for nose inhalability can be found in the literature even though the data are more scarce (Hinds 1999) and among other models a logistic function has been proposed which describes better than the ICRP66 model the nasal inhalability curve from calm air (Menache et al. 1995).

The inhalability at air velocity 1 and 5 m/s versus the particle diameter is shown in Fig. 9.13. The majority of the PM$_{10}$ particles enter the human respiratory

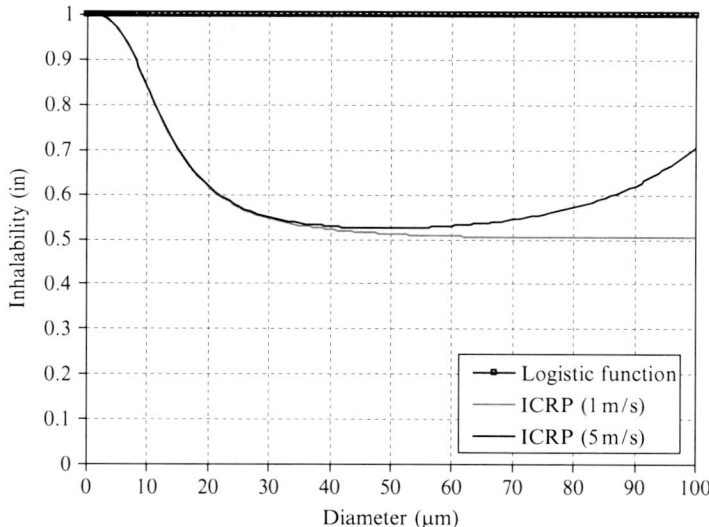

Fig. 9.13 Particle inhalability at wind velocity 1 and 5 m/s using the ICRP model in relation to particle diameter and a logistic function

9.4.3 Calculation of Particle Deposition in the Respiratory Tract

The deposition of particles in the human respiratory tract as analyzed previously is dependent on the location inside the body, the gender, anatomical characteristics of individuals and inhalation route. The determination of the deposition is a difficult task even for well specified particles (spherical with the same density and diameter) since it requires several calculations and the use of a computer. The methodology for the calculation of particle deposition and removal from the human respiratory tract which is presented here is based on the HRTM (Human Respiratory Tract Model) model. The semi-empirical equations for the deposition of particles are adopted from the ICRP (International Commission on Radiation Protection) (ICRP 1994).

In the HRTM model the respiratory system is divided into five regions, two extrathoracic (ET_1 and ET_2), two thoracic (BB and bb) and the alveolar region (AI) as described in the current chapter. Specifically, the deposition fraction in each region of the respiratory tract during the inhalation and exhalation processes is calculated assuming that the respiratory tract can be described as a series of filters from which is passing the air and a fraction of the particles suspended in the air volume are deposited with a deposition yield DF (see Fig. 9.14). The increase of particle size due to hygroscopicity is not taken into account and therefore the deposition is calculated in relation to the particle size.

The particle deposition in the five regions of the respiratory tract can be calculated using a computer model (e.g. LUDEP model which is based on the ICRP publication) which solves a complex system of equations that include particle characteristics, anatomical data and air flow characteristics. However, for an average person under an average activity (average parameter values) the deposition of particles can be described with the following empirical equations (Hinds 1999):

For the *extrathoracic region:*

$$DF_{ET} = IF \left(\frac{1}{1+\exp(6,84+1,183\ln d_p)} + \frac{1}{1+\exp(0,924-1,885\ln d_p)} \right) \quad (9.23)$$

where, IF is the inhalability at air velocity 0 m s^{-1} and d_p is the particle diameter (μm).

For the *tracheobronchial region:*

$$DF_{TB} = \left(\frac{0,00352}{d_p} \right) \left[\exp\left(-0,234(\ln d_p + 3,40)^2\right) + 63,9\exp\left(-0,819(\ln d_p - 1,61)^2\right) \right]$$

$$(9.24)$$

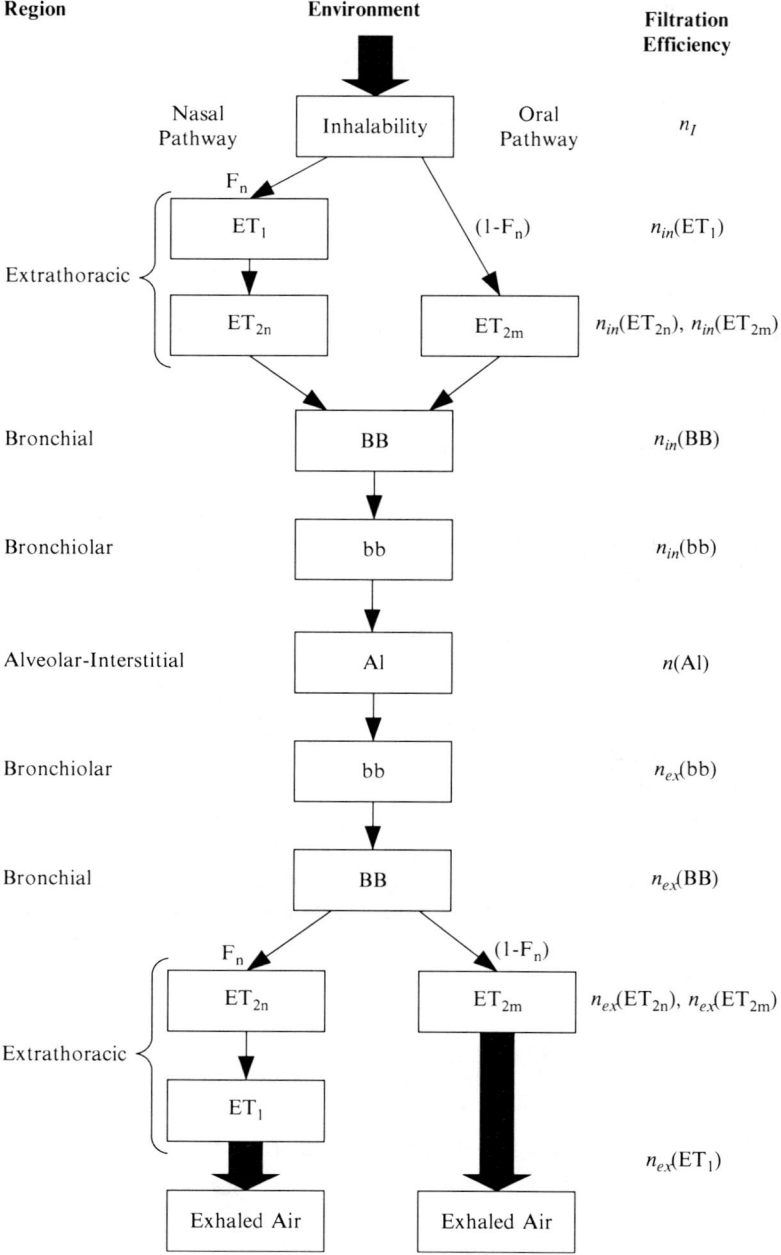

Fig. 9.14 Deposition of particles in the human respiratory tract during a continuous breathing cycle by transport through a series of filters. The two pathways for intake are the nasal pathway with fractional airflow F_n and the oral pathway with fractional airflow $(1-F_n)$. The subscripts "in" and "ex" of the filtration efficiency, η, represent the inhalation and exhalation phases of the breathing cycle, respectively (Adapted from ICRP 1994)

9.4 Particulate Matter Dose Through Inhalation

For the *alveolar region*:

$$DF_{AL} = \left(\frac{0,0155}{d_p}\right)\left[\exp\left(-0,416(\ln d_p + 2,84)^2\right) + 19,11\exp\left(-0,482(\ln d_p - 1,362)^2\right)\right] \quad (9.25)$$

The total deposition fraction can be expressed as the summation of the fractions as:

$$DF = IF\left(0,0587 + \frac{0,911}{1 + \exp(4,77 + 1,485 \ln d_p)} + \frac{0,943}{1 + \exp(0,508 - 2,58 \ln d_p)}\right) \quad (9.26)$$

The above equations can be used for calculation of the deposition fraction for particles with diameters between 0.001 and 100 μm and the results agree with the ICRP model calculations with a deviation of ± 0.03%. The above expressions can be used for calculation of the deposition fraction also for non-spherical particles under the condition of using the aerodynamic diameter for particles larger than 0.5 μm and the equivalent geometric diameter for particles smaller than 0.5 μm.

9.4.4 Particle Clearance in the Human Respiratory Tract

The deposited particles in the RT undergo further clearance (Gradon et al. 1996). The clearance depends on the particle solubility, its chemical composition and size as well as on physiological parameters (metabolic rate, health condition, age, gender).

Inhaled pollutants deposit in the airways by various deposition mechanisms, and subsequently interact with the epithelium depending on their reactivity and solubility. The amount and site of PM deposition and of the clearance rate are affected by:

- Particle geometric and physicochemical characteristics (size, shape, charge, density, chemical composition),
- The morphology and inflammatory state of the respiratory tract,
- Physiological and respiration parameters (breathing pattern, ventilation rate, exercise level, metabolic rate, health condition, body posture, age, gender),
- Environmental conditions (temperature, humidity, ambient concentration).

The particle clearance from the RT is a natural defence mechanism of the human body. The clearance translocates the particles to the gastrointestinal (GI) region, the regional lymph nodes (LN) via lymphatic channels and into the blood by the absorption mechanism (ICRP 1994). The clearance rate by each route depends on the region in which particles were deposited, the physicochemical properties of the particles (e.g. solubility, particle size, chemical composition) and their mass/number concentration.

The clearance mechanisms of deposited particles in the RT may be performed by two mechanisms which act competitively (particle transport and absorption into blood). These are also the main mechanisms used in the ICRP model to study particle clearance:

1. *Particle transport*: Particles are transported to the GI tract and the lymph nodes and from one part of the RT to another. The transport is performed with the mucus layer or uptake by macrophages. At the upper RT additional mechanisms such as coughing contribute also to particle removal.
2. *Absorption into blood*: The movement of materials from the deposited particles into blood occurs with the dissociation of particles and the absorption into blood. This mechanism refers also to the movement of ultrafine particles directly into the blood (ICRP 1994).

There are different clearance mechanisms which dominate at each part of the RT. In particular:

9.4.4.1 Upper Respiratory Tract (ET1 and ET2 Regions)

In the upper respiratory tract and especially in the posterior nasal passages, insoluble particles are cleared by mucus flow (USEPA 2004). The mucus is transported to pharynx and swallowed. Otherwise, particles can be also removed by sneezing. In the anterior part of the respiratory tract particles are mainly removed by extrinsic means (nose blowing) (ICRP 1994). The mucus transport velocity in the anterior part is small (close to 2 mm/h). The time that is needed for particles to be transported from the anterior to the posterior part is close to 10–20 min (USEPA 2004). Soluble particles in the nasal passages are absorbed to the blood through the epithelium.

9.4.4.2 Tracheobrobcial Region (Bb and Bb Regions)

This region is covered by mucus and mucociliary clearance to the GI and pharynx is the main mechanism for particle transport. In addition, insoluble particles can be also removed by macrophage and the soluble particles can be transported into blood. In this region the mucus transport is reduced from the trachea region to the regions deeper in the RT. The clearance from this region is quick (close to 24 h) but some insoluble particles may stay for time periods longer than 24 h (USEPA 2004).

9.4.4.3 Alveolar–Interstitial Region (Al Region)

In the alveolar region particles are removed with (1) absorption to blood and (2) phagocytosis from the macrophage and sequential transport to the mucus layer and transport to the GI region or transport to the blood through the lymph system and the interstitium (see Fig. 9.15).

9.4 Particulate Matter Dose Through Inhalation

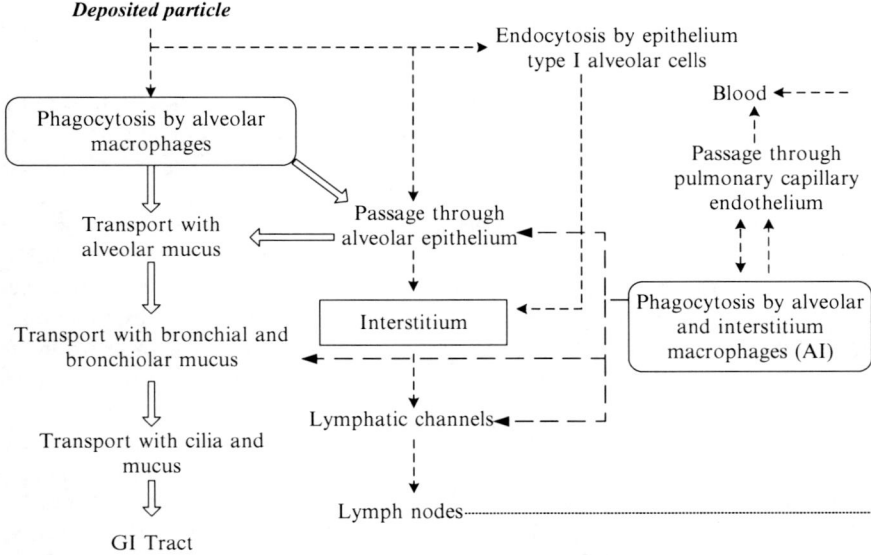

Fig. 9.15 Mechanisms for particle clearance from the alveolar region for insoluble particles (Adapted from USEPA 2004)

Lung retention experimental data for insoluble particles are scarce but data show a high retention time with some material remaining in the lungs for decades (ICRP 1994). The clearance mechanism can be described as a multiphase process with each phase corresponding to a separate clearance mechanism with high retention times. The ICRP model (ICRP 1994) describes the retention in the alveolar region as a two-component exponential function, with about 30% of particles having a half-time of 30 days and the remaining a half-time of several hundred days.

The clearance through the mucus layer is dependent on the particle size and deposition area. Macrophages are effective for the phagocytosis of particles except of particles of asbestos and silica. These particles produce an inflammation to the epithelium cells and the macrophage which have an effect on the phagocytosis rate (Donaldson et al. 2000). An increased dose in the alveolar region (particle mass/number) has also an effect on the retention time (USEPA 2004). In general, the clearance from the alveolar region follows an exponential decay and can be represented by three separate rates, the fast (AI_1 with a rate close to $0.02 \, d^{-1}$) with the transport of macrophage to the mucus layer, the medium (AI_2 with a rate close to $0.001 \, d^{-1}$) with the transport of macrophage through the interstitium and the slow one (AI_3 with a rate close to $0.0001 \, d^{-1}$) which acts with the particle breakup and further absorption and transport (ICRP 1994).

Finally, the available dose due to particle exposure and inhalation can be calculated as the product of the mean concentration of particles in air, the mean breathing frequency and the exposure time. Therefore, the delivered dose rate (D_{app} – $\mu g \, h^{-1}$) from the inhalation of polydisperse particles is given by:

$$D_{app} = \sum_{i=1}^{n} B(a_i IF_{0,i} C_{A,i}) n_{i,j} \qquad (9.27)$$

where, $IF_{0,i}$ is the particle's inhalability, $C_{A,i}$ is the particle concentration in air (μg m^{-3}), B is the breathing frequency (m^3 h^{-1}), α_i is mass fraction of particles and $n_{i,j}$ is the retention of particles with size i to the region j of the RT. The term i refers to the particle mean diameter.

The recommended approach for the clearance mechanism in the RT includes calculations of particle clearance in the lung following deposition for soluble, readily absorbed into blood, and relatively insoluble and nontoxic solid particles (ICRP 1994). Particles deposited in the respiratory tract are cleared by three main routes: by absorption into blood, to the gastrointestinal tract via the pharynx and to lymph nodes via lymphatic channels. The material travels in the respiratory tract following the course in Fig. 9.16.

Each region of the respiratory tract is divided into a combination of compartments cleared with a different constant rate so that the overall clearance approximates the time-dependent clearance behaviour of the region. Clearance from ET$_1$ to the environment is not considered in the model because the material deposited there is removed by extrinsic means (nose blowing, wiping); a reference value for residence time is 1 d^{-1} as proposed in the ICRP model. Particle residence times and clearance rates for mechanical transport are given for each compartment in the ICRP Publication 66 and are presented in Table 9.3.

In the model there are no calculations included for the transport of particles from the ET1 region to the environment since this rate is dependent on parameters which

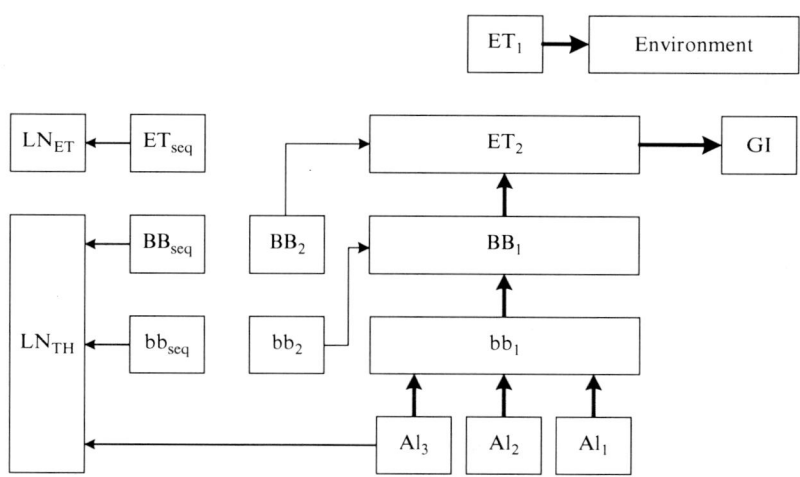

Fig. 9.16 Model of particle movement in the different regions of the RT, where each region consists of specific parts which describe the time dependent transport rate (ICRP 1994). Details on the different regions are given in Table 9.3.

9.4 Particulate Matter Dose Through Inhalation

Table 9.3 Rates of clearance transport of particles together with their half time in the different regions of RT

Clearance Rates				
Pathway	From	To	Rate (d^{-1})	Half-time
$m_{1,4}$	AI_1	bb_1	0.02	35 d
$m_{2,4}$	AI_2	bb_1	0.001	700 d
$m_{3,4}$	AI_3	bb_1	0.0001	7,000 d
$m_{3,10}$	AI_3	LN_{TH}	0.00002	–
$m_{4,7}$	bb_1	BB_1	2	8 h
$m_{5,7}$	bb_2	BB_1	0.03	23 d
$m_{6,10}$	bb_{seq}	LN_{TH}	0.01	70 d
$m_{7,11}$	BB_1	ET_2	10	100 min
$m_{8,11}$	BB_2	ET_2	0.03	23 d
$m_{9,10}$	BB_{seq}	LN_{TH}	0.01	70 d
$m_{11,14}$	ET_2	GI tract	100	10 min
$m_{12,13}$	ET_{seq}	LN_{ET}	0.001	700 d

Table 9.4 Deposition fraction of particles at different compartments of the RT (Adapted from International Commission on Radiological Protection ICRP 1994)

Deposition site	Compartment	Fraction of deposit in region assigned to compartment
ET_2	ET_2	0.9995
	ET_{seq}	0.0005
BB	BB_1	0.993-f_s
	BB_2	f_s
	BB_{seq}	0.007
bb	bb_1	0.993-f_s
	bb_2	f_s
	bb_{seq}	0.007
Al	AI_1	0.3
	AI_2	0.6
	AI_3	0.1

cannot be predicted such as the frequency and duration of coughing. The deposition fraction of particles at different compartments is given in Table 9.4.

The fraction of the initial deposit assigned to each compartment is given in the table where $f_s = 0.5$, for $d_{ae} \leqslant 2.5\sqrt{\rho/\chi}\mu m$, and $f_s = 0.5 \exp\left(-0.63\left(d_{ae}\sqrt{\chi/\rho} - 2.5\right)\right)$, for $d_{ae} > 2.5\sqrt{\rho/\chi}\mu m$.

Absorption into blood is assumed to occur at the same rate in the regions considered. Dissolution of particles and uptake into blood following deposition can occur with variable rates dependent on the aerosol's physical and chemical characteristics. However, absorption behavior of particles has been classified in three main categories; fast, moderate and slow for particles that take days, weeks or years to dissociate and enter the veins. Particle dissolution rates used in the model are according to the ICRP classification for fast and slow cleared particles and are presented in Table 9.5.

Table 9.5 Absorption rates used in the ICRP model for fast and slow clearance (International Commission on Radiological Protection ICRP 1994)

ICRP Publication 30 Classification	D (days)	W (weeks)	Y (years)
Type of absorption behavior:	F (fast)	M (moderate)	S (slow)
Fraction dissolved rapidly, f_r	1	0.1	0.001
Approximate dissolution rate:			
Rapid (d^{-1}), s_r	100	100	100
Slow (d^{-1}), s_s	–	0.005	0.0001
Model parameters			
Initial dissolution rate (d^{-1}), s_p	100	10	0.1
Transformation rate (d^{-1}), s_{pt}	0	90	100
Final dissolution rate (d^{-1}), s_t	–	0.005	0.0001
Fraction of material bound, f_b	0	0	0
Uptake rate of bound material (d^{-1}), s_b	–	–	–

Table 9.6 Values of the coefficient w_R

Type and spectrum of energy	Coefficient of radiation balance w_R
Photons, of different energy level	1
Electrons and positrons of different energy level	1
Radiation γ, β και x	1
Neutrons (energy E in MeV)	$5 + 17e^{-2\ln(2E)/6}$
Particles α, heavy nucleus	20

The model approach, as mentioned before, does not estimate clearance for particles exhibiting moderate absorption behavior. In addition retention for slow cleared particles is estimated with the assumption that all particles become dissolute slowly and that the fraction of material readily absorbed is zero.

The particle absorption into blood can be differentiated in two separate stages: (1) the breakup into substances which can be absorbed into blood and (2) the direct absorption into blood.

Retention in each region is given by the mass balance equation $\frac{dR_i(t)}{dt} = -\lambda_i(t)R_i(t)$ where $\lambda_i(t)$ is the instantaneous clearance rate of the deposit in compartment i and $R_i(t)$ the retained mass after time t. The rate is the sum of mechanical movement and absorption into blood rates $\lambda_i(t) = m_i(t) + s_i(t)$. Transport equations for particles are solved numerically per minute (matrix method) to determine retention. The relationship between target tissues and clearance components is given in Table 9.6. Retention results are assigned to the source components.

9.4.5 Particle Deposition Measurements

The respiratory tract deposition has been studied in the literature using various aerosolized dusts in human and animal lungs as well as in lung airway replicas. Recent measurements on particle deposition in human airways have been taken by

9.4 Particulate Matter Dose Through Inhalation

several authors. An excellent overview of past measurement aerosol deposition studies is presented by USEPA (2004).

Chan and Lippmann (1980) used monodispersed Fe_2O_3 particles with mass median diameter larger than 2 μm and density of 2.56 g/cm^3 for in vivo deposition determination in 26 healthy nonsmokers and for in vitro measurements in hollow casts. The tidal volume was approximately 1,000 cm^3. A mechanistic dosimetry model was used to simulate the deposition measurements from Chan and Lippmann (1980) (Lazaridis et al. 2001). In this dosimetry model the Aerosol General Dynamic Equation is solved numerically during inhalation, using a discrete-nodal point method for describing the particle size distribution. This model incorporates explicitly the mechanisms of nucleation, condensation, coagulation, convection and deposition of gases and particles, as well as a module for considering gas phase reactions.

To compare model predictions with experimental data it was assumed that the tidal volume was inspired in 1 s, corresponding to an air flow rate of 1,000 cm^3/s. This is relatively high air flow, corresponding to nonsedentary breathing conditions under which respired particles undergo enhanced inertial deposition. The model predicts the evolution of the size distribution and composition of inhaled particles and their deposition characteristics for each generation of human airways. The model has modular structure and the user has the flexibility to include or exclude specific physical processes in a particular simulation. Predictions of the dosimetry model compared with the experimental data for tracheobronchial deposition of Chan and Lippmann (1980) are presented in Fig. 9.18. Model predictions were obtained for three typical bifurcation angles θ, the only parameter not rigidly fixed in Weibel's lung model. Typical deposition conditions correspond to bifurcation angles of 45° (typical deposition conditions), 65° (maximum deposition conditions), and 25° (minimum deposition conditions). Fig. 9.17 shows that particle deposition data in the tracheobronchial region seems bound by the theoretical results for the most and least favorable deposition conditions. Varying the bifurcation angles affects only inertial deposition, which takes place mainly in the proximal airways. Therefore, alveolar deposition is less affected by these morphological variations (Fig. 9.18). Overall, the natural variability of the lung morphology, which is transformed into model uncertainty with respect to the bifurcation angles, significantly affects the particle deposition profile along the tracheobronchial airways. Indeed, deposition of large particles in the upper airways can double under adverse conditions (*i.e.* larger surfaces for impaction in the carinal ridges). In contrast, deposition of small particles, which are less affected by inertial impaction, is less affected by the variation in branching angles.

Figs. 9.19 and 9.20 show model predictions of deposition in the tracheobronchial and the alveolar regions respectively, vs data from different studies (Lazaridis et al. 2001). The experimental data were obtained by tracking radioactive labelled, poorly soluble particles of diameter larger than 0.1 μm. For example there have been measurements of total and regional deposition of monodisperse aerosols in the RT of healthy subjects who breathed through the mouth in an upright position. Radioactively labeled polystyrene particles with aerodynamic diameters ranging

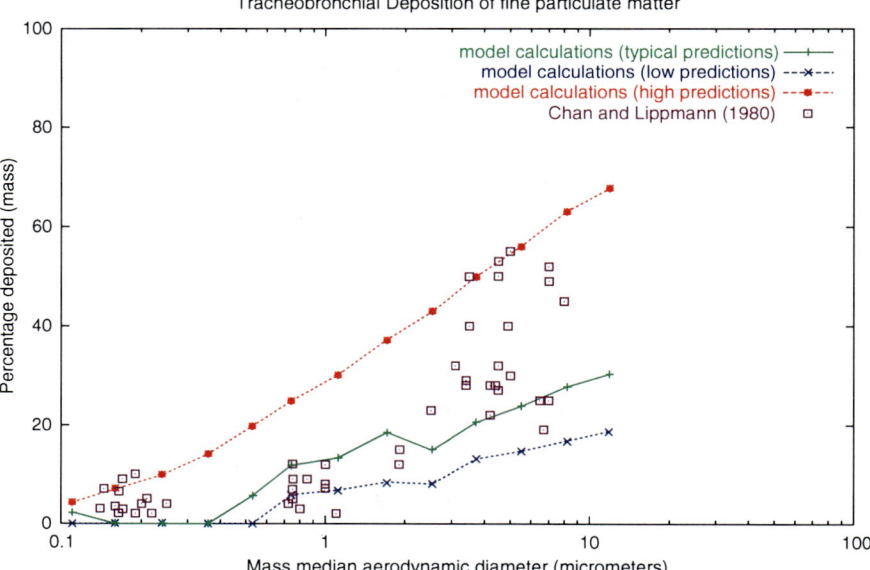

Fig. 9.17 Comparison of model predictions with the experimental data for tracheobronchial deposition from Chan and Lippmann (1980)

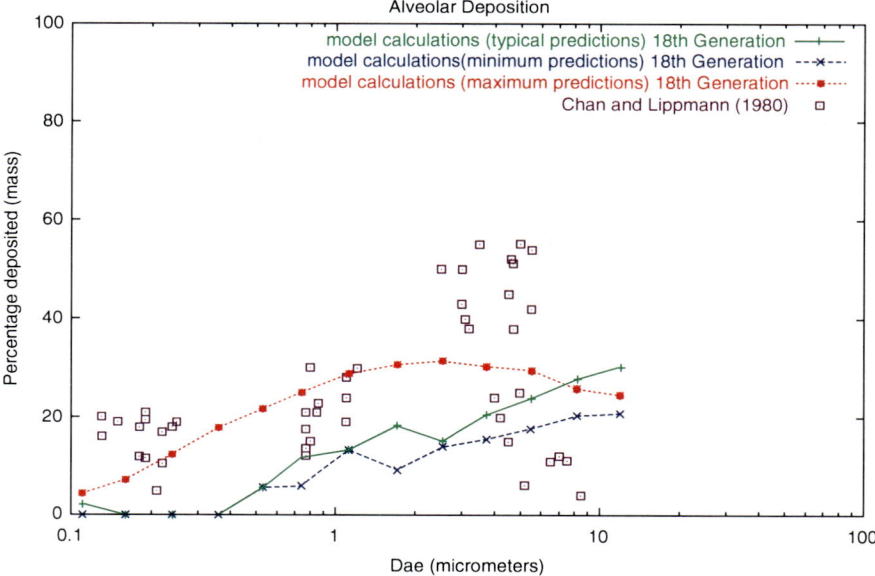

Fig. 9.18 Comparison of model predictions with the experimental data for alveolar deposition from Chan and Lippmann (1980)

9.5 Application: Internal Dose from Radon Inhalation

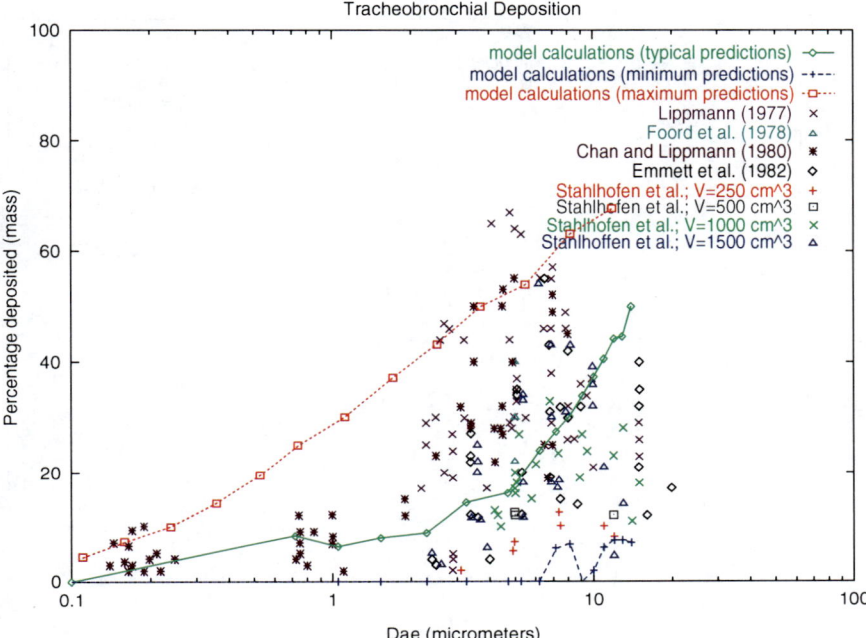

Fig. 9.19 Predicted tracheobronchial deposition vs experimental data (Adapted from Lazaridis et al. 2001)

between 3.5 and 10.0 μm and density of 1,060 kg/m^3 were used. The data show that total deposition rate increased with increasing particle size.

In vivo measurements were based on the amount of radioactivity retained in the lung as a function of time, where the "fast-cleared" and "slow-cleared" deposition fractions correspond to deposition in the ciliated tracheobronchial and alveolar regions, respectively (USEPA 2004). Model parameters include air flow of 1,000 cm^3 s^{-1}, log-normally distributed inert aerosol with an initial 1 μm count median diameter, and a 1.7 geometric standard deviation. Although experimental data are quite scattered, owing both to nonconsistency in methodology and to intersubject variability (USEPA 2004), the theoretical predictions are in fair agreement with the general deposition pattern in the tracheobronchial region.

9.5 Application: Internal Dose from Radon Inhalation

An application of the calculation of internal dose of inhaled radon daughters with internal radiation is presented. The human dose after exposure to radon (^{222}Rn) is originated by the daughters with short half-life time which adhere to the respiratory system after inhalation.

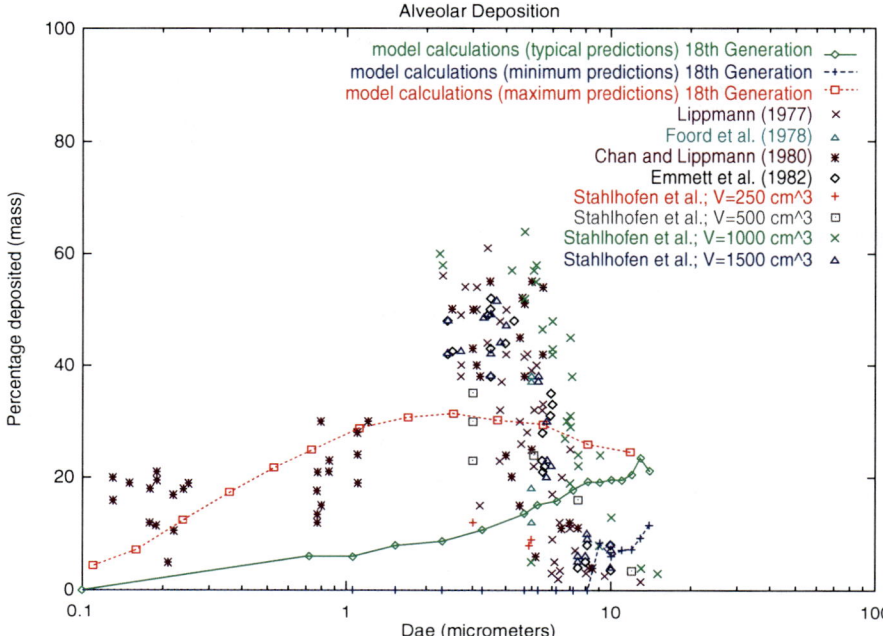

Fig. 9.20 Predicted alveolar deposition vs experimental data (Adapted from Lazaridis et al. 2001)

The harmfulness of radiation is dependent on its range of influence (α, β, or γ radiation), its absorption from the tissues and the type of tissue. The energy which is absorbed per unit mass of tissue which received the radiation dose is called an absorbed dose. $D_{T,R}$ is given as the ratio of the mean energy received from the radiation by the tissue to its mass. The unit for the absorbed dose in the International System (SI) is the Gray (Gy) which is equal to the energy of 1 Joule which is absorbed at 1 kg of tissue (1 Gy = 1 J kg^{-1}).

Since the body response after radiation is dependent on the radiation type the absorbed dose is weighted also based on the radiation type. The equivalent dose ($H_{T,R}$,) which relates the biological response of the absorbed dose with the radiation type is calculated by multiplying with a coefficient of radiation balance (w_R):

$$H_{T,R} = w_R D_{T,R} \tag{9.28}$$

When the radiation spectrum is composed from energy types with different values of w_R, the total equivalent dose (H_T) is given by the expression:

$$H_T = \sum_R w_R D_{T,R} \tag{9.29}$$

The values of the coefficient w_R are dependent on the energy type emitted from the radionuclei and are given in Table 9.6.

9.5 Application: Internal Dose from Radon Inhalation

The unit for the equivalent dose in the International System (SI) is the Sievert (Sv) which relates the absorbed dose in human tissues with the biological response since each radiation type results in a different biological response even in the case that the absorbed dose is equal (1 Sv = 1 J kg^{-1}).

The body response is dependent on the tissue sensitivity. The function which relates the equivalent dose with the tissue sensitivity is the effective dose. Effective dose is the equivalent radiation γ dose for the whole body which has to be applied to a human in order to have the same health risk as the actual applied radiation.

The effective dose is a unit for the human health risk assessment independently from the radiation type, the radiation conditions and tissue targeted. The measurement unit in SI is also the Sievert (Sv).

The regulated dose limits are given as a bounded effective dose for a time period of 50 years for adults and 70 years for children. The bounded effective dose ($H_T(t)$) is the sum of the equivalent bounded doses in tissues multiplied by the coefficient w_T. In more detail the dose $H_T(t)$ is given as an integral for a time period t (years) of the equivalent dose rate in tissue which a human receives after the radiation exposure:

$$H_T(t_0) = \int_{t_0}^{t_0+t} \dot{H}_T(t, t_0) dt \qquad (9.30)$$

where t_0 is the age of the radiation effect and $\dot{H}_T(t)$ is the rate of an equivalent dose given to the tissue at time t.

Finally the effective dose is given as a sum of the weighted equivalent doses from internal and external radiation to all tissues and organs of the body

$$E = \sum_T w_T H_T = \sum_T w_T \sum_R w_R D_{T,R} \qquad (9.31)$$

where $D_{T,R}$ is the absorbed average dose given to a tissue or organ T due to radiation R, and w_R is the weighting radiation factor and w_T the weighting tissue factor for the tissue or organ T. The coefficient w_T is a dimensionless number which is used for computation of the equivalent dose to a tissue or organ. The values of w_T are given in Table 9.7.

The limit for the effective dose of professional exposure is 20 mSv during a year and 100 mSv during a period of five consecutive years.

In the case of radon and its daughter radionuclides the effective dose is calculated from the expression

$$E = h \times W \qquad (9.32)$$

where h is a conversion coefficient for the daughter nuclides and W is the exposure to a potential energy α (J/m^3) (the energy which is emitted from the radioactive decay of nuclides, including decays of radon and thoron to ^{210}Pb and ^{208}Pb respectively). The exposure W to potential energy α is the product of the energy α and the duration of exposure and is measured at J h/m^3. The conversion

Table 9.7 Values for the tissue weighting factor (w_T) (International Commission on Radiological Protection ICRP 1994)

Tissue or organ	Tissue weighting factor (w_T)
Tissues of the respiratory tract	
Extrathoracic airways (ET1) (anterior nose)	0.001
Extrathoracic region ET2 (posterior nasal passages, larynx, pharynx, and mouth)	0.998
Trachea and main bronchi	0.1665
cells of branches of the bronchi	0.1665
Bronchial - bronchiolar	0.333
Alveolar region of the respiratory tract	0.333
Lymph nodes of the extrathoracic region	0.001
Lymph nodes of the tracheobronchial region	0.001
Organs in the body	
lungs	0.12
skin	0.01
Bones surface	0.01
red marrow	0.12
genitals	0.20
colon	0.12
gastric	0.12
bladder	0.05
breast	0.05
liver	0.05
esophagus	0.05
thyroid	0.05
Other organs (adreanal, brain, upper intestine, kidney, muscle, pancreas, spleen, uterus)	0.05

Table 9.8 Exposure sources to radiation and mean effective dose

Source	Effective dose[a]
Cossmic radiation	~2
Radiation γ from the Earth's surface	~2
Ingestion of radionuclides: Daughter radon nuclei	~4
Rest of nuclides	~3
Radiation deposition on soil	<0.4
Roentgen rays (x)	≤3
Nuclear power stations ς	~0.4

[a] The effective dose is given in rem (roentgen equivalent man) and refers to a period of 70 years. The rem expresses the radiation quantity which results to the same biological effect as 1 roentgen of radiation X.

coefficient h is measured at Sv m^3/J h and its value for radon is 1 for houses and 1.1 for working places, whereas for the thoron it is 0.5 for working places.

The effective dose during a human life (70 years period) from radon and other sources is given in Table 9.8.

9.5 Application: Internal Dose from Radon Inhalation

The radon dose is dependent on its concentration in relation to the daughter nuclides in the particulate and gaseous phase. After their formation the daughter nuclides ^{14}Pb, ^{214}Bi, ^{218}Po tend to accumulate to internal surfaces indoors and onto particles in air.

In the determination of risk, the concentration of daughter nuclides of radon is expressed in the unit WL (working level), which means concentration of nuclides with small half-life time per 1 liter of air which emits potential α energy equal to 1.3×10^5 MeV or 2.1×10^{-5} J/m^3. At indoor places 1 pCi/L radon corresponds to 0.004 WL. The accumulated human exposure to radon is expressed at WLM (working level month) which is equal to 170 WL h or 3.54 mJ h/m^3.

The conversion coefficients indoors after exposure to radionuclides of radon and thoron are given in Table 9.9. The values have been calculated assuming that the effective dose is equal to 0.4 and that the person works 2,000 h per year.

Example. *Calculate the annual effective dose from internal radiation that a worker experiences inside a place with mean Thoron (^{220}Rn) and radon (^{222}Rn) concentrations equal to 50 Bq/m^3 and 635 Bq/m^3 respectively.*

The annual exposure of the worker to Thoron is (2,000 h × 50 Bq/m^3 =) 1×10^5 Bq h/m^3. This exposure corresponds (Table 9.10) to effective dose

$$E = 32 nSv/Bqh/m^3 \times 10^5 Bqh/m^3 = 3,2 mSv$$

Accordingly for the radon the annual effective dose is equal to = 1.4 mSv/(mJh/m^3) × 635 Bq/m^3 × 4.45 × 10^{-3} (mJh/m^3)/(Bq/m^3) = 4 mSv

Therefore the total annual effective dose which the worker receives is equal to 7.2 mSv or 720 mrem. This is double the average annual effective dose which a

Table 9.9 Conversion coefficients for the calculation of the effective dose of radon and thoron

Radon (^{222}Rn)	Conversion coefficient
Effective dose per exposure unit	1.4 mSv per 1mJh/m^3 or 5 mSv per 1 WLM
Annual exposure per concentration unit	4.45 × 10^{-3} mJh/m^3 per 1 Bq/m^3
Thoron(^{220}Rn)	
Effective dose per exposure unit	32 nSv per 1 Bq h/m^3

Table 9.10 Comparison of indoor air pollution limits (ppm) in working places with US national limits in outdoor air for specific gasses

	Indoors			Outdoors	
Gas	8 h (ppmv)⁺	15 mins (ppmv)⁺	Maximum allowed concentration (ppmv)	Natiional air quality standards (US) – NAAQS (ppmv)	Air quality limits in California (ppmv)
CO	35	–	200	9.5 (8 h)	9 (8 h)
NO$_2$	–	1	–	0.053 (yearly)	0.25 (1 h)
O$_3$	0.1	0.3	–	0.08 (8 h)	0.09 (1 h)
SO$_2$	2	5	–	0.14 (24 h)	0.05 (24 h)

The limit values are given by NIOSH (2000).

human receives from various radiation sources. However, this dose is lower than the annual limit value of 50 mSv which is adopted for health protection of workers.

9.6 Health Effects from Air Pollutants

The health impact assessment of air pollution is an extremely complicated problem because it involves a series of independent and dissimilar variables, ranging from social factors to atmospheric physics and chemistry, environmental toxicology, and even climatology and geography. Clearly, there is a great deal of variability in outdoor and also indoor air concentration/composition, depending on the particular conditions in each enclosed environment and the susceptibility of each individual. Therefore, one should distinguish between the multiplicity of scales in addressing the problem, ranging from the person-oriented scale to the population scale, and from the cell toxicity scale to the scale of health end-points.

Human exposure and risk modeling in recent years has focused on the development of "person-oriented" (i.e. anthropocentric) "holistic" approaches and models, that aim to account for total ("cumulative" and "aggregate") exposures of individuals, and of populations consisting of such individuals, to co-occurring stressors (e.g. mixtures of contaminants present in the air inhaled and in the water and food digested by the person). At the end of the chain of events from pollutant emission to health effects, one needs to deal with the toxicology of air pollution. Indeed, the toxicology of air pollution, both outdoor and indoor, is exceedingly complex. There are different types of pollutants and great variation in individual susceptibility to their effects at low, environmentally relevant concentrations. The toxicological characteristics of most chemicals at high concentrations are irrelevant to their behaviour and effects as air pollutants. Therefore, human health effects occurring from exposure to air pollutants are better characterised for populations than for individual patients. As is well-known, the statistically-based approach is taken in epidemiological analyses. These analyses have shown that exposure to ambient air particulate matter (PM) is associated with pulmonary and cardiovascular diseases and cancer. A current trend is to attempt to associate indoor air PM with asthma and allergies, or even ophthalmology diseases. However, it has been difficult to identify a direct link between the physico-chemical features of the indoor-generated PM and respiratory allergies, asthma and health effects in general. On the other hand, there exists a large body of work addressing the problem on an individual basis (non-statistically), by focusing on specific indoor microenvironments.

The mixture of air pollutants which are potentially harmful to humans is presented in Fig. 9.21.

The current modelling approaches focus on individual persons, real or "virtual", with well-defined physiological, socioeconomic, etc. attributes, and take into account how the detailed activities of these persons in space and time affect their "microenvironments" and their corresponding exposures to stressors, as well as

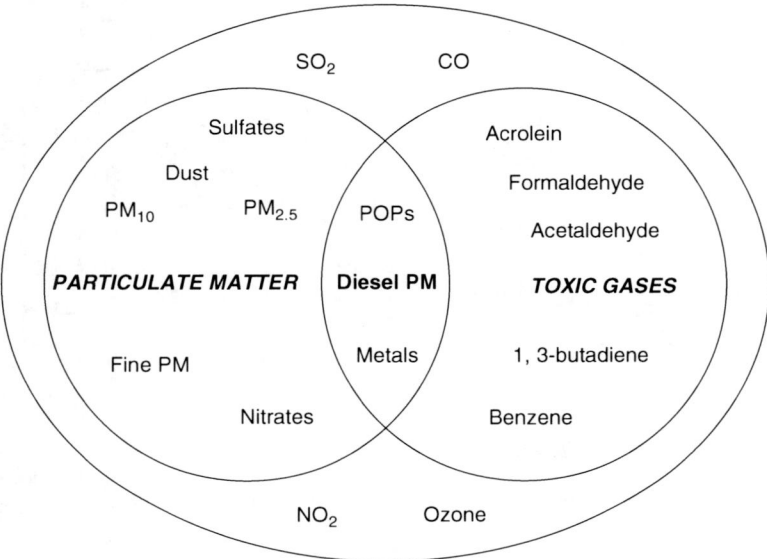

Fig. 9.21 The air pollution mixture

how they affect the physiological processes determining a biologically relevant dose (e.g. inhalation rates; metabolic rates, etc.). Along this line of philosophy, Georgopoulos and Lioy (2006) presented an overview of the MENTOR modeling system for implementing a consistent source-to-dose modeling approach for human exposure and dose assessment from global/regional to local, neighborhood, and eventually personal resolution. MENTOR provides "tools" that link regional and local information with microenvironmental conditions and human activities, as shown schematically in Fig. 9.22 (adapted from concepts in Georgopoulos and Lioy (2006) and USEPA (2004)). A subsequent important step was to complement the above sequence with the dose-to-biological effect chain. This is based on computational methods and tools for modeling the interactions of biomolecules and bionetworks with environmental agents at the genomic/transcriptomic, proteomic, metabomomic, cytomic, and physiomic levels with the objective of improving methods for risk analysis. Such an integrated system is the MENTOR/DORIAN system, which is intended to facilitate the consistent multiscale, source-to-outcome modeling following exposures to contaminants for both individuals and populations. The modelling of the whole source-to-response continuum is made through a cascade of models, schematically shown in Fig. 9.23.

In the whole sequence of events from pollutant emissions to health effects there are studies which involve scientific disciplines ranging from air pollution experts in modelling and measurements to toxicologists. For example at the cellular and molecular level, studies focus on the oxidative stress which leads to damage to the cell membrane, organelles such as mitochondria, lysosomes and nucleus and oxidation of cellular DNA. This complicated interplay is illustrated in Fig. 9.24.

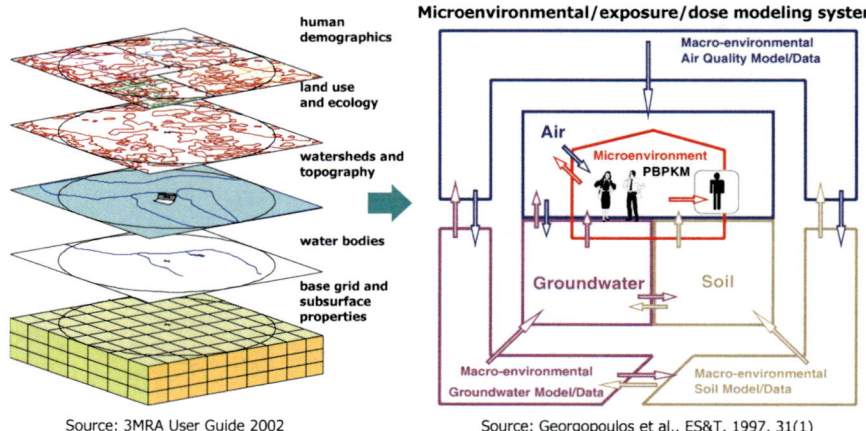

Fig. 9.22 MENTOR supports source-to-dose modeling over multiple scales, down to resolutions relevant to exposures of individuals considering atmospheric or multimedia contaminants; it links environmental models and databases with exposure/dose models (Adapted from Georgopoulos and Lioy 2006)

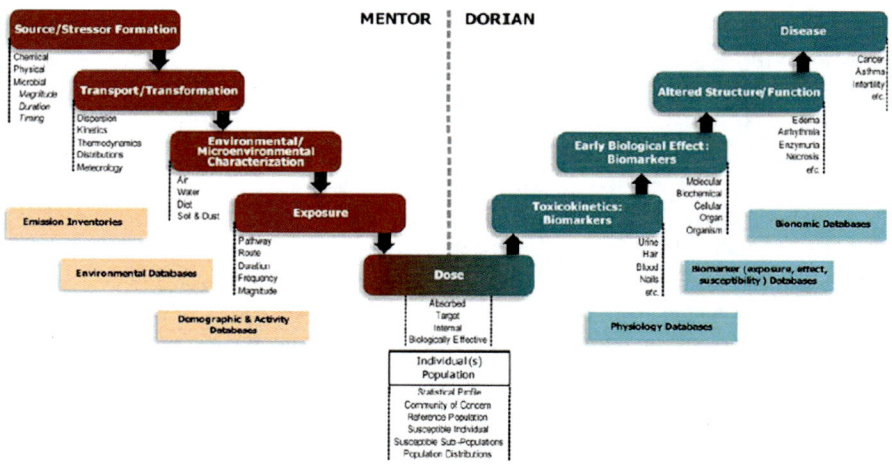

Fig. 9.23 Schematic of the modeling of the whole source-to-response chain in the MENTOR/DORIAN system (Adapted from Georgopoulos and Lioy 2006)

The health effects arising from air pollution is dependent on exposure conditions and duration, the pollution mixture and the exposure pathway. Human exposure arises both from anthropogenic and natural pollutant sources. Besides air pollution there are other exposure pathways for toxic chemical compounds to reach the human body, arising from food and water uptake as well as from dermal absorption.

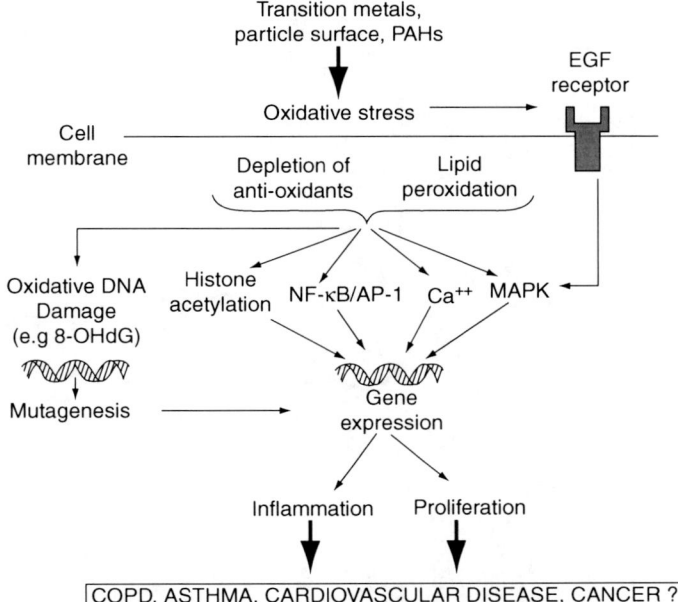

Fig. 9.24 A simplified scheme of cellular and molecular pathways involved in the oxidative stress mediated transformation of cell physiology that eventually leads to clinical symptoms (Adapted from Donaldson et al. 2000)

For example the household glass (plates, glasses) which is used daily may result in human exposure to lead.

An extreme example of the effect of air pollution on the death rate in an urban area is shown in Fig. 9.25. A considerable increase of the death rate is correlated with an increase in the concentration of particulate matter and sulphur dioxide. Several similar observations in urban areas resulted in a series of epidemiological studies for the correlation between air pollution and population mortality.

It is generally accepted that no chemical component is totally safe at high concentrations or extensive exposure. For determination of the limits of exposure or acceptable dose concentrations, it is necessary to gather data about the effects of toxic components on human health, which data however are seldom available. The data are usually limited to toxicity experiments with animals and epidemiological studies of outdoor air from specific chemical components. Air pollution standards have been adopted mainly for outdoor air as described in Chapter 1. Table 9.10 presents a comparison of indoor pollution levels (working places) for specific gaseous components

Metals that are included in emitted aerosols and gases from different athropogenic activities such as metallurgical fumes (Cd, Cr, As, Ni) affect the heart and olfactory canals. In addition metals such as Cd, As and Ni can be carried by proteins and transferred to the human tissues and mainly to the kidney.

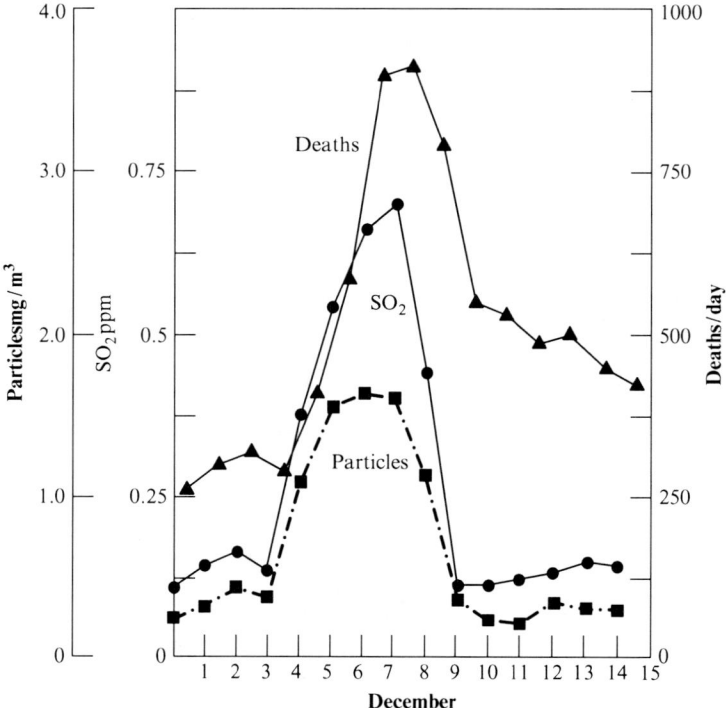

Fig. 9.25 Particle and sulphur dioxide concentration during December 1950 in London and association with number of deaths

Arsenic affects the function of kidney, liver and spleen. The health effects of specific metals are shown in Table 9.11.

Of course specific metals (such as Co, Cu, Fe, Mn, Mo, Se, Al etc.) are necessary to the human body in trace quantities. On the contrary, overdoses of these metals result in significant health effects.

For medical purposes, radioactive components have been used that have the tendency to accumulate in specific tissues. For example iodine (^{231}I) and strontium (^{90}Sr) accumulate to the thyroid gland and bones respectively. Fig. 9.26 shows the accumulation areas of the main gaseous and particulate matter pollutants in the human body.

9.7 Health Effects from Exposure to Particulate Matter

In the indoor environment humans are exposed to suspended particles which are produced indoors or infiltrate from the outdoor air. In Tables 9.12 and 9.13 are given estimates for the risk in public health from particulate matter as reflected in the scientific literature.

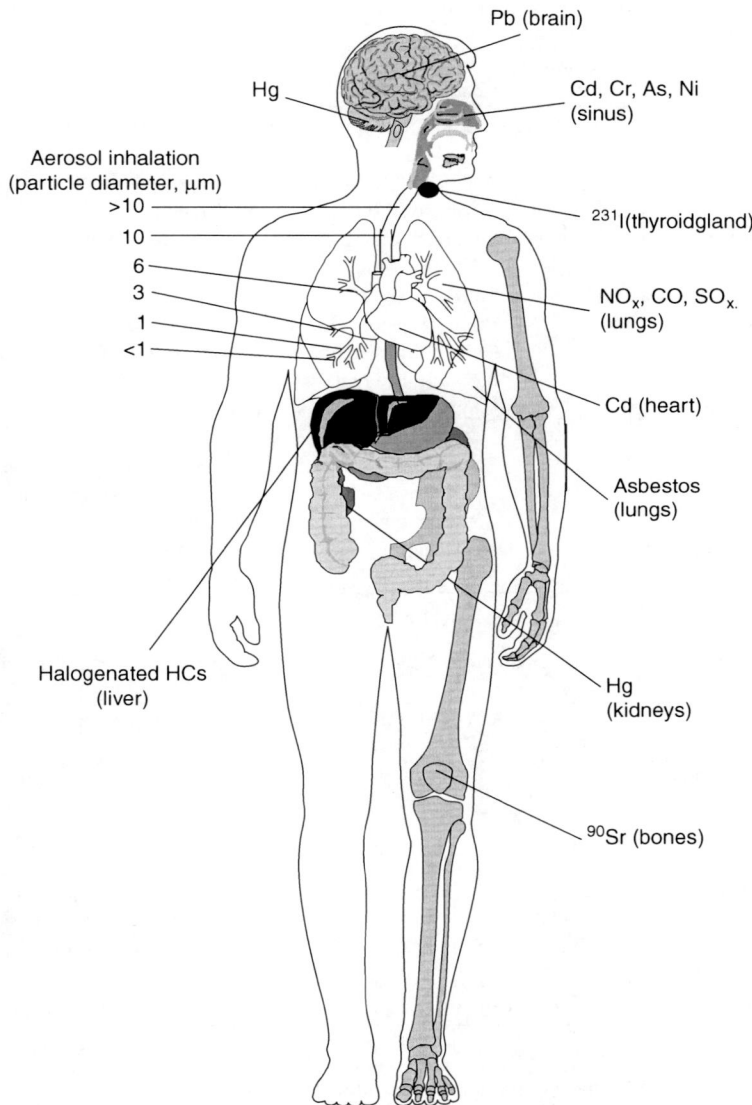

Fig. 9.26 The influence in the human body of several gaseous and aerosol components and the areas (organs and tissues) of accumulation (Adapted from Turco 2002)

Several chemical components of particulate matter, such as metals and PAHs, may be toxic to the human body. Specifically the PAHs result in mutagenic effects on humans and cancer.

Table 9.11 Metals which are directly emitted in air from industrial units and their health effects (Adapted from Turco 2002)

Metal	Concentration (ppmm)	Health effect
Arsenic (As)	0.5	Cancer of the lungs, liver, and skin; teratogenenic; poisonous in large doses
Cadmium (Cd)	0.2	Accumulation in the kidneys, lungs, and heart; symptoms like Wilson's disease; 50 ppmm fatal within 1 h; carcinogenic
Chromium (Cr)	1.0	Skin rashes, lung cancer (after continued exposure); carcinogenic
Iron (Fe)	10.0	Siderosis, or red lung disease
Lead (Pb)	0.15	Brain damage; red blood cell anemia; paralysis of limbs
Manganese (Mn)	5.0	Aching limbs and back; drowsiness; loss of bladder control; nasal bleeding
Mercucy (Hg)	0.05	Central nervous system attack; tremors and neuropsychiatric disturbance
Nickel (Ni)	1.0	Skin rashes, cancer of the sinus and lungs (after continued exposure); exposure to 0.001 ppmm of nickel carbonyl leads to nausea, vomiting, and possible death
Vanadium (V)	0.5	Acute spasm of the bronchi; emphysema
Zinc (Zn)	5.0	Fever, muscular pain, nausea, and vomiting

Table 9.12 Relative risk to human health related to an increase of the PM_{10} and $PM_{2.5}$ concentrations by 10 $\mu g/m^3$

Health effects	Relative risk for $PM_{2.5}$ (95% confidence limit)	Relative risk for PM_{10} (95% confidence limit)
Breathing difficulty		1.0305 (1.0201–1.0410)
Coughing		1.0356 (1.0197–1.0518)
Symptoms to the lower respiratory tract		1.0324 (1.0185–1.0464)
Change of the maximum breathing flux		−0,13 (−0.17 to −0.09%)
Hospital admission with respiratory problems		1.0080 (1.0048–1.0112)
Mortality	1.015 (1.011–1.019)	1.0074 (1.0062–1.0086)

Table 9.13 Expected number of persons (of one million population) having health problems after an exposure of 3 days with a PM_{10} concentration of 50 $\mu g/m^3$ or 100 $\mu g/m^3$

Health index	Number of persons affected from an air quality episode of duration of 3 days with PM_{10} concentration:	
	50 $\mu g/m^3$	100 $\mu g/m^3$
Morbidity	4	8
Hospital admission with breathing problems	3	6
Use of bronchodilators (days)	4,683	10,514
Person days with increased symptoms	5,185	11,267

9.7 Health Effects from Exposure to Particulate Matter

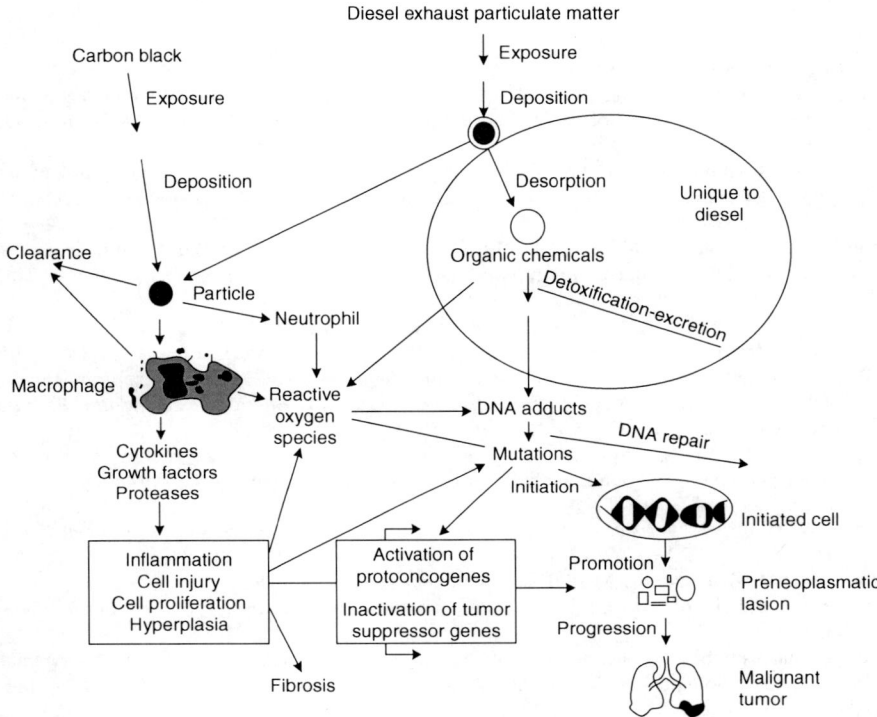

Fig. 9.27 Proposed mechanism for cancer occurance in rats which are exposed to elevated levels of diesel particles (Adapted from USEPA 2004)

As an example Fig. 9.27 presents a proposed mechanism by which exposure to diesel particles in rats leads to lung cancer. This mechanism is proposed after many years of laboratory research (USEPA 2004).

Problems

9.1 *What is the portion of particles with diameter of 5 μm which are inhaled and are deposited in the upper respiratory tract ?*

9.2 *An adult person performs a mild exercise and inhales particles with diameter 4 μm and density 1 g cm^{-3}. What percentage of those particles is deposited in the upper respiratory tract? What percentage of particles are deposited in the alveolar region if the particles' diameter is equal to 3 μm and their density 4 g cm^{-3}? How different is the dose in the alveolar region for particles with diameter 30 μm and density 2 g cm^{-3};*

References

ACGIH. (1997). *Threshold limit values and biological exposure indices.* Cincinnati: ACGIH.

Balashazy, I., Hofmann, W., & Lorine, M. (1999). Relative contributions of individual deposition mechanisms to total aerosol deposition in human airways. *Journal of Aerosol Science, 30,* S729–S730.

Chan, T. L., & Lippmann, M. (1980). Experimental measurements and empirical modelling of the regional deposition of inhaled particles in humans. *American Industrial Hygiene Association Journal, 41,* 399–409.

Donaldson, K., Stone, V., Gilmour, P. S., Brown, D. M., & Macnee, W. (2000). Ultrafine particles: Mechanisms of lung injury. *Philosophical Transactions of the Royal Society London A, 358,* 2741–2749.

EPA (1992). *Guidelines for Exposure Assessment.* Report EPA/600/Z-92/001.

Georgopoulos, P. G., & Lioy, P. J. (2006). From a theoretical framework of human exposure and dose assessment to computational system implementation: The Modeling Environment for Total Risk Studies (MENTOR). *Journal of Toxicology and Environmental Health. Part B: Critical Reviews, 9*(6), 457–483.

Gradon, L., Pratsinis, E., Prodgorski, A., Scott, S., & Panda, S. (1996). Modeling retention of inhaled particles in rat lungs including toxic and overloading effects. *Journal of Aerosol Science, 27*(3), 487–503.

Hinds, W. C. (1999). *Aerosol technology: Properties, behavior, and measurement of airborne particles* (2nd ed.). New York: Wiley.

Housiadas, C., & Lazaridis, M. (2010). Inhalation dosimetry modeling. In M. Lazaridis & I. Colbeck (Eds.), *Human exposure to pollutants via dermal absorption and inhalation.* Dordrecht: Springer.

International Commission on Radiological Protection (ICRP). (1994). Human respiratory tract model for radiological protection. ICRP publication 66. *Annals of the ICRP, 24*(4), 482. Pergamon, Oxford.

International Commission on Radiological Protection (ICRP). (2002). Guide for the practical application of the ICRP human respiratory tract model. *Annals of the ICRP, 32*(1–2), 1–312. Pergamon, Oxford.

Menache, M. G., Miller, F. J., & Raabe, O. G. (1995). Particle inhalability curves for humans and small laboratory animals. *The Annals of Occupational Hygiene, 39,* 317–328.

Finlay, W. H. (2005). *The mechanics of inhaled pharmaceutical aerosols.* London: Academic.

Lazaridis, M., Broday, D. M., Hov, Ø., & Georgopoulos, P. G. (2001). Integrated exposure and dose modeling and analysis system. 3. Deposition of inhaled particles in the human respiratory tract. *Environmental Science & Technology, 35,* 3727–3734.

Martonen, T. B., Rosati, J. A., & Isaacs, K. K. (2005). Modeling deposition of inhaled particles. In L. S. Ruzer & N. H. Harley (Eds.), *Aerosols handbook: Measurement, dosimetry and health effects* (pp. 113–155). New York: CRC Press.

National Council on Radiation Protection and Measurement. (1994). *Deposition, retention and dosimetry of inhaled radioactive substances.* Report S. C. 57–2, NCRP, Bethesda, MD.

Stahlhofen, W., Gebhart, J., Heyder, J., & Scheuch, G. J. (1983). *Journal of Aerosol Science, 14,* 186–188.

Stahlhofen, W., Gebhart, J., Heyder, J., & Scheuch, G. J. (1986). *Journal of Aerosol Science, 17,* 333–336.

Turco, R. P. (2002). *Earth under siege. From air pollution to global change.* Oxford: Oxford University Press.

USEPA (2004). *Air quality criteria for particulate matter*, US Environmental Protection Agency, Research Triangle Park, NC, USA (Report EPA/600/P-99/002aF and bF).

Weibel, E. R. (1963). *Morphometry of the human lung.* New York: Academic.

Vincent, J., Mark, D., Miller, B. G., Armbruster, L., & Ogden, J. L. (1990). Aerosol inhalability at higher windspeeds. *Journal of Aerosol Science, 21,* 577–586.

Appendix A

Units and Physical Constants

Table A.1 Useful parameters, constants and conversion factors

Symbol	Value	Unit	Description
Constants			
k	0.4 (0.35–0.41)		Von Karman's constant
a_k	1.53–1.68		Kolmogorov's constant
Geophysical parameters			
g	9.8	$m \times s^{-2}$	Gravitational acceleration
r_e	6.37×10^6	m	Average radius of Earth
ω	7.27×10^{-5}	$radians \times s^{-1}$	Rate of Earth's rotation
	2π radians / 24 h		
f_c	$(1.46 \times 10^{-4}) \sin(\varphi)$	s^{-1}	Coriolis parameter (φ is the latitude)
$P_{inertial}$	$12/\sin(\varphi)$	h	Inertial pressure
Radiation parameters			
S	1,370	$W\ m^{-2}$	Sun's radiation
	1,113	$K\ m\ s^{-1}$	(it is used the air density at sea level)
σ_{SB}	5.67×10^{-8}	$W\ m^{-2}\ K^{-4}$	Stefan-Boltzmann's constant
Parameters of air			
Values at the sea level and stable atmosphere			
ρ_{SL}	1.225	$kg\ m^{-3}$	Air density
	0.01225	$mb\ s^2\ m^{-2}$	
	0.001225	$kPa\ s^2\ m^{-2}$	
P	101.325	kPa	Pressure
	1013.25	mb	
	1.013×10^5	$N\ m^{-2}$	
	82,714	$m^2\ s^{-2}$	
T	288.15	K	Temperature
	15	°C	
μ	$1,789 \times 10^{-5}$	kg	Dynamic molecular viscosity
ν	1.461×10^{-5}	$m^2\ s^{-1}$	Kinematic molecular viscosity
k_θ	2.53×10^{-2}	$W\ m^{-1}\ K^{-1}$	Molecular thermal conductivity
ν_θ	2.06×10^{-5}	$m^2\ s^{-1}$	Molecular thermal diffusivity
C_{pd}	1004.67	$J\ kg^{-1}\ K^{-1}$	Dry air thermal conductivity at stable pressure
	1004.67	$m^2\ s^{-2}\ K^{-1}$	

(continued)

Table A.1 (continued)

Symbol	Value	Unit	Description
C_p	$C_{pd}(1+0.84q)$	$m^2\ s^{-2}\ K^{-1}$	Wet thermal conductivity at stable pressure
R	287.04	$J\ K^{-1}\ kg^{-1}$	Gas constant for dry air
	287.04	$m^2\ s^{-2}\ K^{-1}$	
	2,87	$mb\ K^{-1}\ m^3\ kg^{-1}$	
	0,287	$kPa\ K^{-1}\ m^3\ kg^{-1}$	

Conversion Units of Energy

$(kJ\ mol^{-1}) \times 0.2390 = kcal\ mol^{-1}$
$\qquad \times 0.0104 = eV$
$\qquad \times 83.59 = cm^{-1}$
$(kcal\ mol^{-1}) \times 4.184 = kJ\ mol^{-1}$
$\qquad \times 0.04336 = eV$
$\qquad \times 349.8 = cm^{-1}$
$(cm^{-1}) \times 1.196 \times 10^{-2} = kJ\ mol^{-1}$
$\qquad \times 2.859 \times 10^{-3} = kcal\ mol^{-1}$
$\qquad \times 1.204 \times 10^{-4} = eV$
$(eV) \times 96.49 = kJ\ mol^{-1}$
$\qquad \times 23.06 = kcal\ mol^{-1}$
$\qquad \times 8.066 \times 10^3 = cm^{-1}$
Energy $(kcal\ mol^{-1}) = 2.859 \times 10^4 / \lambda\ (nm)$
Energy $(kJ\ mol^{-1}) = 1.196 \times 10^5 / \lambda\ (nm)$

Fundamental Physical Constants

Velocity of light	$c =$	$2.9979 \times 10^8\ m\ s^{-1}$
Planck's constant	$h =$	$6.62608 \times 10^{-34}\ J\ s$
Boltzmann's constant	$k =$	$1.3807 \times 10^{-23}\ J\ K^{-1}$
Avogadro number	$N =$	$6.0221 \times 10^{23}\ mol^{-1}$
Gas constant	$R =$	$8.314\ J\ K^{-1}\ mol^{-1}$
	$=$	$0.082058\ L\ atm\ K^{-1}\ mol^{-1}$

Conversion Factors for Light Absorption Coefficients

Units at logarithmic scale with base e or 10
$(cm^2\ molecule^{-1}) \times 2.69 \times 10^{19} = (atm\ at\ 273\ K)^{-1}\ (cm^{-1})$
$\qquad \times 2.46 \times 10^{19} = (atm\ at\ 298\ K)^{-1}\ (cm^{-1})$
$\qquad \times 3.24 \times 10^{16} = (Torr\ at\ 298\ K)^{-1}\ (cm^{-1})$
$\qquad \times 6.02 \times 10^{20} = (L\ mol^{-1}\ cm^{-1})$
$(atm\ at\ 298\ K)^{-1}\ (cm^{-1}) \times 4.06 \times 10^{-20} = cm^2\ molecule^{-1}$
$\qquad \times 1.09 = (atm\ at\ 273\ K)^{-1}\ (cm^{-1})$
$(L\ mol^{-1}\ cm^{-1}) \times 4.46 \times 10^{-2} = (atm\ at\ 273\ K)^{-1}\ (cm^{-1})$
$\qquad \times 4.09 \times 10^{-2} = (atm\ at\ 298\ K)^{-1}\ (cm^{-1})$
$\qquad \times 5.38 \times 10^{-5} = (Torr\ at\ 298\ K)^{-1}\ (cm^{-1})$
$\qquad \times 1.66 \times 10^{-21} = (cm^2\ molecule^{-1})$

(continued)

(continued)

Conversion of the logarithmic base and units
(cm^2 $molecule^{-1}$), base e × 1.17×10^{19} = (atm at 273 K)$^{-1}$ (cm^{-1}), base 10
× 1.07×10^{19} = (atm at 298 K)$^{-1}$ (cm^{-1}), base 10
× 1.41×10^{16} = (Torr at 298 K)$^{-1}$ (cm^{-1}), base 10
× 2.62×10^{20} = (L mol^{-1} cm^{-1}), base 10
(L mol^{-1} cm^{-1}), base 10 × 3.82×10^{-21} = cm^2 $molecule^{-1}$, base e
× 0.103 = (atm at 273 K)$^{-1}$ (cm^{-1}), base e
× 9.42×10^{-2} = (atm at 298 K)$^{-1}$ (cm^{-1}), base e
(atm at 273 K)$^{-1}$ (cm^{-1}), base 10 × 8.57×10^{-20} = cm^2 $molecule^{-1}$, base e
× 51.6 = L mol^{-1} cm^{-1}, base e
(Torr at 298 K)$^{-1}$ (cm^{-1}), base 10 × 7.11×10^{-17} = cm^2 $molecule^{-1}$, base e
× 4.28×10^4 = L mol^{-1} cm^{-1}, base e
(atm at 298 K)$^{-1}$ (cm^{-1}), base 10 × 9.35×10^{-20} = cm^2 $molecule^{-1}$, base e
× 2.51 = (atm at 273 K)$^{-1}$ (cm^{-1}), base e

Factors of Units Conversion for Reactions in the Gaseous Phase

Concentrations$^\alpha$

1 mol L^{-1} = 6.02×10^{20} molecules cm^{-3}
1 ppm = 2.46×10^{13} molecules cm^{-3} = $40.9 \times (MB)^\beta$ μg m^{-3}
1 ppb = 2.46×10^{10} molecules cm^{-3} = $0.0409 \times (MB)$ μg m^{-3}
1 ppt = 2.46×10^7 molecules cm^{-3} = $(4.09 \times 10^{-5}) \times (MB)$ μg m^{-3}
1 atm at 298 K = 4.09×10^{-2} mol L^{-1} = 2.46×10^{19} molecules cm^{-3}
1 Torr = 133.3 Pa = 1.333 mbar

Constant of reaction velocity of second order
cm^{-3} $molecule^{-1}$ s^{-1} × 6.02×10^{20} = L mol^{-1} s^{-1}
ppm^{-1} min^{-1} × 4.08×10^5 = L mol^{-1} s^{-1}
ppm^{-1} min^{-1} × 6.77×10^{-16} = cm^{-3} $molecule^{-1}$ s^{-1}
atm^{-1} s^{-1} × 4.06×10^{-20} = cm^{-3} $molecule^{-1}$ s^{-1}

Constant of reaction velocity of third order
cm^6 $molecule^{-2}$ s^{-1} × 3.63×10^{41} = L^2 mol^{-2} s^{-1}
ppm^{-2} min^{-1} × 9.97×10^{12} = L^2 mol^{-2} s^{-1}
ppm^{-2} min^{-1} × 2.75×10^{-29} = cm^6 $molecule^{-2}$ s^{-1}

$^\alpha$ The concentrations at ppm, ppb and ppt refer to standard conditions pressure of 1 atm and temperature of 25 C, where 1 atm = 760 Torr = 1.01325×10^5 Pa = 1.01325 bar.
$^\beta$ MB = molecular weight of the chemical components

Index

A

Absolute humidity, 102–103
Absorption, 14, 17, 20, 33, 36–42, 45, 48, 50, 52, 95, 99, 111, 116, 155, 162, 164, 171, 185, 218–222, 243, 258, 306, 309–310, 316–320, 333–338, 342, 348
Accumulation mode, 13, 14, 172–175
Aerosol, 4, 10, 11, 13–14, 43, 44, 48, 49, 63, 151, 152, 154, 165, 166, 169–198, 208, 234, 243, 257, 258, 261–263, 269, 291, 297, 323, 324, 329, 337–339, 341, 349, 351
Air density, 10, 11, 15, 22, 27–28, 30, 55, 111, 145
Air exchange rate, 256, 262, 265, 288, 296, 297, 299–300
Air masses, 5, 18, 19, 22, 23, 26, 70, 77–86, 101, 106, 128, 131–132, 138–140, 225
Air temperature, 20, 22, 24, 25, 63, 78, 96, 97, 100, 101, 104, 105, 107, 114–116, 143, 186, 194, 214–216, 300
Aitken mode, 172, 173
Alveolar region, 257, 311, 314, 319, 323, 325–327, 331, 333–335, 339, 341, 344, 353
Aquatic chemistry in the atmosphere, 162–163
Asbestos, 62, 257, 260, 284, 285, 335, 351
Atmosphere's height, 6
Atmospheric aerosols, 13–14, 169–198
Atmospheric diffusion, 202, 205, 217–220, 234
Atmospheric stability, 23, 70–77, 209, 216–217, 224
Autumnal equinox, 93, 94

B

Beaufort scale, 126, 127
Bioaerosols, 197–198, 256–258, 260, 295–296
Biogenic emissions, 243–245
Boundary layer, 4, 16–20, 69, 70, 86–89, 91, 115, 128, 131, 133, 135, 136, 144, 145, 154, 155, 208, 209, 218, 236, 237, 246, 249
Breathing frequency, 315, 325, 327, 328, 335, 336
Buoyant plume, 212, 214

C

Carbohemoglobin, 283, 284
Carbon chemical components, 156–157
Carbon monoxide, 11, 60, 61, 153, 157, 160, 239–241, 248, 256, 257, 259, 260, 283–284, 293, 294
Chemistry of hydrocarbons, 161
Circular force, 126
Climate, 2, 3, 6, 39, 47, 49–53, 68, 94, 97, 141, 156, 170, 239
Climate change, 4, 52–53, 171, 247
Cloud cover, 3, 49, 50, 73, 74, 81, 99, 100, 152, 230
Clouds, 8, 13, 17–19, 25, 42, 48–50, 68, 72, 75, 78, 79, 82–84, 93, 94, 96, 99–101, 106–111, 126, 137, 140, 144, 165, 166, 171, 174, 250
Coagulation, 13, 111, 172, 174, 194, 196, 257, 324, 339
Coarse particles, 13, 14, 60, 173–175, 178–181, 247, 285, 297, 324, 327, 331
Cold front, 26, 81–85, 106
Condensation, 13, 14, 23, 75, 82, 83, 85, 95, 101, 102, 105–107, 111, 144, 171, 172, 174, 186, 187, 191, 193–195, 246, 247, 257, 261, 323, 339
Continuous winds, 140–141
Coriolis force, 126, 128–135, 139–141, 145

359

Count median diameter (CMD), 177, 341
Covariance coefficient, 89
Cunningham correction factor, 177

D
Daily thermal width, 96, 97
Dermal absorption, 258, 306, 309–310, 316–319, 348
Dew point, 104–107
Diffusional deposition, 325
Diffusion stage, 194
Dispersion coefficient, 88, 89, 209, 210
Dose-response functions, 315–322
Dry atmospheric air, 11
Dry deposition, 13, 174, 236
Dry vertical adiabatic lapse rate, 23

E
Earth's radiation, 4, 5, 14, 25, 36–37, 41, 42, 57
Earth's rotation, 5, 10, 24, 32, 126, 128–130, 139, 140, 145
Eddy diffusivity, 205
Elongated high, 123, 124
Elongated low, 123, 124
Equations of circulation, 145–148
Euler description, 202–205
Evaporation, 10, 11, 13, 17, 50, 78, 95, 101, 102, 104, 105, 111, 174, 187, 188, 193–195, 261, 277, 285, 290
Exosphere, 14, 21–22
Expected dose, 316, 317
Extrathoracic airways, 311, 314, 344

F
Fine particles, 173, 181, 286
First law of thermodynamics, 75
Formaldehyde, 61, 258, 259, 270–272, 274, 275, 287–290, 293, 294, 347
Friction force, 126, 131–136
Friction velocity, 92, 135, 148, 247
Frontal inversion, 26
Fronts, 20, 26, 57, 77–86, 106, 139, 140, 258

G
Gaussian model, 74, 201–232
General circulation, 24, 137–144
Geostrophic wind, 18–20, 85, 132–133, 135
Gravitational settling, 13, 175, 214, 297, 325
Greenhouse effect, 45–47, 53, 152

H
Half life time, 275–277, 279, 282, 341, 345
Halogen chemical components, 157
Health effects, 61, 68, 69, 158, 182, 197, 258, 259, 284, 289, 290, 307, 319–321, 324, 346–353
Heat capacity, 17, 51, 70, 97–99, 142
Heavy metals, 239, 285–286
Heterogeneous nucleation, 186, 192–193
Heteromolecular nucleation, 186
Heterosphere, 11, 14
Holland plume rise, 214
Homogeneous nucleation, 186, 192, 193
Homomolecular nucleation, 186
Human dose, 320, 341
Human exposure, 34, 69, 170, 171, 234, 235, 256, 258–260, 262, 268, 277, 279, 284, 289, 294, 305–353
Human respiratory tract, 171, 177, 178, 260, 293, 294, 309, 310, 312, 328, 329, 331–338
Humidity, 3, 10, 11, 20, 24, 26, 70, 77, 78, 82–84, 91, 100–111, 120, 121, 144, 177, 186, 193, 194, 257, 260, 262, 266, 271, 274, 318, 333

I
ICRP model, 325, 330, 333–336, 338
Indoor air quality, 256–262, 288, 296
Inertial impaction, 322–325, 339
Infiltration rate, 297–300
Inhalability, 314, 324, 328–330, 336
Inhalation exposure, 310–315
Intake, 62, 258, 306, 307, 309, 315–316, 318–320, 329, 332
Interception, 322–325
Intermediate stage, 213
Internal dose, 69, 259, 307, 309, 316–321, 341–346
Ionosphere, 22
Isobaric curves, 122–124

K
Kirchhoff's law, 34

L
Lagrange description, 203–206
Land breeze, 141–143
Law of Buys-Ballot, 135–136

Law of ideal gasses, 24, 27–29, 55, 71, 203
Law of Stefan-Boltzmann, 35–36, 48
Laws of radiation, 33–36
Length scale, 68, 69
Logarithmic canonical distribution, 178, 179

M
Magnetosphere, 22
Mass median aerodynamic diameter, 177, 340
Mass median diameter (MMD), 177, 339
Mean absolute normalized gross error (MANGE), 238
Mean arithmetic diameter, 179
Mean bias, 238
Mean error, 238
Mean kinetic energy, 91
Mean normalized bias (MNB), 238
Mean value, 16, 58, 86–88, 115, 122, 203, 289, 328
Mesopause, 14, 15, 21
Mesosphere, 14, 15, 21
Microenvironmental model, 279, 297, 298, 347
Mie scattering, 43, 44
Milankovitch cycles, 53
Mixing layer, 17–19, 55, 77
Mixing ratio, 24, 30, 31, 55, 56, 103, 155
Model of single volume, 235–237
Moist vertical adiabatic lapse rate, 23–24
Mountain breezes, 141, 143–144

N
Nitrogen chemical components, 156
Nitrogen oxides, 2, 11, 105, 156, 158–160, 164, 180, 239, 242, 247, 248, 257, 260, 264–266, 294
Nocturnal boundary layer, 19–20
Normalized covariance, 89
Nucleation mode, 13, 14, 171, 172
Nucleation rate, 186, 190–192

O
Occluded front, 84–86
Organic aerosols, 180, 182–185, 243
Ozone, 2, 9, 12, 20, 21, 31, 40–42, 45, 50, 58–62, 68, 152, 153, 157–161, 163–166, 243, 257, 259, 260, 262–265, 269–271, 273–275, 324, 347
Ozone hole, 14, 68, 165

P
Particle clearance, 333–338
Particulate matter, 2–4, 58–60, 62, 68, 157–159, 170, 175, 183, 185–196, 239, 242, 243, 247, 249, 256–258, 260, 261, 276, 293, 322–341, 346, 349–353
Pasquill class, 74
Pasquill stability class, 74, 211, 214
Penetration, 34, 99, 225, 276, 297, 302, 306, 307, 312
Pesticides, 257, 258, 261, 263, 290
Photochemical cycle nitrogen oxides, 158–160
Photochemical cycle of ozone, 158–160
Photochemistry, 158–162, 324
Photosynthesis, 3, 8, 9
Photosynthetically active radiation (PAR), 243
≠$$$-Pinene, 244, 268, 269, 271, 273, 274
Planck's law, 35, 37
Plume rise, 212–216, 224, 225, 227
Polar front, 79–80, 140
Polychloric diphenyls, 260
Polycyclic aromatic hydrocarbon (PAH), 61, 260, 291
Potential temperature, 22, 55, 112, 148, 216
Precipitation, 3, 13, 101, 103, 107–111
Pressure gradient force, 124–126, 128, 132–135, 143
Pressure gradient wind, 134–135
Pressure high (Anticyclone), 16, 123–124
Pressure low (Cyclon), 16, 122–123
Pressure saddle point, 123, 124
Pressure vertical changes, 120–121
Primary organic carbon, 184–185
Probability density, 87, 88, 203, 204, 206, 218

R
Radiation, 8, 9, 14, 18, 25, 31–53, 58, 70, 75, 92–96, 98, 99, 105–107, 157–160, 163–166, 171, 243, 244, 246, 247, 262, 275, 276, 279, 281, 282, 300, 309, 325, 338, 341–346
Radon, 11, 256, 257, 260, 275–283, 341–346
Radon inhalation, 341–346
Radon isotopes, 277–279
Rayleigh scattering, 43–44
Reflection, 6, 22, 34, 38, 48–50, 218, 219, 221, 222
Regional deposition, 322, 325, 339
Relative humidity, 3, 24, 26, 83, 101–106, 177, 186, 193, 194, 257, 262
Residual layer, 17, 19
Residual volume, 314, 315, 327

Respiratory tract, 171, 177, 178, 197, 260, 282, 289, 293, 294, 309, 310, 312, 320, 322–338, 344, 352
Reynolds stress, 91, 92
Richardson's number, 216, 217

S

Scattering, 14, 38, 39, 42–44, 52, 171, 246
Sea breeze, 141–144
Sea salt emissions, 245–246
Secondary organic carbon, 185
Single cell models, 138–140, 251
Size distribution of aerosols, 171–180
Soot, 2, 182–184
Specific humidity, 103
Stability class, 74, 211, 214
Stability conditions, 70–77, 209, 211, 212, 214, 215, 217, 223–225
Standard atmosphere, 29–31, 70
Stationary front, 84–86
Stratopause, 20–21
Stratosphere, 14, 16, 20–21, 151, 157, 163–166
Subsidence inversion, 26, 75
Sulphur chemical components, 155, 174
Summer solstice, 93
Sun's radiation, 6, 9, 14, 18, 22, 24, 32–34, 36–45, 47–50, 53, 73, 74, 93–96, 99, 121, 152, 158, 159, 163, 265
Supersaturation, 24, 186, 192
Surface inversion, 25, 26, 75

T

Temperature daily variability, 96–98
Temperature inversion, 18, 25–27, 75–78, 96, 117, 224
Terpenes, 174, 180, 243, 244, 269–271, 273
Thermal stage, 212–213
Thermosphere, 14, 15, 21
Three cell models, 138–140
Threshold dose, 321

Tidal volume, 314, 315, 339
Tobacco smoke, 197, 291, 293–294
Total exposure, 308
Tracheo-bronchial region, 311, 314, 322, 326, 327, 331, 339, 341, 344
Transition probability density, 218
Troposphere, 11, 14–17, 20, 21, 23, 25, 26, 28–30, 68, 69, 101, 120, 151–155, 157–162, 247
Turbulence, 17–20, 29, 68, 86–92, 107, 114, 145, 152, 205, 209, 212, 213, 216–218, 223, 224
Turbulent kinetic energy, 87, 91

U

Ultrafine particles, 172, 174, 175, 297, 325, 334
Uptake, 306–309, 315–317, 319, 334, 337, 338, 348

V

Valley breezes, 141, 143–144
Vernal equinox, 93, 94
Vital capacity (VC), 314, 315, 327
Volatile organic compounds (VOCs), 157, 158, 174, 197, 239, 240, 243, 257, 258, 260, 262, 267–271, 274, 287, 290, 298, 301

W

Warm front, 20, 26, 82–84, 86
Water vapour, 8, 24, 41–42, 48, 51, 79, 101–106, 111, 162, 195
Wave cyclone, 84–86
Weibel model, 312, 314, 339
Wet deposition, 13, 19, 101, 110, 114, 181, 264
Wien's first law, 35
Wien's second law, 35
Winter solstice, 94